EARTH ENGINE & GEEMAP
GEOSPATIAL DATA SCIENCE WITH PYTHON

QIUSHENG WU

LOCATE PRESS

Credits & Copyright

Earth Engine and Geemap
Geospatial Data Science with Python

by Qiusheng Wu

Published by Locate Press

COPYRIGHT © 2023 LOCATE PRESS INC.
ISBN: 978-1-7387675-1-9

Direct permission requests to info@locatepress.com or mail:
Locate Press Inc., Suite 433, 113-437 Martin St.
Penticton, BC, Canada, V2A 5L1

Publisher Website https://locatepress.com
Book Website https://locatepress.com/book/gee

Version: 7c8ed2f (2023-08-15)

Contents

Preface

Introduction

Google Earth Engine[1] (GEE) is a cloud computing platform that offers access to a vast data catalog[2] of satellite imagery and geospatial datasets. The platform has gained immense popularity in the geospatial community in recent years and has played a significant role in empowering numerous environmental applications at local, regional, and global scales. GEE provides users with both JavaScript and Python APIs for making computational requests to the Earth Engine servers. While the GEE JavaScript API has robust documentation[3] and an interactive IDE (i.e., GEE JavaScript Code Editor[4]), the GEE Python API has relatively limited functionality for visualizing results interactively, and there is a lack of documentation. To fill this gap, the geemap Python package was created, which is built upon several open-source Python libraries, such as the earthengine-api[5], folium[6], ipyleaflet[7], and ipywidgets[8]. Geemap allows users to analyze and visualize Earth Engine datasets interactively within a Jupyter environment with minimal coding.

This book takes a hands-on approach to help users get started with the GEE Python API and geemap. It begins with the basics of geemap, including creating and customizing interactive maps. Then, users learn to load cloud-based Earth Engine datasets and local geospatial datasets onto the interactive maps. As readers progress through the chapters, they will explore practical examples of using geemap to visualize and analyze Earth Engine datasets and learn how to export data from Earth Engine. Additionally, the book covers more advanced topics, such as building and deploying interactive web apps with Earth Engine and geemap.

Who this book is for

This book is designed for students, researchers, and data scientists who want to explore Google Earth Engine using the Python ecosystem of diverse libraries and tools. Regardless of whether you are a new or experienced user of the Earth Engine JavaScript API, this book is suitable for you.

What this book covers

Chapter 1 *Introducing GEE and geemap*: provides an introduction to using the GEE Python API and setting up a Python environment for using geemap.

Chapter 2 *Creating Interactive Maps*: teaches readers how to create and customize interactive maps using various plotting backends.

---

[1] Google Earth Engine: earthengine.google.com

[2] data catalog: tiny.geemap.org/nkuh

[3] documentation: tiny.geemap.org/mpxn

[4] GEE JavaScript Code Editor: code.earthengine.google.com

[5] earthengine-api: tiny.geemap.org/xzvx

[6] folium: tiny.geemap.org/isxr

[7] ipyleaflet: tiny.geemap.org/mclp

[8] ipywidgets: tiny.geemap.org/xzmw

Chapter 3 *Using Earth Engine Data*: covers the basic data types of Earth Engine and teaches how to search for and load Earth Engine datasets onto an interactive map.

Chapter 4 *Using Local Geospatial Data*: teaches how to load local vector and raster datasets onto an interactive map and how to download data from OpenStreetMap.

Chapter 5 *Visualizing Geospatial Data*: covers various tools and techniques for visualizing geospatial data, such as split-panel maps, linked maps, and timeseries inspector. This chapter also covers creating map elements such as color bars, legends, and labels.

Chapter 6 *Analyzing Geospatial Data*: covers statistical methods and machine learning techniques for analyzing geospatial data, such as zonal statistics, unsupervised classification, supervised classification, and accuracy assessment.

Chapter 7 *Exporting Earth Engine Data*: teaches how to export vector and raster data from Earth Engine and how to download thousands of image chips from Earth Engine within a few minutes.

Chapter 8 *Making Maps with Cartoee*: teaches how to create publication-quality maps using the cartoee module. This chapter covers plotting Earth Engine data on the map and customizing map projections.

Chapter 9 *Creating Timelapse Animations*: provides practical examples of using geemap to create timelapse animations from satellite and aerial imagery, such as Landsat, GOES, MODIS, and NAIP.

Chapter 10 *Building Interactive Web Apps*: teaches how to build interactive web apps with Earth Engine and geemap from scratch. This chapter also covers deploying web apps to the cloud for public access.

Chapter 11 *Earth Engine Applications*: covers various Earth Engine applications, such as surface water mapping, forest cover change analysis, flood mapping, and global land cover mapping.

To get the most out of this book

This book assumes that you have some basic knowledge of Python. This means that you should be comfortable with the basic Python syntax, including variables, lists, dictionaries, loops, and functions. If you are familiar with creating lists and defining variables, and have worked with for loops before, you have sufficient Python knowledge to get started.

To get the most out of this book, it is highly recommended that you type the code yourself in a Jupyter environment, such as Jupyter notebook, JupyterLab, or Google Colab. This will aid your understanding of the code and the material in the book.

Download Jupyter notebook examples

The Jupyter notebook examples for this book can be downloaded from the book's GitHub repository at `https://github.com/giswqs/geebook`. If there are any updates to the code, they will be made available in this repository.

Conventions used

There are a number of text conventions used throughout this book.

Code in text: Indicates code words in text, folder names, filenames, file extensions, pathnames, URLs, etc. Here is an example: Set `EARTHENGINE_TOKEN` as a system environment variable to your Earth Engine API key.

Python code blocks are set as follows:

```python
import geemap
# Create an interactive map
Map = geemap.Map(center=[40, -100], zoom=4)
Map
```

Bold: Indicates a new term, an important word, or words that you see onscreen. For instance, words in menus or dialog boxes appear in bold. Here is an example: Click on **Advanced settings** to set EARTHENGINE_TOKEN as an environment variable.

Get in touch

We welcome feedback from our readers.

Questions: If you have any questions about the materials covered in the book, go to the book repository discussion board[9] to ask questions and share ideas.

Errata: Although we have made every effort to ensure the accuracy of the book content, errors may occur. If you find any mistakes in the book, please report them to us. Please go to the book repository and submit an issue[10].

Acknowledgments

This book is based upon work partially supported by the National Aeronautics and Space Administration (NASA) under Grant No. 80NSSC22K1742 issued through the Open Source Tools, Frameworks, and Libraries 2020 Program[11].

Some of the text and code examples in this book are adapted from the Earth Engine User Guides[12]. The Earth Engine team deserves credit for their excellent work on developing the Earth Engine platform and comprehensive user guides. I would like to give a special thank you to Khalil Misbah for designing the geemap logo[13].

I would like to express my sincere gratitude to the reviewers who provided valuable feedback on this book, including Ellen Brock, Jake Gearon, and Emma Izquierdo-Verdiguier. Your insights and suggestions helped me to improve the quality of my work and make it more accessible to readers. Your time and effort are greatly appreciated.

I also want to acknowledge the efforts of my editor and publishing team, who worked tirelessly to bring this book to fruition. Your support and guidance throughout the process were invaluable, and I could not have done it without you.

Finally, I would like to thank my family and friends for their unwavering support and encouragement. Your love and belief in me kept me going during the challenging times, and I am forever grateful.

---

[9] discussion board: bit.ly/geebook-qa

[10] submit an issue: bit.ly/geebook-issues

[11] Open Source Tools, Frameworks, and Libraries 2020 Program: bit.ly/3RVBRcQ

[12] Earth Engine User Guides: tiny.geemap.org/iweo

[13] geemap logo: tiny.geemap.org/gzul

1. Introducing Earth Engine and Geemap

1.1 Introduction

Google Earth Engine[1] is a widely used cloud-computing platform in the geospatial community. It features a multi-petabyte data catalog of satellite imagery and geospatial datasets, enabling users to easily and efficiently visualize, manipulate, and analyze geospatial data. Built upon the Earth Engine Python API[14] and open-source mapping libraries, geemap[15] makes it much easier to analyze and visualize Earth Engine datasets in a Jupyter environment. Since its initial release in April 2020, geemap has become the most popular Python package for interactive analysis and visualization of Earth Engine data.

This chapter covers the fundamentals of Geospatial Data Science, Google Earth Engine, and geemap. We will walk through the process of setting up a conda environment and installing geemap. Additionally, we will explore how to utilize geemap with Google Colab without needing to install anything on your device. Finally, we will provide some useful resources for further learning about Earth Engine and geemap beyond the scope of this book.

1.2 What is Geospatial Data Science

Before introducing geospatial data science, we need to understand what **data science** is. The term "data science" has gained a lot of attention during the past decade, along with related terms such as **big data**, **data analytics**, and **machine learning**. According to Google Trends[16] , the online search interest over time in "data science" has experienced a rapid increase since 2016 (see Fig. 1.1). When we googled "data science", 4.1 billion records were returned, compared to 7.6 billion on "big data", 2.6 billion on "machine learning", and 1.9 billion on "data analytics". Interestingly, the interest in "big data" has been decreasing since 2018, while the interests in "data science" and "machine learning" continue to increase.

Okay, so what is **data science**? Data science is a broad term that encompasses many areas of interest. From a high-level perspective, data science is the science of data or the study of data (Cao et al., 2017). From the disciplinary perspective, data science is an interdisciplinary field that uses scientific methods, processes, algorithms, and systems to extract knowledge and insights from noisy, structured and unstructured data, and apply knowledge and actionable insights from data across a broad range of application domains (Dhar et al., 2013).

Geospatial data science is a discipline within data science that specifically focuses on the spatial component of data. It brings forth theories, concepts and applications that are specific to geographic data in the realm of data science (Hassan et al., 2019). A good example of geospatial data science is NOAA's analysis of spatial and temporal datasets (e.g., satellite imagery, weather data, and climate models) to provide hurricane forecasts using statistics, machine learning, and mathematical models (Eftelioglu et al., 2017).

---

[14]Earth Engine Python API: tiny.geemap.org/ciny

[15]geemap: geemap.org

[16]Google Trends: bit.ly/40lfnpW

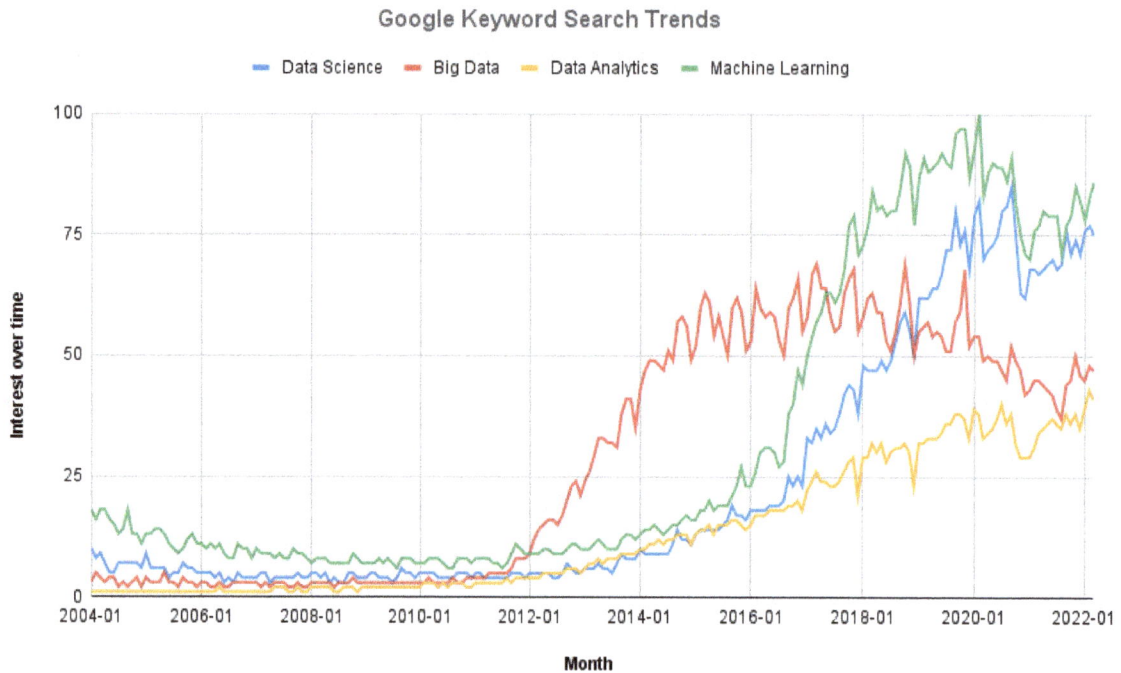

Figure 1.1: Online search interest trends on data science-related keywords by Google as of March 28, 2022. The numbers on the vertical axis represent search interest relative to the highest point on the chart for the given region (worldwide) and time (2004-2022). A value of 100 is the peak popularity for the term. A value of 50 means that the term is half as popular. A score of 0 means there was not enough data for this term.

1.3   What is Google Earth Engine

Google Earth Engine (GEE) is a cloud computing platform with a multi-petabyte data catalog[2] of satellite imagery and geospatial datasets (Gorelick et al., 2017). During the past few years, GEE has become very popular in the geospatial community, and it has empowered numerous environmental applications at local, regional, and global scales (Amani et al., 2020; Boothroyd et al., 2020; Tamiminia et al., 2020; Wu et al., 2019). Since GEE became publicly available in 2010, there has been an exponential growth in the number of peer-reviewed journal publications empowered by GEE (see Fig. 1.2). Based on the most recent bibliometric analysis, there are 1,077 peer-reviewed journal publications with the word "Google Earth Engine" in the title and 2,969 publications with the word "Google Earth Engine" in either the title or abstract. In 2022, the number of publications with "Google Earth Engine" in the title or abstract reached 1,150, which is more than a 280-fold increase from the year 2014 with only 4 publications.

To use Earth Engine, you must first sign up for an Earth Engine account[17] (Fig. 1.3). You cannot use Google Earth Engine unless your application has been approved. Once you receive the application approval email, you can log in to the Earth Engine JavaScript Code Editor[18] to get familiar with the JavaScript API.

---

[17]sign up for an Earth Engine account: tiny.geemap.org/plpk

[18]Earth Engine JavaScript Code Editor: code.earthengine.google.com

The number of journal publications empowered by Google Earth Engine

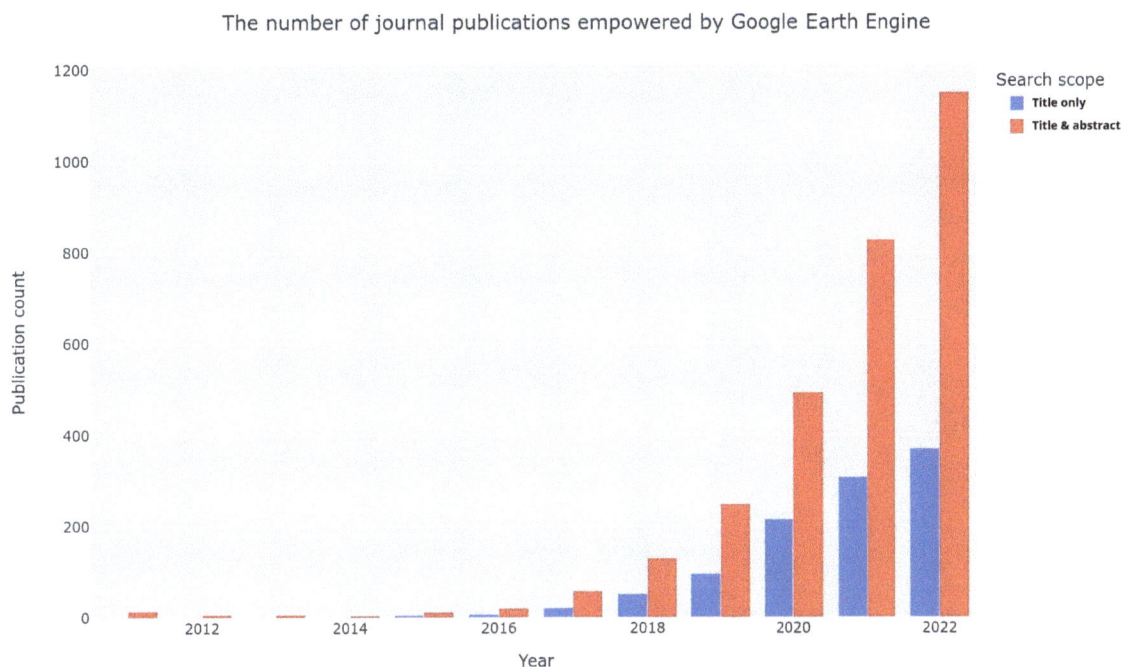

Figure 1.2: The number of journal publications empowered by Google Earth Engine.

Figure 1.3: Signing up for an Earth Engine account.

1.4 What is geemap

GEE provides users with both JavaScript and Python APIs for making computational requests to the Earth Engine servers. While the GEE JavaScript API has robust documentation[3] and an interactive IDE (i.e., GEE JavaScript Code Editor[4]), the GEE Python API has relatively limited functionality for visualizing results interactively, and there is a lack of documentation. The **geemap** Python package was created to fill this gap (Wu, 2020). It is built upon a number of open-source Python libraries, such as the earthengine-api[5] , folium[6] , ipyleaflet[7] , and ipywidgets[8] packages. Geemap enables users to analyze and visualize Earth Engine datasets interactively within a Jupyter environment with minimal coding (see Fig. 1.4).

Geemap is intended for students and researchers who would like to utilize the Python ecosystem of diverse libraries and tools to explore Google Earth Engine. It is also designed for existing GEE users who would like to transition from the GEE JavaScript API to the Python API. Geemap provides an interactive graphical user interface for converting GEE JavaScript projects to Python scripts without coding. It can save users a lot of time and effort by providing a simple interface for exploring and visualizing Earth Engine datasets.

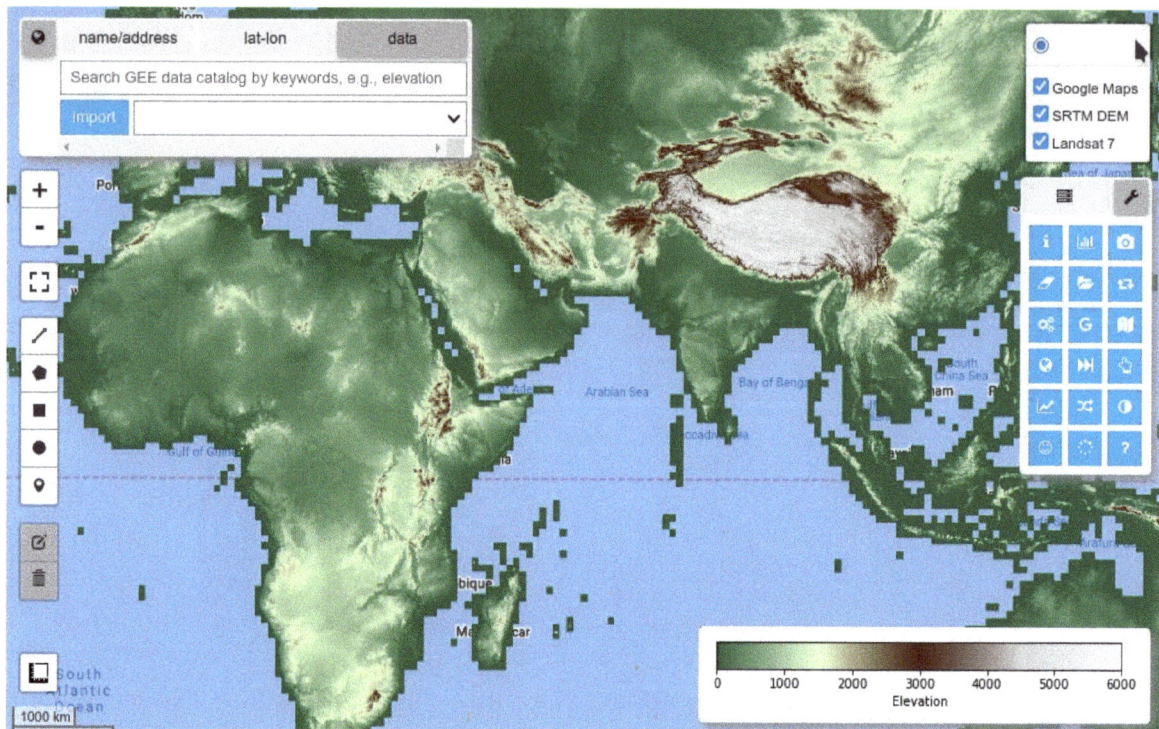

Figure 1.4: The geemap graphical user interface built upon ipyleaflet and ipywidgets.

1.5 Installing geemap

The geemap package has some optional dependencies, such as GeoPandas[19] and localtileserver[20]. These optional dependencies can sometimes be a challenge to install, especially on Windows. Therefore, we advise you to closely follow the recommendations below to avoid installation problems. Note that installing geemap will automatically install all of its dependencies, including the earthengine-api package. Therefore, you do not need to install the earthengine-api package separately.

Installing with conda

To install geemap and its dependencies, we recommend you use the conda[21] package and environment manager. This can be obtained by installing the Anaconda Distribution[22] (a free Python distribution for data science), or through Miniconda[23] (minimal distribution only containing Python and the conda package manager). Also see the installation docs[24] for more information on how to install Anaconda or Miniconda locally.

Geemap is available on the conda-forge[25] Anaconda channel, a community effort that provides conda packages for a wide range of software. Creating a new conda environment to install geemap is not

---

[19]GeoPandas: geopandas.org

[20]localtileserver: tiny.geemap.org/twyv

[21]conda: conda.io/en/latest

[22]Anaconda Distribution: tiny.geemap.org/shvw

[23]Miniconda: tiny.geemap.org/xpyp

[24]installation docs: tiny.geemap.org/lrzx

strictly necessary, but given that some geemap dependencies might have a version conflict with other geospatial packages in an existing conda environment, it is a good practice to start fresh by installing geemap and its dependencies in a clean environment for your project. The following commands create a new conda environment named gee and install geemap in it:

```
conda create -n gee python
conda activate gee
conda install -c conda-forge geemap
```

First, open the **Anaconda Prompt** or **Terminal** and type "conda create -n gee python". Press **Enter** to create a new conda environment named gee (see Fig. 1.5).

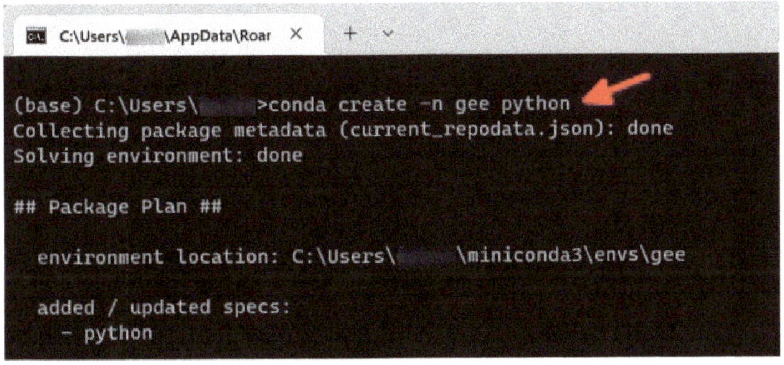

Figure 1.5: Creating a new conda environment named 'gee'.

Next, activate the new conda environment by typing "conda activate gee" and press **Enter**. Then, install geemap into the environment we just activated by typing "conda install -c conda-forge geemap" and press **Enter** (see Fig. 1.6).

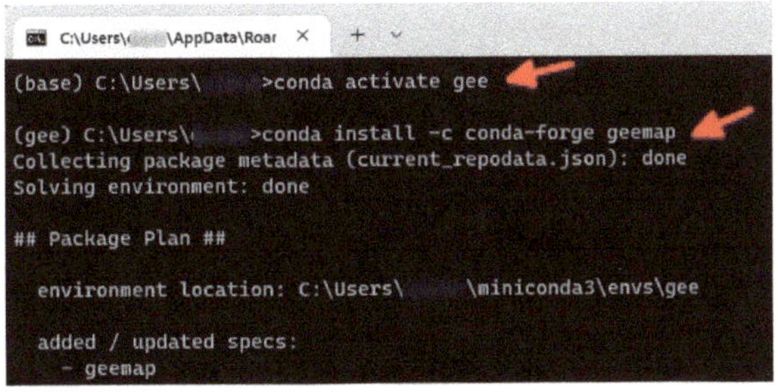

Figure 1.6: Activating the new conda environment and installing geemap.

Geemap has a list of optional dependencies specified in the requirements_all.txt[26] , such as GeoPandas, localtileserver, osmnx[27] , rioxarray[28] and rio-cogeo[29]. It can be a bit cumbersome to install these optional dependencies individually, but luckily these optional dependencies are available through the pygis[30] Python package which can be installed with a single command.

Since pygis has many dependencies, it might take a while for conda to resolve dependencies. Therefore, we highly recommend you to install Mamba[31] , a fast, robust, and cross-platform package manager.

---

[25]conda-forge: tiny.geemap.org/ucpm

[26]requirements_all.txt: tiny.geemap.org/qvgt

[27]osmnx: github.com/gboeing/osmnx

[28]rioxarray: github.com/corteva/rioxarray

[29]rio-cogeo: github.com/cogeotiff/rio-cogeo

[30]pygis: pygis.gishub.org

Mamba is a re-write of conda that significantly increases the speed of resolving and installing packages. It runs on Windows, macOS, and Linux, and is fully compatible with conda packages and supports most of conda's commands. The following commands install Mamba and pygis:

```
conda install -c conda-forge mamba
mamba install -c conda-forge pygis
```

To install Mamba, type "conda install -c conda-forge mamba" and press **Enter** (see Fig. 1.7).

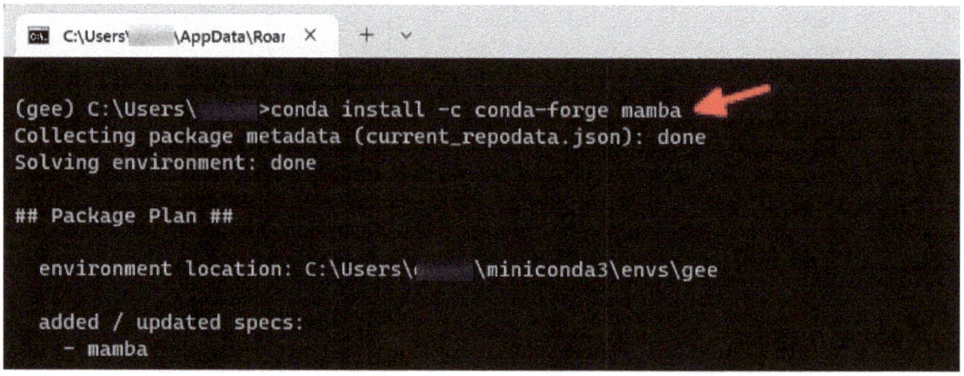

Figure 1.7: Installing the Mamba package manager.

Once Mamba is installed in a conda environment, you can then simply replace any conda command with mamba. For example, to install pygis, type "mamba install -c conda-forge pygis" and press **Enter** (see Fig. 1.8).

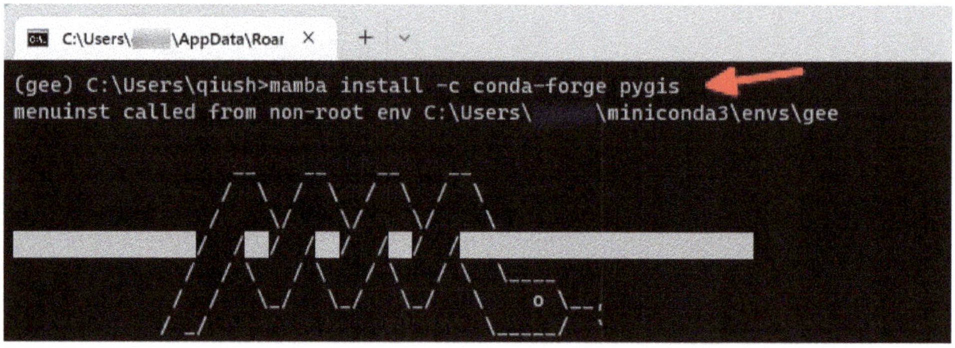

Figure 1.8: Installing optional dependencies of geemap through the pygis package.

Congratulations! You have successfully installed geemap and its dependencies. We will dive into geemap in the next chapter.

Installing with pip

Geemap is also available on PyPI[32]. It can be installed with pip using the following command:

```
pip install geemap
```

All optional dependencies of geemap are listed in requirements_all.txt[26] , which can be installed using one of the following:

---

[31]Mamba: github.com/mamba-org/mamba

[32]PyPI: pypi.org/project/geemap

- `pip install geemap[extra]`: installing extra optional dependencies listed in requirements_extra.txt.
- `pip install geemap[all]`: installing all optional dependencies listed in requirements_all.txt.
- `pip install geemap[backends]`: installing keplergl, pydeck, and plotly.
- `pip install geemap[lidar]`: installing ipygany, ipyvtklink, laspy, panel, pyntcloud[LAS], pyvista, pyvista-xarray, and rioxarray.
- `pip install geemap[raster]`: installing geedim, localtileserver, rio-cogeo, rioxarray, netcdf4, and pyvista-xarray.
- `pip install geemap[sql]`: installing psycopg2 and sqlalchemy.
- `pip install geemap[apps]`: installing gradio, streamlit-folium, and voila
- `pip install geemap[vector]`: installing geopandas and osmnx.

Installing from source

You may install the latest development version by cloning the GitHub repository with Git[33] and using pip to install from the local directory:

```
git clone https://github.com/gee-community/geemap
cd geemap
pip install .
```

It is also possible to install the latest development version directly from the GitHub repository with:

```
pip install git+https://github.com/gee-community/geemap
```

Upgrading geemap

If you have installed geemap before and want to upgrade to the latest version, you can run the following command in your terminal:

```
pip install -U geemap
```

If you use conda, you can update geemap to the latest version by running the following command in your terminal:

```
conda update -c conda-forge geemap
```

To install the development version from GitHub directly within a Jupyter notebook without using Git, run the following code in a Jupyter notebook and restart the kernel to take effect:

```
import geemap

geemap.update_package()
```

Using Docker

Geemap is also available on Docker Hub[34].

To use geemap in a Docker container, you first need to install Docker[35]. Once Docker is installed, you can pull the latest geemap image from Docker Hub by running the following command in your terminal:

---

[33]Git: git-scm.com

[34]Docker Hub: hub.docker.com/r/giswqs/geemap

[35]Docker: docs.docker.com/get-docker

```
docker run -it -p 8888:8888 giswqs/geemap:latest
```

1.6   Creating a Jupyter notebook

Let's activate the conda environment created in the previous section:

```
conda activate gee
```

Next, launch JupyterLab by typing the following commands in the **Terminal** or **Anaconda Prompt**:

```
jupyter lab
```

JupyterLab will open as a new tab in the browser. Click the **Python 3** icon in the top left corner of the JupyterLab **Launcher** window (see Fig. 1.9) or go to **File -> New -> Notebook** to create a new notebook. Select the newly created notebook in the JupyterLab File Browser tab and press **F2** to rename the notebook, e.g., **chapter01.ipynb**.

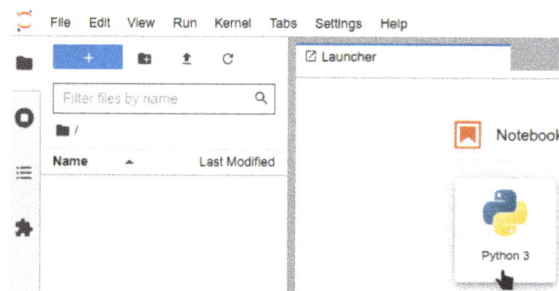

Figure 1.9: The JupyterLab user interface.

Jupyter notebook has two modes: **Edit mode** and **Command mode**. The Edit mode allows you to type into the cells like a normal text editor. The Command mode allows you to edit the notebook as a whole, but not type into individual cells. Jupyter notebook has many keyboard shortcuts (Yordanov et al., 2017). Here are some commonly used shortcuts. Note that the shortcuts are for Windows and Linux users. For Mac users, replace `Ctrl` with `Command`.

Shortcuts in both modes:

- `Shift + Enter`: run the current cell, select below
- `Ctrl + Enter`: run selected cells
- `Alt + Enter`: run the current cell, insert below
- `Ctrl + S`: save and checkpoint

While in command mode (press `Esc` to activate):

- `A`: insert cell above
- `B`: insert cell below
- `X`: cut selected cells
- `C`: copy selected cells
- `V`: paste cells below
- `Y`: change the cell type to Code
- `M`: change the cell type to Markdown
- `P`: open the command palette

While in edit mode (press `Enter` to activate):

- Esc: activate the command mode
- Tab: code completion or indent
- Shift + Tab: show tooltip

1.7 Earth Engine authentication

You need to authenticate Earth Engine before using it. The package for the Earth Engine Python API is called earthengine-api[5] , which should have been automatically installed by the geemap package as described in Section 1.5. Enter the following script into a code cell of a Jupyter notebook and press Shift + Enter to execute:

```
import ee

ee.Authenticate()
```

After running the above script, a new tab will open in the browser asking you to sign in to your Earth Engine account. After signing in, you will be asked to authorize the Google Earth Engine Authenticator. If this is the first time you are authenticating Earth Engine, click **CHOOSE PROJECT** to select a Cloud Project to use for Earth Engine (see Fig. 1.10).

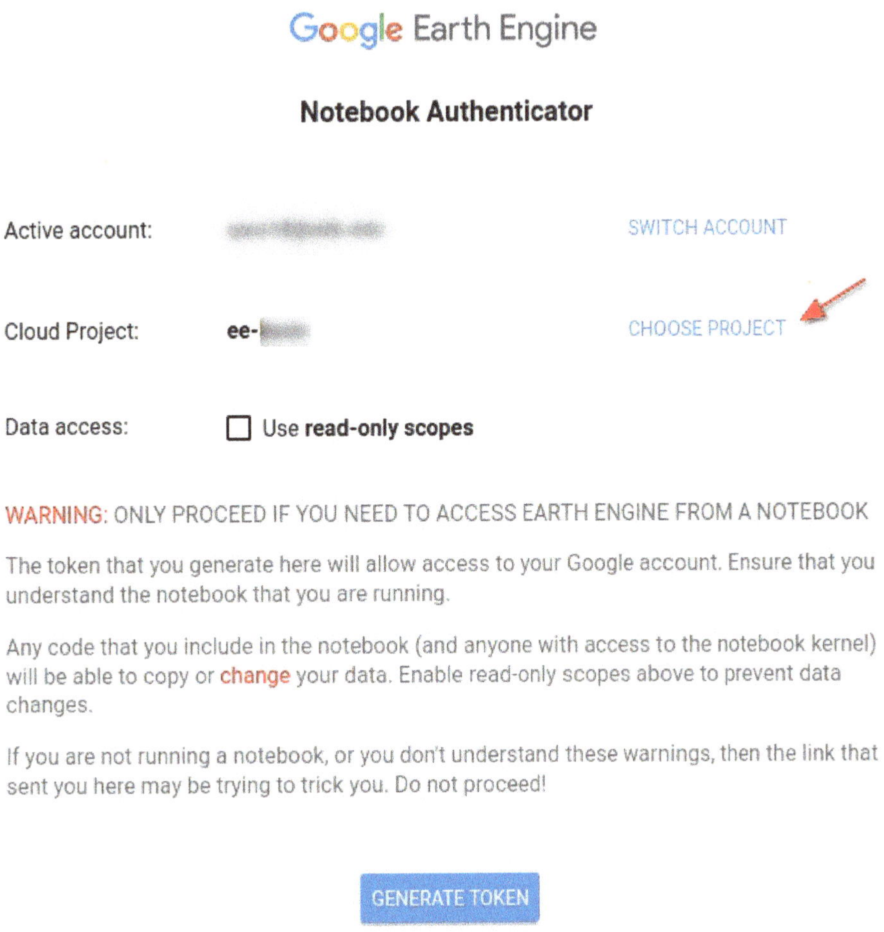

Figure 1.10: Earth Engine Notebook Authenticator.

You can either choose an existing Cloud Project or create a new one. If you choose to create a new

Cloud Project, enter a project name, e.g., ee-your-username and click the blue **SELECT** button to create a new Cloud Project. If a red warning message appears at the bottom of the page, click on the **Cloud Terms of Service** link to accept the terms of service and then click the **SELECT** button again (see Fig. 1.11).

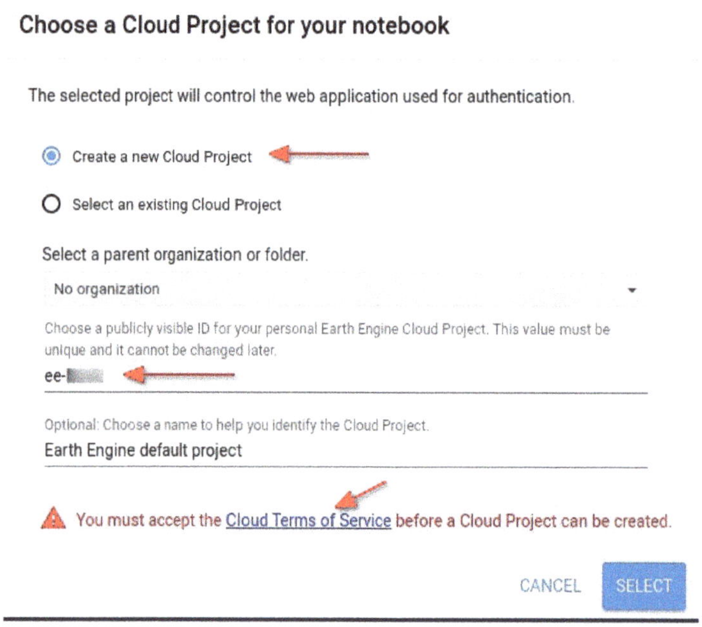

Choose a Cloud Project for your notebook

The selected project will control the web application used for authentication.

⦿ Create a new Cloud Project ⟵

◯ Select an existing Cloud Project

Select a parent organization or folder.

No organization   ▾

Choose a publicly visible ID for your personal Earth Engine Cloud Project. This value must be unique and it cannot be changed later.

ee-▮▮▮ ⟵

Optional: Choose a name to help you identify the Cloud Project.

Earth Engine default project

⚠ You must accept the Cloud Terms of Service before a Cloud Project can be created.

                                                                CANCEL   SELECT

Figure 1.11: Creating a new Cloud Project.

After selecting a Cloud Project, click the **GENERATE TOKEN** button to generate a new token. You will be asked to choose your Earth Engine account for the Notebook Client (see Fig. 1.12).

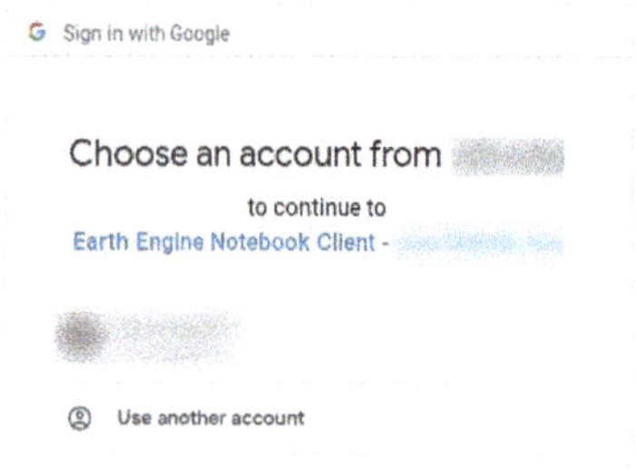

G Sign in with Google

Choose an account from ▮▮▮

to continue to
Earth Engine Notebook Client - ▮▮▮

▮▮▮▮▮▮

ⓧ Use another account

Figure 1.12: Choosing an account for the Earth Engine Notebook Client.

Click the **Allow** button to allow the Notebook Client to access your Earth Engine account (see Fig. 1.13).

An authentication code will be generated and displayed on the page. Copy the authorization code and paste it into the notebook cell asking for the verification code. Press **Enter** and the Successfully saved authorization token message should appear beneath the authorization code you entered (see Fig. 1.14).

Congratulations! You have successfully authenticated Earth Engine for use in your Jupyter notebook. In general, authentication for local installations is a one-time step that generates a persistent authorization token stored on a local computer. The token can be found in the following file path depending on

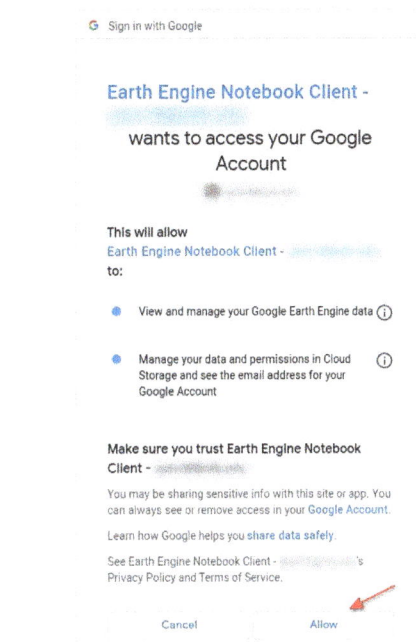

Figure 1.13: Choosing an account for the Earth Engine Notebook Client.

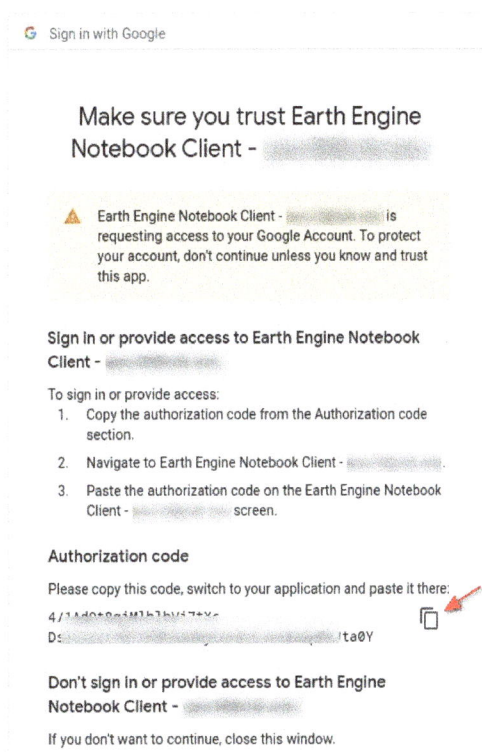

Figure 1.14: Copying the authentication code.

your operating system. Note that you might need to show the hidden directories on your computer in order to see the .config folder under the home directory.

```
Windows: C:\\Users\\USERNAME\\.config\\earthengine\\credentials
Linux: /home/USERNAME/.config/earthengine/credentials
MacOS: /Users/USERNAME/.config/earthengine/credentials
```

Once Earth Engine is authenticated, you can run the following script to initialize Earth Engine for a new Python session.

```
ee.Initialize()
```

In general, you will need to initialize Earth Engine for each new Python session, i.e., whenever you open a Jupyter notebook or Python script and want to use Earth Engine. Fortunately, geemap can automatically initialize Earth Engine for you when creating an interactive map, which will be covered in the next chapter. In other words, you rarely need to run ee.Initialize() explicitly.

1.8   Using Google Colab

If you have difficulties installing geemap on your computer, you can try out geemap with Google Colab[36] without installing anything on your machine. Google Colab is a free Jupyter notebook environment that runs entirely in the cloud. Most importantly, it does not require a setup and the notebooks that you create can be simultaneously edited by your team members - just like the way you edit documents in Google Docs!

Click 01_introduction.ipynb[37] to launch the notebook in Google Colab.

Next, press **Ctrl + /** to uncomment the following line to install geemap:

```
# %pip install geemap
```

After geemap has been installed successfully, type the following code in a new cell:

```
import geemap

Map = geemap.Map()
Map
```

Follow the on-screen instructions to authenticate Earth Engine. After that, you should be able to see the interactive map displayed beneath the code cell (see Fig. 1.15).

1.9   Using geemap with a VPN

When using geemap through a VPN, it's important to use "geemap.set_proxy(port=your-port-number)" to connect to Earth Engine servers (Fig. 1.16). Failure to do so may result in a connection timeout issue.

```
import geemap

geemap.set_proxy(port='your-port-number')
Map = geemap.Map()
Map
```

---

[36]Google Colab: colab.research.google.com
[37]01_introduction.ipynb: tiny.geemap.org/ch01

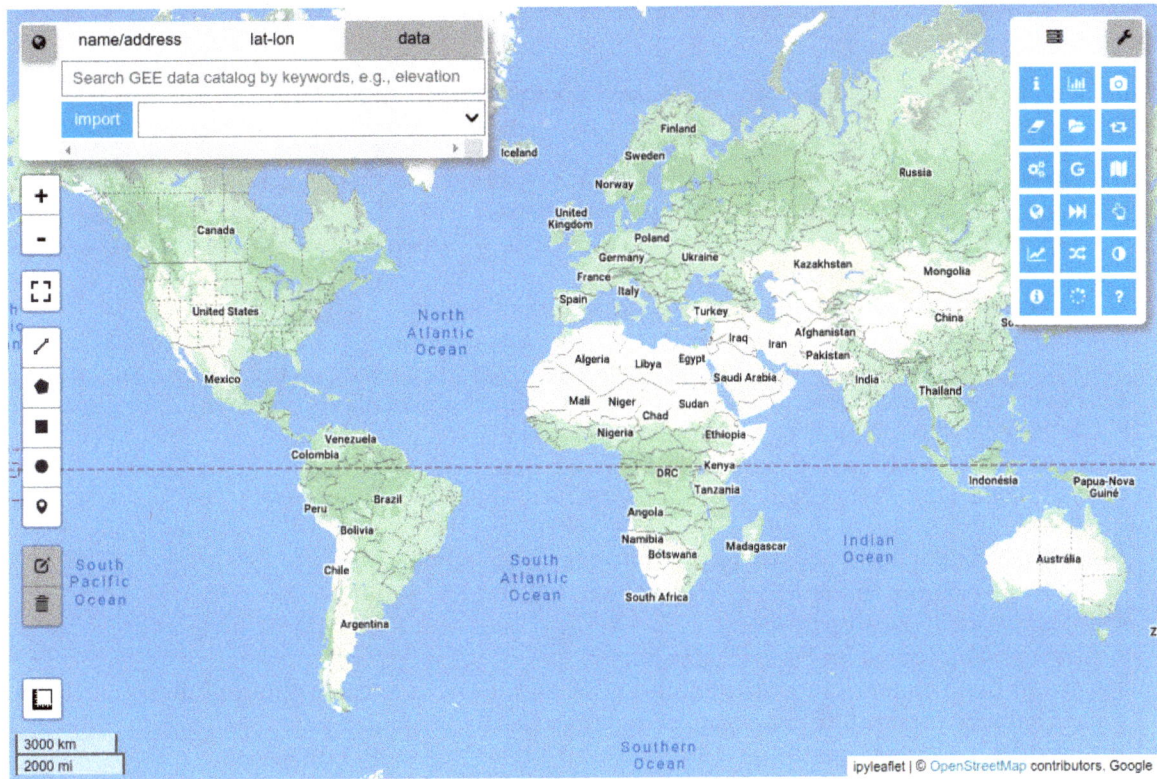

Figure 1.15: The interactive map displayed in Google Colab.

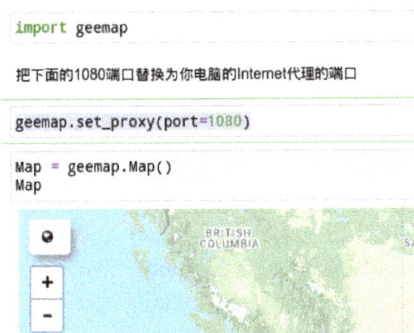

Figure 1.16: Using geemap with a VPN.

1.10   Key features of geemap

Below is a partial list of geemap features. Please check the geemap API Reference[38] and tutorials[39] for more details.

- Convert Earth Engine JavaScript projects to Python scripts and Jupyter notebooks.
- Display Earth Engine data layers on interactive maps.
- Support Earth Engine JavaScript API-styled functions in Python, such as Map.addLayer(), Map.setCenter(), Map.centerObject(), Map.setOptions().
- Visualize Earth Engine vector and raster data without coding.
- Retrieve Earth Engine data interactively using the Inspector tool.
- Creating interactive plots from Earth Engine data by simply clicking on the map.
- Convert data between the GeoJSON and Earth Engine FeatureCollection formats.
- Use drawing tools to interact with Earth Engine data.
- Use shapefiles with Earth Engine without having to upload data to one's GEE account.
- Export data in the Earth Engine FeatureCollection format to other formats (i.e., shp, csv, json, kml, kmz).
- Export Earth Engine Image and ImageCollection as GeoTIFF.
- Extract pixels from an Earth Engine Image into a 3D numpy array.
- Calculate zonal statistics by group.
- Add a custom legend for Earth Engine data.
- Convert Earth Engine JavaScript projects to Python code from directly within a Jupyter notebook.
- Add animated text to GIF images generated from Earth Engine data.
- Add colorbar and images to GIF animations generated from Earth Engine data.
- Create satellite timelapse animations with animated text using Earth Engine.
- Search places and datasets from Earth Engine Data Catalog.
- Use the timeseries inspector to visualize landscape changes over time.
- Export Earth Engine maps as HTML files and PNG images.
- Search Earth Engine API documentation within Jupyter notebooks.
- Import Earth Engine assets from personal Earth Engine accounts.
- Publish interactive GEE maps directly within a Jupyter notebook.
- Add local raster datasets (e.g., GeoTIFF) to the map.
- Support Cloud Optimized GeoTIFF (COG) and SpatioTemporal Asset Catalog (STAC).
- Perform image classification and accuracy assessment.
- Extract pixel values interactively and export data as shapefile and CSV.
- Visualize land cover change with Sankey diagrams.
- Load vector data from a PostGIS server.
- Create publication-quality maps with cartoee.

1.11   Summary

In this chapter, we began by covering the fundamentals of Geospatial Data Science, Google Earth Engine, and geemap. We then provided guidance on setting up a conda environment for installing geemap and its dependencies. Additionally, we walked through the process of using geemap with Google Colab as a cloud-based alternative to a local installation.

By now, you should have a fully functional conda environment that is ready for working with Earth Engine and geemap. In our next chapter, we will explore geemap in greater depth.

---

[38] API Reference: geemap.org/geemap

[39] tutorials: geemap.org/tutorials

2. Creating Interactive Maps

2.1 Introduction

There are numerous Python packages available for interactive mapping and geospatial analysis (Wu, 2021). However, each package has its own unique API for creating maps and visualizing data, which can be difficult for beginners to navigate. Geemap simplifies the process by providing a unified API interface for creating interactive maps and visualizing data, allowing users to switch between plotting backends with only one line of code.

Throughout this chapter, we will explore how to create interactive maps using one of five different plotting backends. We will also provide practical examples of how to add basemaps to an interactive map. With hundreds of basemaps available, they can be easily added to an interactive map using only a single line of code.

2.2 Technical requirements

To follow along with this chapter, you will need to have geemap and several optional dependencies installed. If you have already followed Section 1.5 - *Installing geemap*, then you should already have a conda environment with all the necessary packages installed. Otherwise, you can create a new conda environment and install pygis[30] with the following commands, which will automatically install geemap and all the required dependencies:

```
conda create -n gee python
conda activate gee
conda install -c conda-forge mamba
mamba install -c conda-forge pygis
```

Next, launch JupyterLab by typing the following commands in your terminal or Anaconda prompt:

```
jupyter lab
```

Alternatively, you can use geemap in a Google Colab cloud environment without installing anything on your local computer. Click 02_maps.ipynb[40] to launch the notebook in Google Colab.

Once in Colab, you can uncomment the following line and run the cell to install pygis, which includes geemap and all the necessary dependencies:

```
# %pip install pygis
```

The installation process may take 2-3 minutes. Once pygis has been installed successfully, click the **RESTART RUNTIME** button that appears at the end of the installation log or go to the **Runtime** menu and select **Restart runtime**. After that, you can start coding.

To begin, import the necessary libraries that will be used in this chapter:

```
import ee
import geemap
```

---

[40]02_maps.ipynb: tiny.geemap.org/ch02

Initialize the Earth Engine Python API:

```
geemap.ee_initialize()
```

If this is your first time running the code above, you will need to authenticate Earth Engine first. Follow the instructions in Section 1.7 - *Earth Engine authentication* to authenticate Earth Engine.

2.3   Plotting backends

Geemap has five plotting backends, including ipyleaflet[7], folium[6], plotly[41], pydeck[42], and kepler.gl[43]. An interactive map created using one of the plotting backends can be displayed in a Jupyter environment, such as Google Colab, Jupyter Notebook, and JupyterLab. By default, `import geemap` will use the ipyleaflet plotting backend.

The five plotting backends supported by geemap do not offer equal functionality. The ipyleaflet plotting backend provides the richest interactive functionality, including the interactive Graphical User Interface (GUI) for loading, analyzing, and visualizing geospatial data interactively without coding. For example, users can add vector data (e.g., GeoJSON, Shapefile, KML, GeoDataFrame) and raster data (e.g., GeoTIFF, Cloud Optimized GeoTIFF [COG]) to the map with a few clicks. Users can also perform geospatial analysis using the WhiteboxTools GUI with 500+ geoprocessing tools directly within the map interface. Other interactive functionality (e.g., split-panel map, linked map, time slider, time-series inspector) can also be useful for visualizing geospatial data. The ipyleaflet package is built upon ipywidgets and allows bidirectional communication between the frontend and the backend, enabling the use of the map to capture user input (QuantStack, 2019).

In contrast, folium has relatively limited interactive functionality. It is meant for displaying static data only. It can be useful for developing interactive web apps when ipyleaflet is not supported. Note that the aforementioned interactive GUI is not available for other plotting backends. Geemap provides a unified API that makes it very easy to switch from one plotting backend to another. To choose a specific plotting backend, use one of the following:

- `import geemap.geemap as geemap`
- `import geemap.foliumap as geemap`
- `import geemap.deck as geemap`
- `import geemap.kepler as geemap`
- `import geemap.plotlymap as geemap`

Ipyleaflet

You can simply use `geemap.Map()` to create an interactive map with the default settings. First, let's import the `geemap` package:

```
import geemap
```

Next, create an interactive map using the ipyleaflet plotting backend. The geemap.Map[44] class inherits the ipyleaflet.Map[45] class. Therefore, you can use the same syntax to create an interactive map as you would with `ipyleaflet.Map`.

---

[41]plotly: plotly.com

[42]pydeck: tiny.geemap.org/ijvg

[43]kepler.gl: tiny.geemap.org/oati

[44]geemap.Map: tiny.geemap.org/bhgy

[45]ipyleaflet.Map: tiny.geemap.org/jhjy

```
Map = geemap.Map()
```

This code creates a new map and assigns it to a new variable named `Map`. To display the map in a Jupyter notebook, simply type the variable name:

```
Map
```

Keep in mind that throughout this book, `Map` is commonly used to refer to the interactive map. It is just a variable name. You can use whatever name you want (e.g., `m`) as long as it complies with the following Python variable names rules:

- A variable name must start with a letter or the underscore character
- A variable name cannot start with a number
- A variable name can only contain alphanumeric characters and underscores (A-z, 0-9, and _)

In general, a Python variable name should be lowercase. The reason we use `Map` rather than `m` is because in the Earth Engine JavaScript API, `Map` is a reserved keyword referring to the interactive map. We want to be consistent with the Earth Engine JavaScript API so that users can have an easier transition to geemap. Users are by no means required to use `Map` as the variable name for the interactive map.

To customize the map, you can specify various keyword arguments, such as `center` ([lat, lon]), `zoom`, `width`, and `height`. The default `width` is `100%`, which takes up the entire cell width of the Jupyter notebook. The `height` argument accepts a number or a string. If a number is provided, it represents the height of the map in pixels. If a string is provided, the string must be in the format of a number followed by `px`, e.g., `600px`.

```
Map = geemap.Map(center=[40, -100], zoom=4, height=600)
Map
```

The default map comes with all the following controls (see Fig. 2.1):

- `attribution_ctrl`
- `data_ctrl`
- `draw_ctrl`
- `fullscreen_ctrl`
- `layer_ctrl`
- `measure_ctrl`
- `scale_ctrl`
- `toolbar_ctrl`
- `zoom_ctrl`

To hide a control, set `control_name` to `False`, e.g., `draw_ctrl=False`.

```
Map = geemap.Map(data_ctrl=False, toolbar_ctrl=False, draw_ctrl=False)
Map
```

You can also set `lite_mode=True` to show only the Zoom Control.

```
Map = geemap.Map(lite_mode=True)
Map
```

To save the map as an HTML file you can call the `save` method:

```
Map.save('ipyleaflet.html')
```

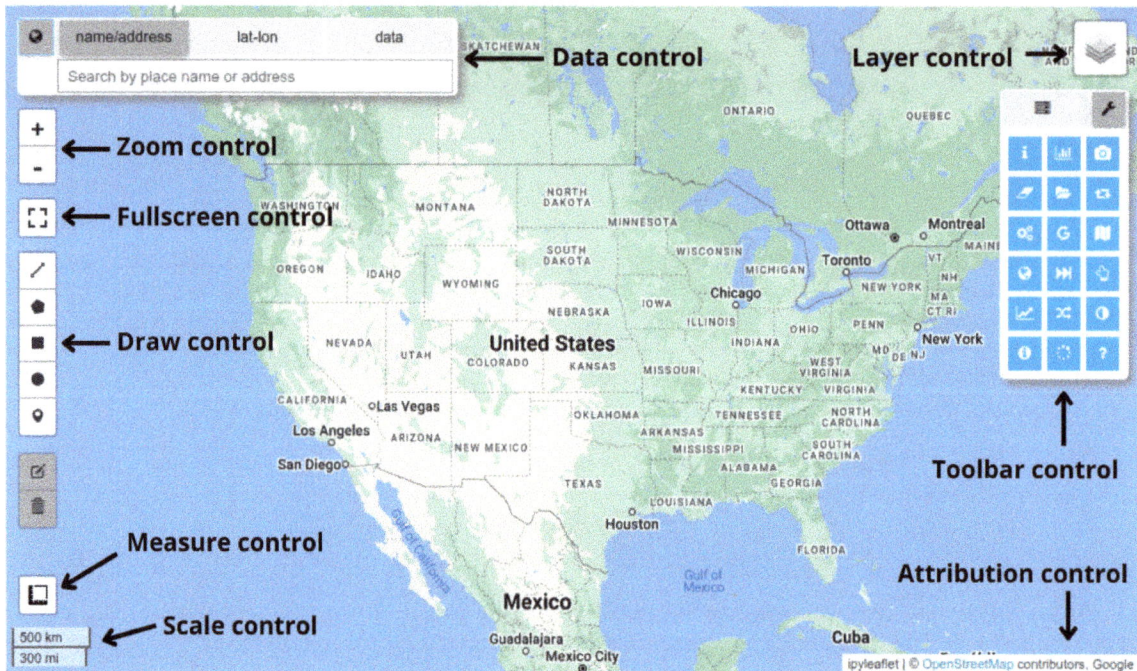

Figure 2.1: A map created using the ipyleaflet plotting backend.

Folium

To create an interactive map using the folium plotting backend, simply import the library as follows:

```
import geemap.foliumap as geemap
```

Then, you can use the same line of code to create and display an interactive map (Fig. 2.2) as you would with the ipyleaflet plotting backend introduced above.

```
Map = geemap.Map(center=[40, -100], zoom=4, height=600)
Map
```

Folium does not support bidirectional communication (QuantStack, 2019). Once the map is created and displayed in a notebook cell, you can't modify the map's properties, e.g., add/remove layers, add controls, change width/height. Also, the data and toolbar controls available in ipyleaflet (see Fig. 2.1) are not supported in folium as folium does not support ipywidgets.

To save the map as an HTML file we can call the save method:

```
Map.save('folium.html')
```

Plotly

To create an interactive map using the plotly plotting backend, simply import the library as follows:

```
import geemap.plotlymap as geemap
```

Then, create and display an interactive map (Fig. 2.3).

```
Map = geemap.Map()
Map
```

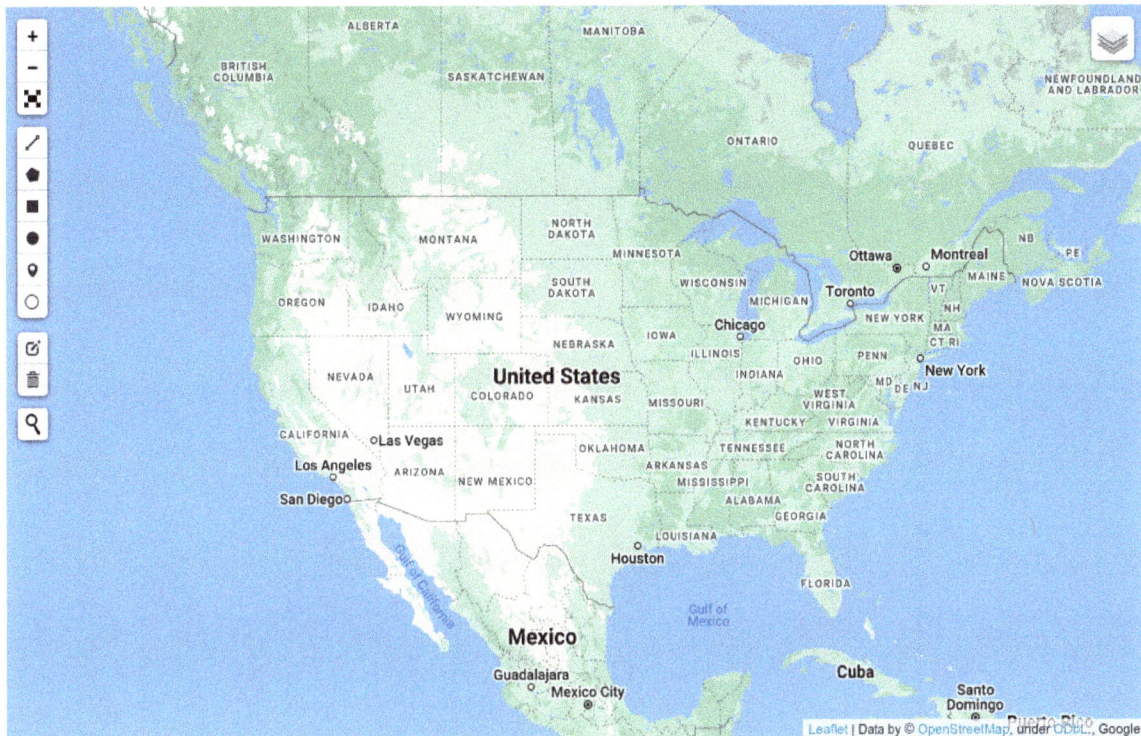

Figure 2.2: A map created using the folium plotting backend.

Note that you might not see the map displayed in Colab as it does not yet support plotly FigureWidget. If you run into an error saying `FigureWidget - 'mapbox._derived' Value Error` (see plotly issue 2570[46]), uncomment the following line and run it.

```
# geemap.fix_widget_error()
```

Pydeck

To create an interactive map using the pydeck plotting backend, simply import the library as follows:

```
import geemap.deck as geemap
```

Then, create and display an interactive map (Fig. 2.4).

```
Map = geemap.Map()
Map
```

KeplerGL

To create an interactive map using the kepler.gl[43] plotting backend, simply import the library as follows:

```
import geemap.kepler as geemap
```

Then, create and display an interactive map (Fig. 2.5).

---

[46]plotly issue 2570: tiny.geemap.org/qqsk

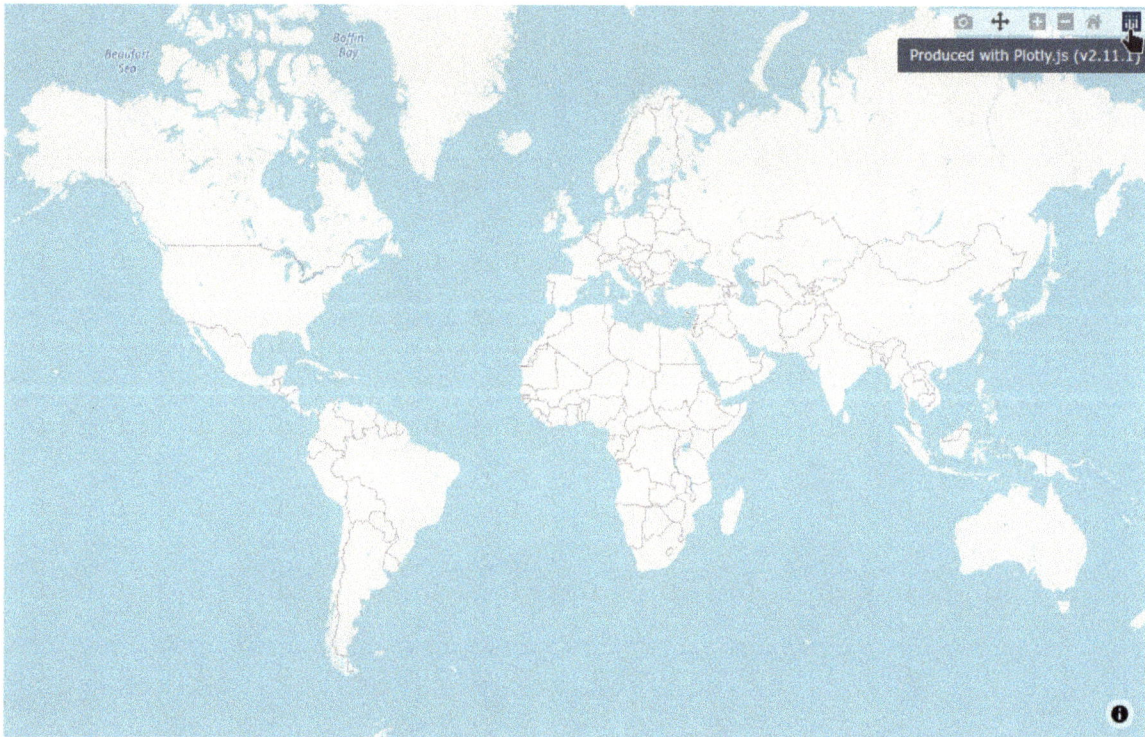

Figure 2.3: A map created using the plotly plotting backend.

Figure 2.4: A map created using the pydeck plotting backend.

```
Map = geemap.Map()
Map
```

Figure 2.5: A map created using the KeplerGL plotting backend.

2.4 Adding basemaps

There are several ways to add basemaps to a map. You can specify the basemap to use in the `basemap` keyword argument when creating the map with the `geemap.Map()` method. Alternatively, you can add basemap layers to the map using the `Map.add_basemap()` method. Geemap has hundreds of built-in basemaps available through xyzservices[47] that can be easily added to the map with only one line of code.

Built-in basemaps

Let's try out some of the built-in basemaps. First, import the geemap library as follows:

```
import geemap
```

Next, create a map by specifying the basemap to use as follows. For example, the `HYBRID` basemap represents the Google Satellite Hybrid basemap (Fig. 2.6).

```
Map = geemap.Map(basemap='HYBRID')
Map
```

You can add as many basemaps as you like to the map. For example, the following code adds the `OpenTopoMap` basemap to the map above (Fig. 2.7):

---

[47]xyzservices: tiny.geemap.org/kqaf

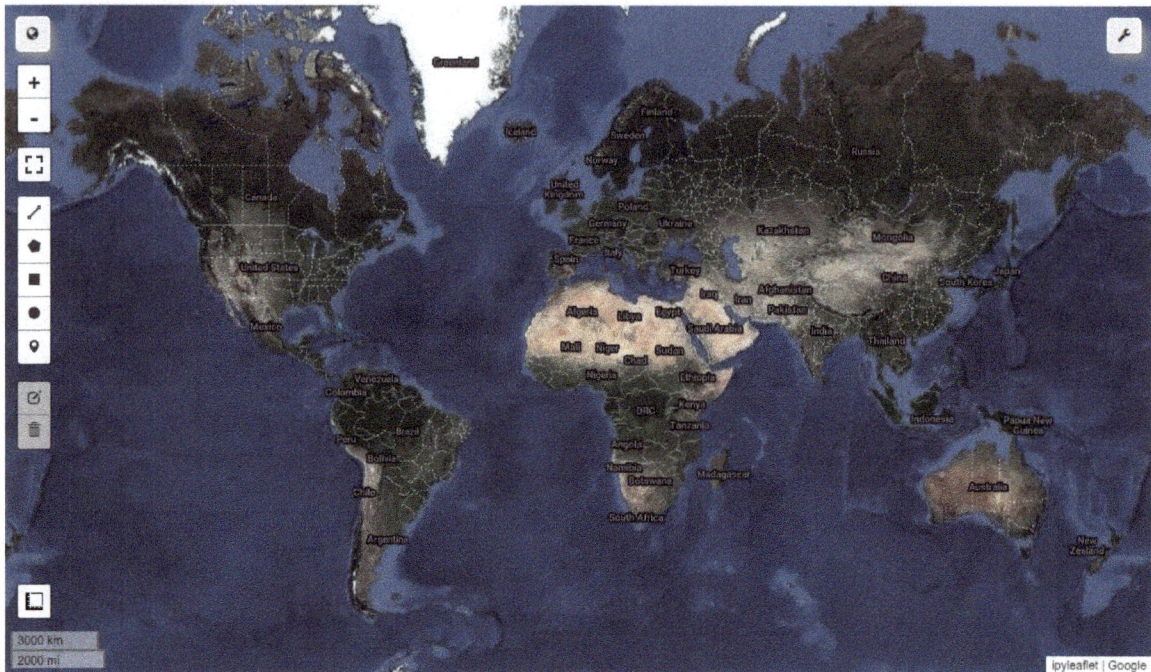

Figure 2.6: The Google Satellite Hybrid basemap.

```
Map.add_basemap('OpenTopoMap')
```

To find out the list of available basemaps:

```
for basemap in geemap.basemaps.keys():
    print(basemap)
```

In total, there are 100+ basemaps available.

```
len(geemap.basemaps)
```

XYZ tiles

You can also add XYZ tile layers to the map using the `Map.add_tile_layer()` method. For example, the following code creates an interactive map and adds the Google Terrain basemap to it:

```
Map = geemap.Map()
Map.add_tile_layer(
    url="https://mt1.google.com/vt/lyrs=p&x={x}&y={y}&z={z}",
    name="Google Terrain",
    attribution="Google",
)
Map
```

WMS tiles

Similarly, you can add WMS tile layers to the map using the `Map.add_wms_layer()` method. For example, the following code creates an interactive map and adds the National Land Cover Database (NLCD) 2019 basemap to it (Fig. 2.8):

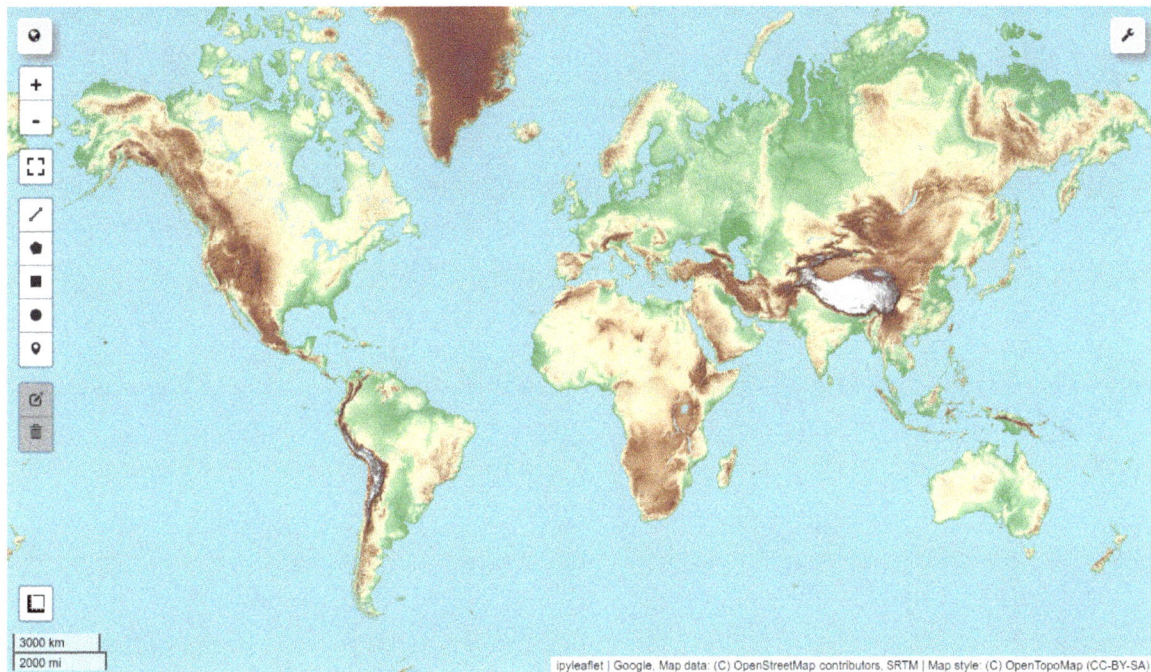

Figure 2.7: The OpenTopoMap basemap.

```
Map = geemap.Map(center=[40, -100], zoom=4)
url = 'https://www.mrlc.gov/geoserver/mrlc_display/NLCD_2019_Land_Cover_L48/wms?'
Map.add_wms_layer(
    url=url,
    layers='NLCD_2019_Land_Cover_L48',
    name='NLCD 2019',
    format='image/png',
    attribution='MRLC',
    transparent=True,
)
Map
```

Want to find out more freely available WMS basemaps? Check out the USGS National Map Services[48]. Once you get a WMS URL, you can follow the example above to add it to the map.

Planet basemaps

Planet Labs[49] provides high-resolution global satellite imagery with a high temporal frequency. The monthly and quarterly global basemaps can be streamed via the XYZ Basemap Tile Service[50] for use in web mapping applications or for visualization purposes. A valid Planet account[51] is required to access the Basemap Tile Service. Once you sign up for a Planet account, you can get your API key by navigating to the Account Settings page[52] as shown in (Fig. 2.9).

---

[48]USGS National Map Services: apps.nationalmap.gov/services

[49]Planet Labs: www.planet.com

[50]XYZ Basemap Tile Service: tiny.geemap.org/skrb

[51]Planet account: tiny.geemap.org/lgen

[52]Account Settings page: tiny.geemap.org/motw

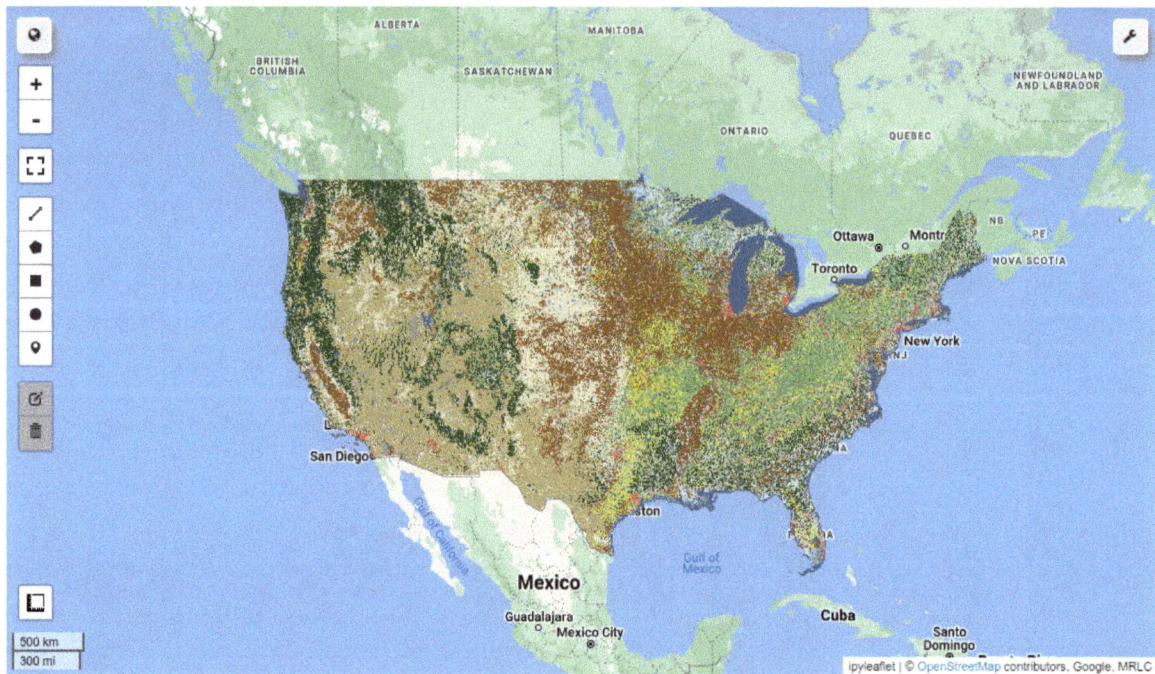

Figure 2.8: The National Land Cover Database (NLCD) 2019 basemap.

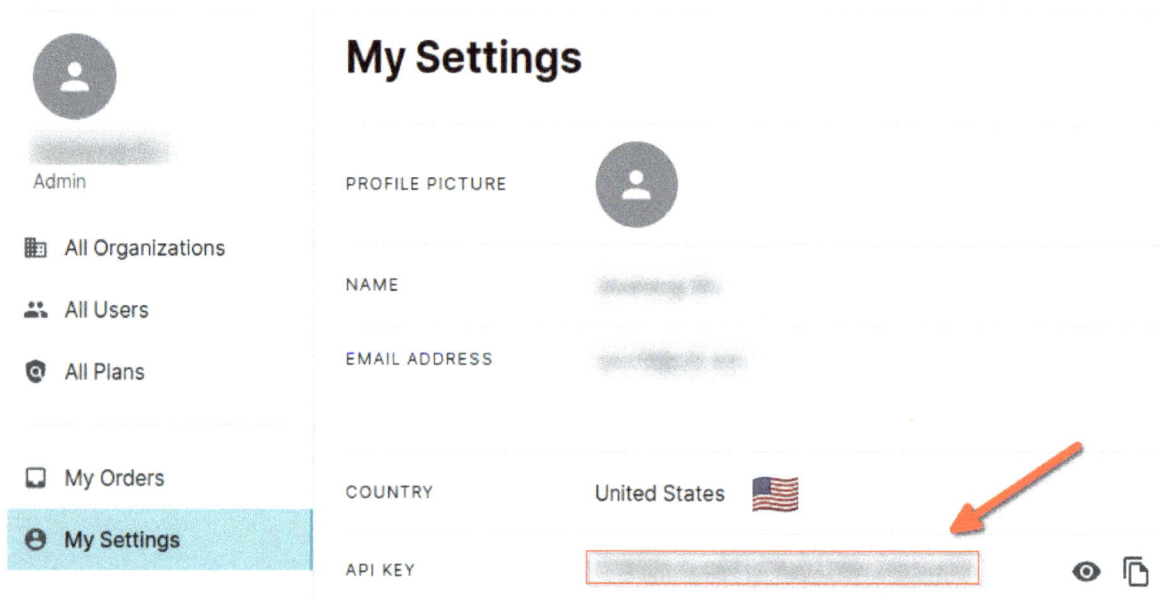

Figure 2.9: Planet account profile page.

Replace `YOUR_API_KEY` below with the API key copied from your Account Settings page shown above. You can create an environment variable called `PLANET_API_KEY` and set it to the API key so that you can access the basemap tiles without having to specify the API key every time.

```
import os

os.environ["PLANET_API_KEY"] = "YOUR_API_KEY"
```

First, let's look at the list of available quarterly basemaps. Planet has quarterly basemaps going back to the first quarter of 2016. The names of the quarterly basemaps are: `Planet_2016q1`, `Planet_2016q2`, ..., `Planet_2022q1`, and so on.

```
quarterly_tiles = geemap.planet_quarterly_tiles()
for tile in quarterly_tiles:
    print(tile)
```

Next, let's look at the list of available monthly basemaps. Planet has monthly basemaps going back to January 2016. The names of the monthly basemaps are: `Planet_2016_01`, `Planet_2016_02`, ..., `Planet_2022_04`, and so on.

```
monthly_tiles = geemap.planet_monthly_tiles()
for tile in monthly_tiles:
    print(tile)
```

To add a monthly basemap to the map, use the `Map.add_planet_by_month()` method as follows:

```
Map = geemap.Map()
Map.add_planet_by_month(year=2020, month=8)
Map
```

To add a quarterly basemap to the map, use the `Map.add_planet_by_quarter()` method as follows:

```
Map = geemap.Map()
Map.add_planet_by_quarter(year=2019, quarter=2)
Map
```

The map should look like Fig. 2.10.

Basemap GUI

Geemap makes it easy for users to add basemaps to their maps without having to write any code. By clicking the "map" icon on the toolbar, the basemap GUI is activated (Fig. 2.11). Simply click the dropdown menu to select the basemap that you want to add. Additionally, if you have created an environment variable called `PLANET_API_KEY` and set it to your Planet API key, the Planet basemaps will also be available to select from the dropdown menu.

To set the Planet API key, use the following code:

```
import os

os.environ["PLANET_API_KEY"] = "YOUR_API_KEY"
```

Then the Planet basemaps will be available via the basemap dropdown menu.

```
Map = geemap.Map()
Map
```

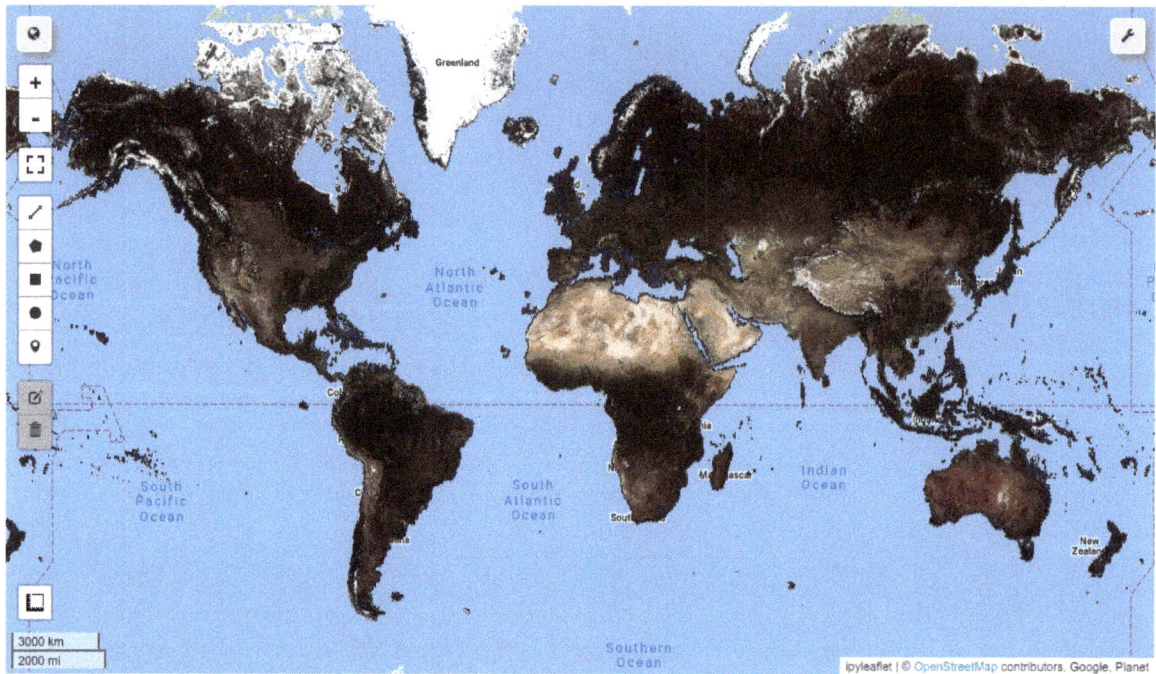

Figure 2.10: The Planet quarterly basemap for second quarter of 2019.

Figure 2.11: The basemap GUI for changing basemaps interactively without coding.

2.5 Summary

In this chapter, we began by introducing the fundamentals of Jupyter notebooks. We then provided hands-on examples of using the five geemap plotting backends to create interactive maps. Additionally, we explored how to incorporate various basemap layers, such as built-in basemaps, XYZ tiles, WMS tiles, and Planet basemaps, into our maps.

By now, you should be proficient in using Jupyter notebook to generate interactive maps and switch between various basemaps seamlessly. That concludes this chapter. In the next chapter, we will delve into Google Earth Engine data.

3. Using Earth Engine Data

3.1 Introduction

Google Earth Engine is a cutting-edge cloud-computing platform that is revolutionizing the field of geospatial analysis. The data types used in Earth Engine differ significantly from traditional data types processed by desktop software.

In this chapter, we will begin by introducing the fundamental Earth Engine data types for storing vector and raster data. We will then dive into the Earth Engine Data Catalog and explore how to search for Earth Engine datasets interactively within a Jupyter environment. Additionally, we will cover how to obtain image metadata and compute descriptive statistics. Lastly, we will teach you how to automatically convert Earth Engine JavaScript code to Python code. By the end of this chapter, you will feel confident in your ability to access Earth Engine data in a Jupyter environment using geemap.

3.2 Technical requirements

To follow along with this chapter, you will need to have geemap and several optional dependencies installed. If you have already followed Section 1.5 - *Installing geemap*, then you should already have a conda environment with all the necessary packages installed. Otherwise, you can create a new conda environment and install pygis[30] with the following commands, which will automatically install geemap and all the required dependencies:

```
conda create -n gee python
conda activate gee
conda install -c conda-forge mamba
mamba install -c conda-forge pygis
```

Next, launch JupyterLab by typing the following commands in your terminal or Anaconda prompt:

```
jupyter lab
```

Alternatively, you can use geemap with a Google Colab cloud environment without installing anything on your local computer. Click 03_gee_data.ipynb[53] to launch the notebook in Google Colab.

Once in Colab, you can uncomment the following line and run the cell to install pygis, which includes geemap and all the necessary dependencies:

```
# %pip install pygis
```

The installation process may take 2-3 minutes. Once pygis has been installed successfully, click the **RESTART RUNTIME** button that appears at the end of the installation log or go to the **Runtime** menu and select **Restart runtime**. After that, you can start coding.

To begin, import the necessary libraries that will be used in this chapter:

```
import ee
import geemap
```

---

[53]03_gee_data.ipynb: tiny.geemap.org/ch03

Initialize the Earth Engine Python API:

```
geemap.ee_initialize()
```

If this is your first time running the code above, you will need to authenticate Earth Engine first. Follow the instructions in Section 1.7 - *Earth Engine authentication* to authenticate Earth Engine.

3.3 Earth Engine data types

Before diving into Earth Engine and using it for geospatial analysis, we first need to understand the data types that are available in Earth Engine. Earth Engine has two fundamental data types: **Image** for raster data (see Fig. 3.1) and **Geometry** for vector data (see Fig. 3.2). The extended data types are **ImageCollection**, **Feature**, and **FeatureCollection**. Keep in mind that Earth Engine objects are server-side objects rather than client-side objects, which means that they are not stored locally on your computer. Similar to video streaming services (e.g., YouTube, Netflix, and Hulu), which store videos/movies on their servers, Earth Engine data is stored on the Earth Engine servers. We can stream geospatial data from Earth Engine on-the-fly without having to download the data, just like how we can watch videos from streaming services using a web browser without having to download the entire video to your computer. This is one of the most powerful features of Earth Engine, which allows us to process massive amounts of geospatial data without having to worry about storage space and computing power. For more information about the differences between client-side and server-side objects, please refer to the Earth Engine documentation - Client vs. Server[54].

The five common Earth Engine data types are:

- **Image**: the fundamental raster data type in Earth Engine.
- **ImageCollection**: a stack or timeseries of images.
- **Geometry**: the fundamental vector data type in Earth Engine.
- **Feature**: a Geometry with attributes.
- **FeatureCollection**: a set of features.

Image

As mentioned above, raster data in Earth Engine are represented as **Image** objects. Images are composed of one or more bands and each band has its own name, data type, scale, mask and projection. Each image has metadata stored as a set of properties.

Loading Earth Engine images

Images can be loaded by passing an Earth Engine asset ID into the `ee.Image` constructor. You can find image IDs in the Earth Engine Data Catalog[55]. For example, to load the NASA SRTM Digital Elevation[56] you can use:

```
image = ee.Image('USGS/SRTMGL1_003')
```

As mentioned above, Earth Engine objects are server-side objects. To inspect the content of an Earth Engine object, simply type its name in a code cell and run the cell. For example, to inspect the image we just loaded:

```
image
```

---

[54]Client vs. Server: tiny.geemap.org/wvez

[55]Earth Engine Data Catalog: tiny.geemap.org/qcsz

[56]NASA SRTM Digital Elevation: tiny.geemap.org/krif

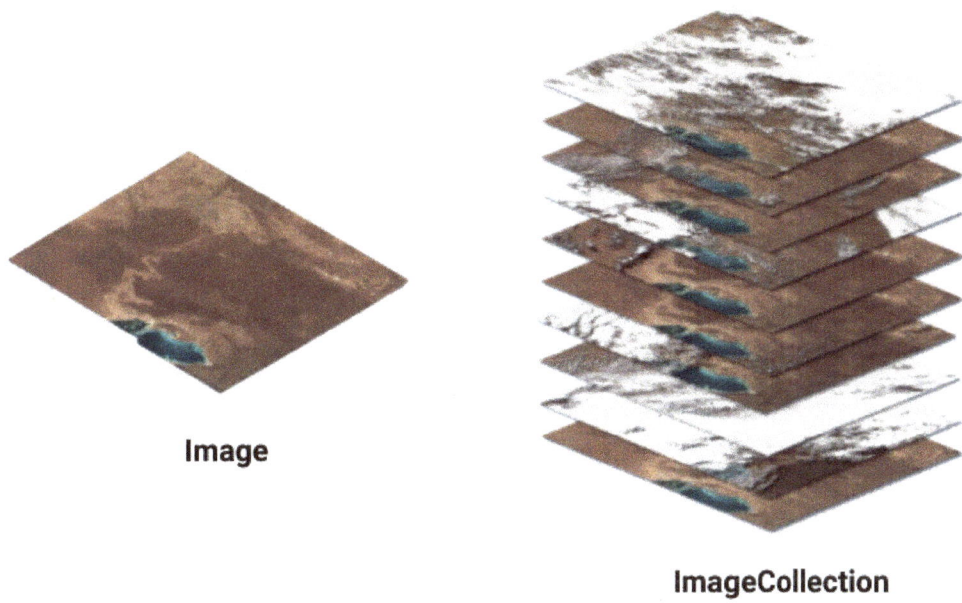

Image

ImageCollection

Figure 3.1: Earth Engine raster data types - Image and ImageCollection.

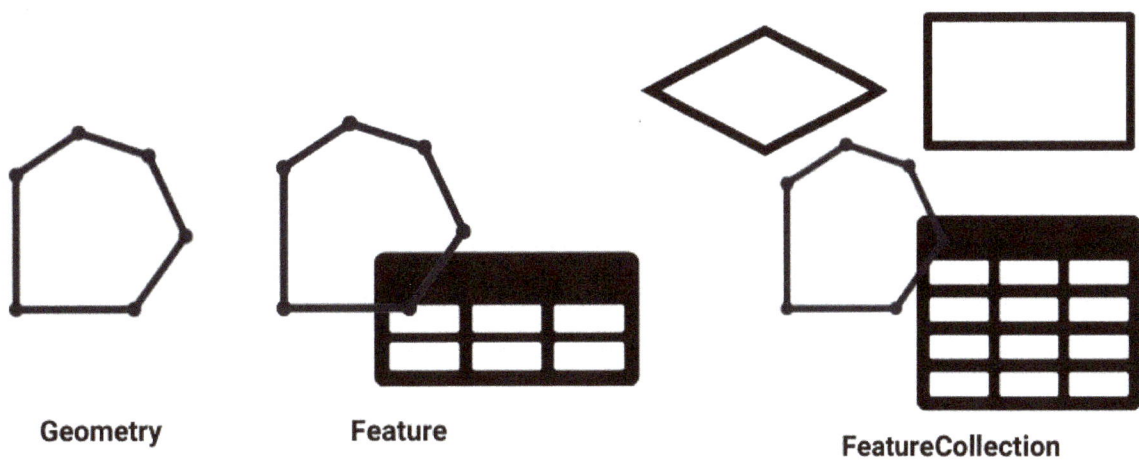

Geometry   **Feature**

FeatureCollection

Figure 3.2: Earth Engine vector data types - Geometry, Feature, and FeatureCollection.

The output should look like Fig. 3.3.

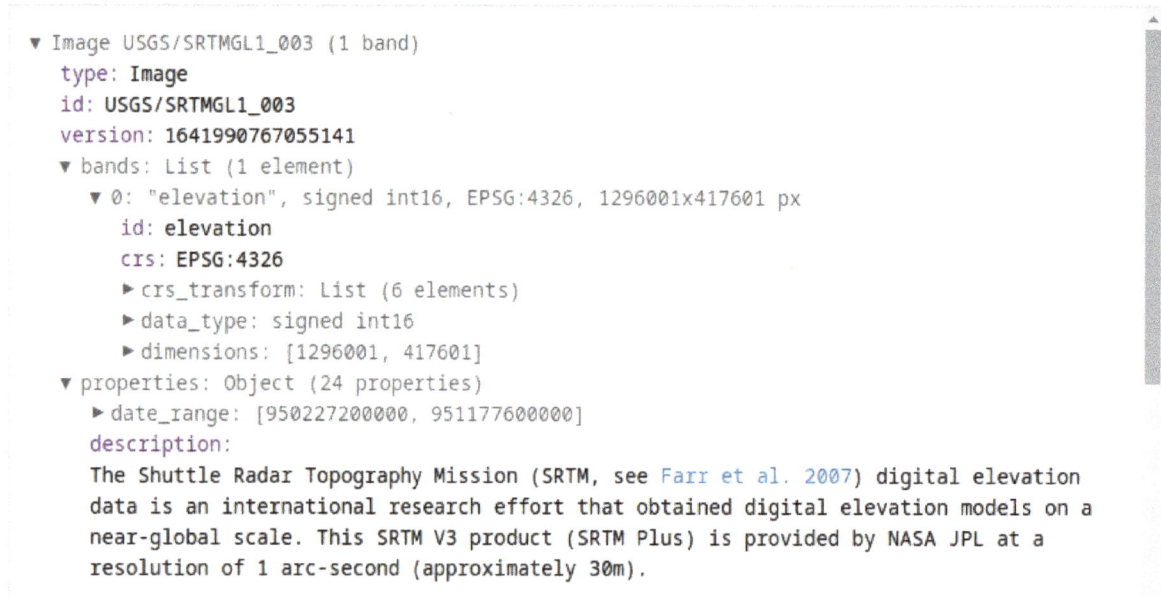

```
▼ Image USGS/SRTMGL1_003 (1 band)
    type: Image
    id: USGS/SRTMGL1_003
    version: 1641990767055141
  ▼ bands: List (1 element)
      ▼ 0: "elevation", signed int16, EPSG:4326, 1296001x417601 px
          id: elevation
          crs: EPSG:4326
        ▶ crs_transform: List (6 elements)
        ▶ data_type: signed int16
        ▶ dimensions: [1296001, 417601]
  ▼ properties: Object (24 properties)
      ▶ date_range: [950227200000, 951177600000]
        description:
        The Shuttle Radar Topography Mission (SRTM, see Farr et al. 2007) digital elevation
        data is an international research effort that obtained digital elevation models on a
        near-global scale. This SRTM V3 product (SRTM Plus) is provided by NASA JPL at a
        resolution of 1 arc-second (approximately 30m).
```

Figure 3.3: Inspecting the metadata of an Earth Engine Image.

The output is a collapsible JSON object, which does not take up much space. You can click on the arrow to expand it. Alternatively, you can directly print any Earth Engine object using the .getInfo() method. However, the output is not a collapsible JSON object. Therefore, the output may be very long and overwhelming. For example:

```
image.getInfo()
```

If you need to inspect the content of an Earth Engine, simply type its name in a code cell and run the cell. Avoid using the .getInfo() method unless you need to convert an Earth Engine object to a client-side object and manipulate its content.

Visualizing Earth Engine images

To visualize an Earth Engine Image, you can use geemap's Map.addLayer()[57] method, similar to the Map.addLayer() method for the JavaScript Code Editor. The Map.addLayer() method accepts parameters as follows:

```
Map.addLayer(ee_object, vis_params={}, name=None, shown=True, opacity=1.0)
```

The description of each parameter for the Map.addLayer() method can be found below:

Parameter	Type	Description	Default
ee_object	Image, ImageCollection, Geometry, Feature, FeatureCollection	The object to add to the map.	required
vis_params	dict	The visualization parameters as a dictionary.	{}

...continued on next page

---

[57]Map.addLayer(): tiny.geemap.org/savn

Parameter	Type	Description	Default
name	str	The name of the layer. Defaults to Layer N.	None
shown	bool	A flag indicating whether the layer should be on by default.	True
opacity	float	The opacity of the layer (0.0 is fully transparent and 1.0 is fully opaque)	1

To change the map visualization effects, you can customize visualization parameters[58] :

Parameter	Description	Type
bands	Comma-delimited list of three band names to be mapped to RGB	list
min	Value(s) to map to 0	number or list of three numbers, one for each band
max	Value(s) to map to 255	number or list of three numbers, one for each band
gain	Value(s) by which to multiply each pixel value	number or list of three numbers, one for each band
bias	Value(s) to add to each pixel	number or list of three numbers, one for each band
gamma	Gamma correction factor(s)	number or list of three numbers, one for each band
palette	List of CSS-style color strings (single-band images only)	comma-separated list of hex strings
opacity	The opacity of the layer (0.0 is fully transparent and 1.0 is fully opaque)	number
format	Either "jpg" or "png"	string

For example, to visualize the NASA SRTM Digital Elevation[56] dataset:

```
Map = geemap.Map(center=[21.79, 70.87], zoom=3)
image = ee.Image('USGS/SRTMGL1_003')
vis_params = {
    'min': 0,
    'max': 6000,
    'palette': ['006633', 'E5FFCC', '662A00', 'D8D8D8', 'F5F5F5'],
}
Map.addLayer(image, vis_params, 'SRTM')
Map
```

The resulting map should look like Fig. 3.4.

Loading Cloud GeoTIFFs

You can use ee.Image.loadGeoTIFF() or geemap.load_GeoTIFF() to load Cloud Optimized GeoTIFFs (COG[59]) hosted in Google Cloud Storage[60] with a URL prefix of gs:// or https://storage.googleapis.com.

---

[58]visualization parameters: tiny.geemap.org/cdqw

[59]COG: www.cogeo.org

[60]Google Cloud Storage: cloud.google.com/storage

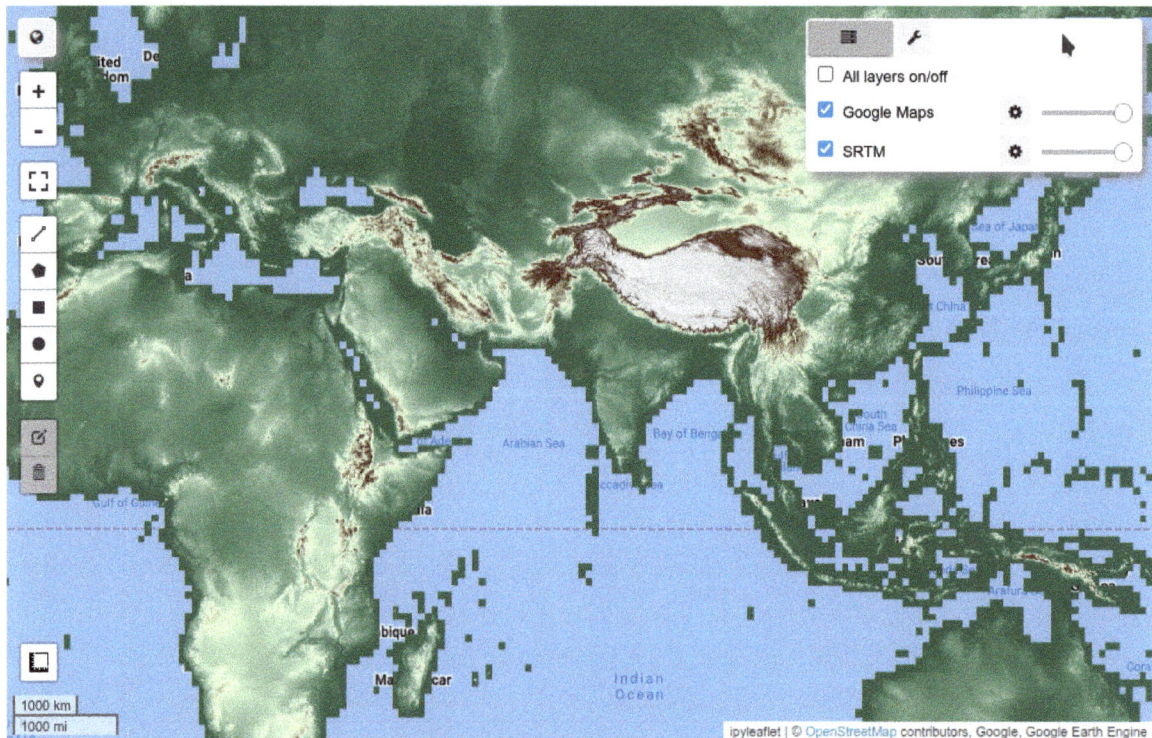

Figure 3.4: Visualizing the NASA SRTM digital elevation data.

For example, the Planet Disaster Data Catalog[61] hosted in Google Cloud Storage (GCS) contains this GeoTIFF[62], which is a 4-band multispectral image of Hurricane Harvey[63] formed in the Atlantic Ocean in August 2017. You can load this image from GCS using geemap.loadGeoTIFF():

```
Map = geemap.Map()
URL = 'https://bit.ly/3aSZ0fH'
image = geemap.load_GeoTIFF(URL)
vis = {
    "min": 3000,
    "max": 13500,
    "bands": ["B3", "B2", "B1"],
}
Map.addLayer(image, vis, 'Cloud GeoTIFF')
Map.centerObject(image)
Map
```

The resulting map should look like Fig. 3.5.

Sometimes, each band of a COG image is stored in a separate file. In this case, you can use geemap.load_GeoTIFFs() to load a list of COG files as a single image.

For example, Google Cloud[64] hosts Landsat data. Each band of a Landsat scene is stored in a separate file. The following example shows how to load three bands of a Landsat scene[65] as an Image. First,

---

[61]Planet Disaster Data Catalog: tiny.geemap.org/uwdn

[62]this GeoTIFF: bit.ly/3aSZ0fH

[63]Hurricane Harvey: tiny.geemap.org/abro

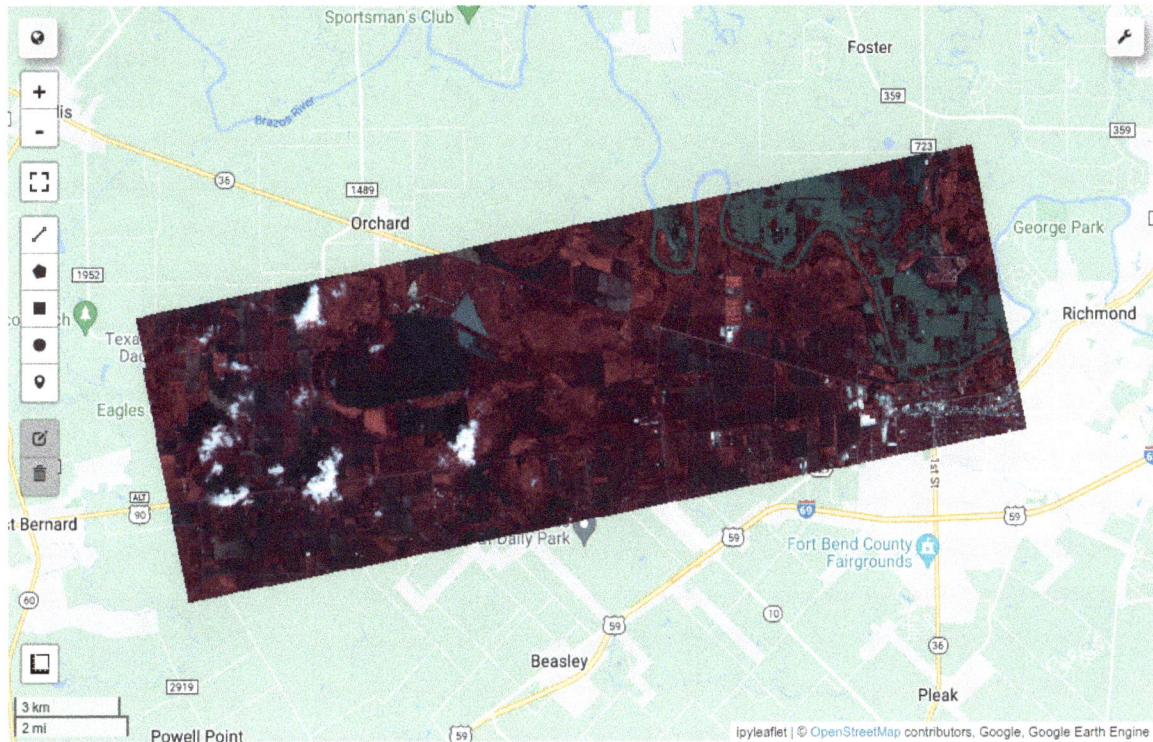

Figure 3.5: Visualizing a Cloud Optimized GeoTIFF (COG) hosted in Google Cloud Storage.

obtain the URL to each band and store it in a variable, respectively:

```
B3 = 'gs://gcp-public-data-landsat/LC08/01/044/034/LC08_L1TP_044034_20131228_20170307_01_T1/
      LC08_L1TP_044034_20131228_20170307_01_T1_B3.TIF'
B4 = 'gs://gcp-public-data-landsat/LC08/01/044/034/LC08_L1TP_044034_20131228_20170307_01_T1/
      LC08_L1TP_044034_20131228_20170307_01_T1_B4.TIF'
B5 = 'gs://gcp-public-data-landsat/LC08/01/044/034/LC08_L1TP_044034_20131228_20170307_01_T1/
      LC08_L1TP_044034_20131228_20170307_01_T1_B5.TIF'
```

Next, put the three variables containing the bands in a list and use `geemap.load_GeoTIFFs()` to load the three bands as an `ImageCollection`, which can be further converted to an `Image` using collection.toBands(). Use the `rename()` function to rename the bands to `Green`, `Red`, and `NIR`. The `selfMask()` function is used to mask out the black nodata area in the image.

```
URLs = [B3, B4, B5]
collection = geemap.load_GeoTIFFs(URLs)
image = collection.toBands().rename(['Green', 'Red', 'NIR']).selfMask()
```

Lastly, add the image to the map and center the map around the image (see Fig. 3.6).

```
Map = geemap.Map()
vis = {'bands': ['NIR', 'Red', 'Green'], 'min': 100, 'max': 12000, 'gamma': 0.8}
Map.addLayer(image, vis, 'Image')
Map.centerObject(image, 8)
Map
```

---

[64]Google Cloud: tiny.geemap.org/yyxb

[65]Landsat scene: tiny.geemap.org/tsrs

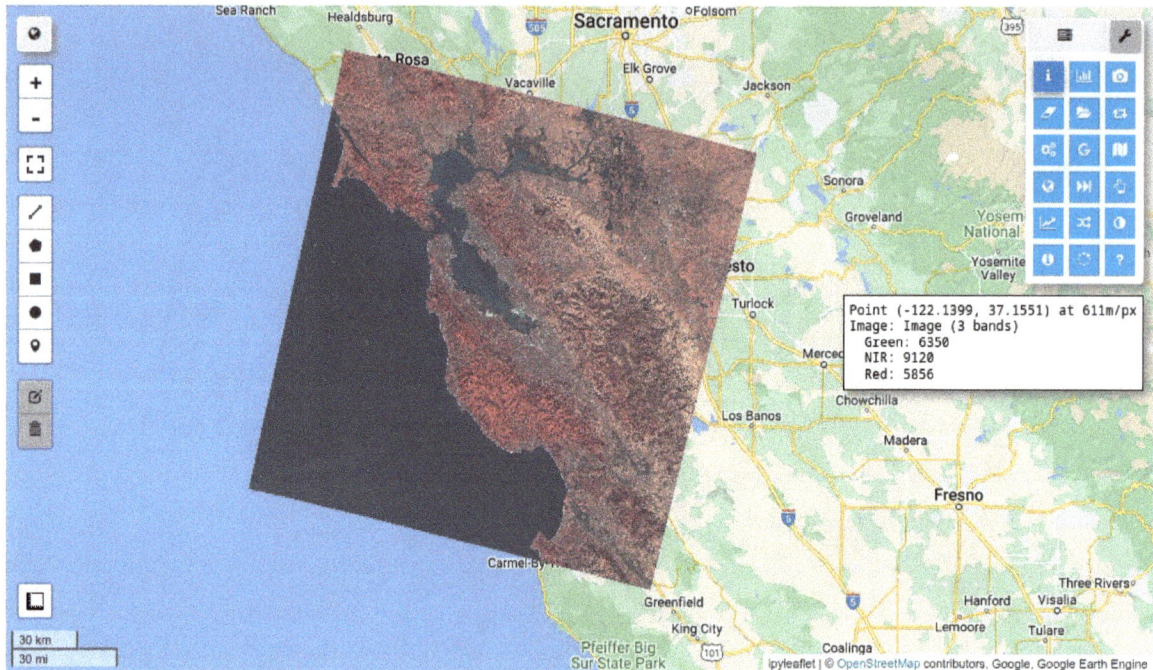

Figure 3.6: Loading a list of Cloud Optimized GeoTIFFs from Google Cloud Storage as an ee.Image.

ImageCollection

Loading image collections

An `ImageCollection` is a stack or sequence of images. An `ImageCollection` can be loaded by passing an Earth Engine asset ID into the `ImageCollection` constructor. You can find the image collection IDs in the Earth Engine Data Catalog[55]. For example, to load the Sentinel-2 surface reflectance[66] image collection:

```
collection = ee.ImageCollection('COPERNICUS/S2_SR')
```

The Sentinel-2 image collection contains millions of images and is too large to be printed. We can inspect the first few images in the collection using the `collection.limit()` method. For example, to inspect the first five images in the collection:

```
collection.limit(5)
```

The output should look like Fig. 3.7.

Visualizing image collections

To visualize an Earth Engine **ImageCollection**, we need to convert an **ImageCollection** to an **Image** by compositing all the images in the collection into a single image representing, for example, the min, max, median, mean or standard deviation of the images. For example, to create a median value image from a collection, use the `collection.median()` method. Let's create a median image from the Sentinel-2 surface reflectance image collection:

---

[66]Sentinel-2 surface reflectance: tiny.geemap.org/xqjn

```
▼ ImageCollection COPERNICUS/S2_SR (5 elements)
  type: ImageCollection
  id: COPERNICUS/S2_SR
  version: 1689560910728390.0
  ▶ bands: []
  ▶ properties: Object (21 properties)
  ▼ features: List (5 elements)
    ▶ 0: Image COPERNICUS/S2_SR/20170328T083601_20170328T084228_T35RNK (23 bands)
    ▶ 1: Image COPERNICUS/S2_SR/20170328T083601_20170328T084228_T35RNL (23 bands)
    ▶ 2: Image COPERNICUS/S2_SR/20170328T083601_20170328T084228_T35RNM (23 bands)
    ▶ 3: Image COPERNICUS/S2_SR/20170328T083601_20170328T084228_T35RNN (23 bands)
    ▼ 4: Image COPERNICUS/S2_SR/20170328T083601_20170328T084228_T35RNP (23 bands)
        type: Image
        id: COPERNICUS/S2_SR/20170328T083601_20170328T084228_T35RNP
        version: 1645343020123264
      ▶ bands: List (23 elements)
      ▶ properties: Object (82 properties)
```

Figure 3.7: Inspecting the first five images in an Earth Engine ImageCollection.

```
Map = geemap.Map()
collection = ee.ImageCollection('COPERNICUS/S2_SR')
image = collection.median()

vis = {
    'min': 0.0,
    'max': 3000,
    'bands': ['B4', 'B3', 'B2'],
}

Map.setCenter(83.277, 17.7009, 12)
Map.addLayer(image, vis, 'Sentinel-2')
Map
```

The resulting map should look like Fig. 3.8.

Filtering image collections

An **ImageCollection** might contain thousands or even millions of images, which might take a while to process and visualize. For example, the Sentinel-2 surface reflectance collection[67] we used earlier contains over 14 million images.

To reduce the number of images to be processed, we can filter the collection so that only a smaller number of images is processed. Earth Engine provides a variety of convenience methods for filtering image collections. Specifically, many common use cases are handled by `collection.filterDate()`, and `collection.filterBounds()`. For general purpose filtering, use `collection.filter()` with an `ee.Filter` as an argument.

The following example demonstrates both convenience methods and the `filter()` method to identify and remove images with high cloud cover from an **ImageCollection**. Specifically, we select all images acquired in 2021 and then filter out images with cloud cover greater than 5% and remove them from the collection:

---

[67]Sentinel-2 surface reflectance collection: tiny.geemap.org/fpad

Figure 3.8: Visualizing an Earth Engine ImageCollection.

```python
Map = geemap.Map()
collection = (
    ee.ImageCollection('COPERNICUS/S2_SR')
    .filterDate('2021-01-01', '2022-01-01')
    .filter(ee.Filter.lt('CLOUDY_PIXEL_PERCENTAGE', 5))
)
image = collection.median()

vis = {
    'min': 0.0,
    'max': 3000,
    'bands': ['B4', 'B3', 'B2'],
}

Map.setCenter(83.277, 17.7009, 12)
Map.addLayer(image, vis, 'Sentinel-2')
Map
```

Since only images with cloud cover less than 5% are selected to create the image composite, the resulting image (Fig. 3.9) looks better than the image without filtering shown in Fig. 3.8.

Geometry

Geometry types

Earth Engine handles vector data with the **Geometry** type. The types of geometries supported by Earth Engine are:

Figure 3.9: Filtering an Earth Engine ImageCollection.

- **Point**: A list of two [x,y] coordinates in a given projection.
- **LineString**: A list of at least two points.
- **LinearRing**: A list of points in the ring, i.e., a closed LineString.
- **Polygon**: A list of rings defining the boundaries of a polygon.
- **Rectangle**: The minimum and maximum corners of the rectangle as a list of two points.
- **BBox**: A bounding box in the format of (west, south, east, north).
- **MultiPoint**: A collection of points.
- **MultiLineString**: A collection of lines.
- **MultiPolygon**: A collection of polygons.

Creating Geometry objects

To create a **Geometry** object, use the `ee.Geometry(coords, proj, geodesic)` constructor. For example:

```
Map = geemap.Map()

point = ee.Geometry.Point([1.5, 1.5])

lineString = ee.Geometry.LineString([[-35, -10], [35, -10], [35, 10], [-35, 10]])

linearRing = ee.Geometry.LinearRing(
    [[-35, -10], [35, -10], [35, 10], [-35, 10], [-35, -10]]
)

rectangle = ee.Geometry.Rectangle([-40, -20, 40, 20])

polygon = ee.Geometry.Polygon([[[-5, 40], [65, 40], [65, 60], [-5, 60], [-5, 60]]])

Map.addLayer(point, {}, 'Point')
```

```
Map.addLayer(lineString, {}, 'LineString')
Map.addLayer(linearRing, {}, 'LinearRing')
Map.addLayer(rectangle, {}, 'Rectangle')
Map.addLayer(polygon, {}, 'Polygon')
Map
```

The resulting map should look like Fig. 3.10.

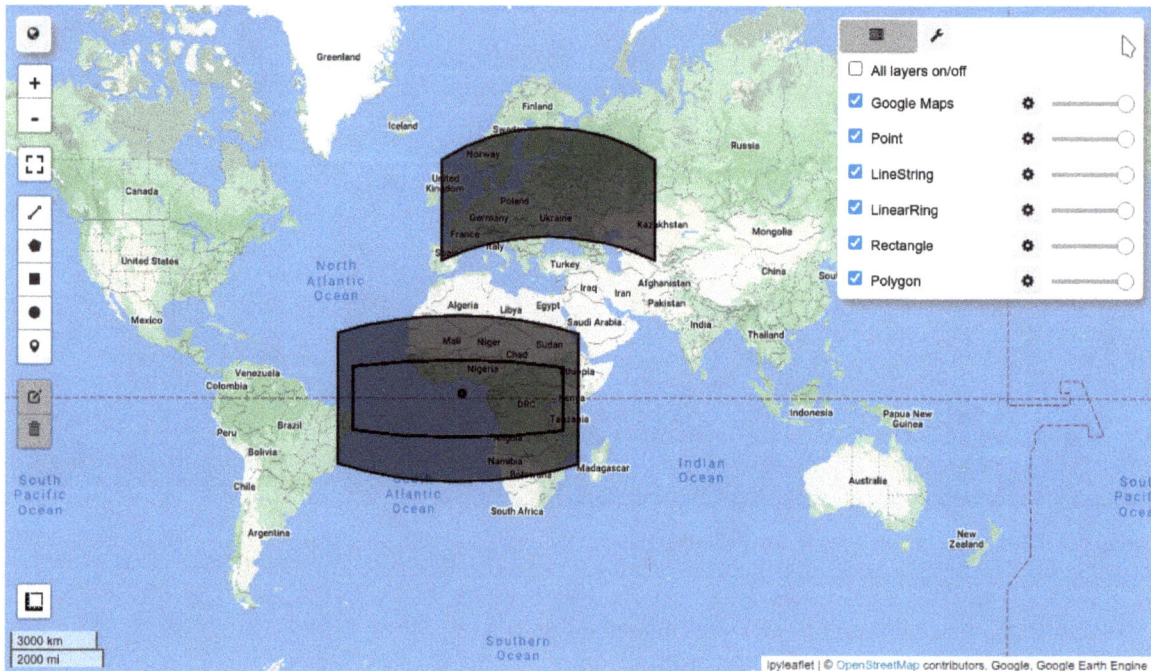

Figure 3.10: Geodesic geometries in Earth Engine.

Keep in mind that all Geometry constructors (except **Point** and **MultiPoint**) have a geodesic parameter, which is set to True by default. Geometry created in Earth Engine is either geodesic (i.e. edges are the shortest path on the surface of a sphere) or planar (i.e. edges are the shortest path in a 2-D Cartesian plane). To create a planar geometry, you need to set the geodesic parameter to False. Since the proj parameter comes before the geodesic parameter, you need to set the proj parameter to None (i.e., 'EPSG:4326') when creating a planar geometry. The following example demonstrates how to create different types of planar geometries:

```
Map = geemap.Map()

point = ee.Geometry.Point([1.5, 1.5])

lineString = ee.Geometry.LineString(
    [[-35, -10], [35, -10], [35, 10], [-35, 10]], None, False
)

linearRing = ee.Geometry.LinearRing(
    [[-35, -10], [35, -10], [35, 10], [-35, 10], [-35, -10]], None, False
)

rectangle = ee.Geometry.Rectangle([-40, -20, 40, 20], None, False)

polygon = ee.Geometry.Polygon(
    [[[-5, 40], [65, 40], [65, 60], [-5, 60], [-5, 60]]], None, False
)
```

```
Map.addLayer(point, {}, 'Point')
Map.addLayer(lineString, {}, 'LineString')
Map.addLayer(linearRing, {}, 'LinearRing')
Map.addLayer(rectangle, {}, 'Rectangle')
Map.addLayer(polygon, {}, 'Polygon')
Map
```

The resulting map should look like Fig. 3.11.

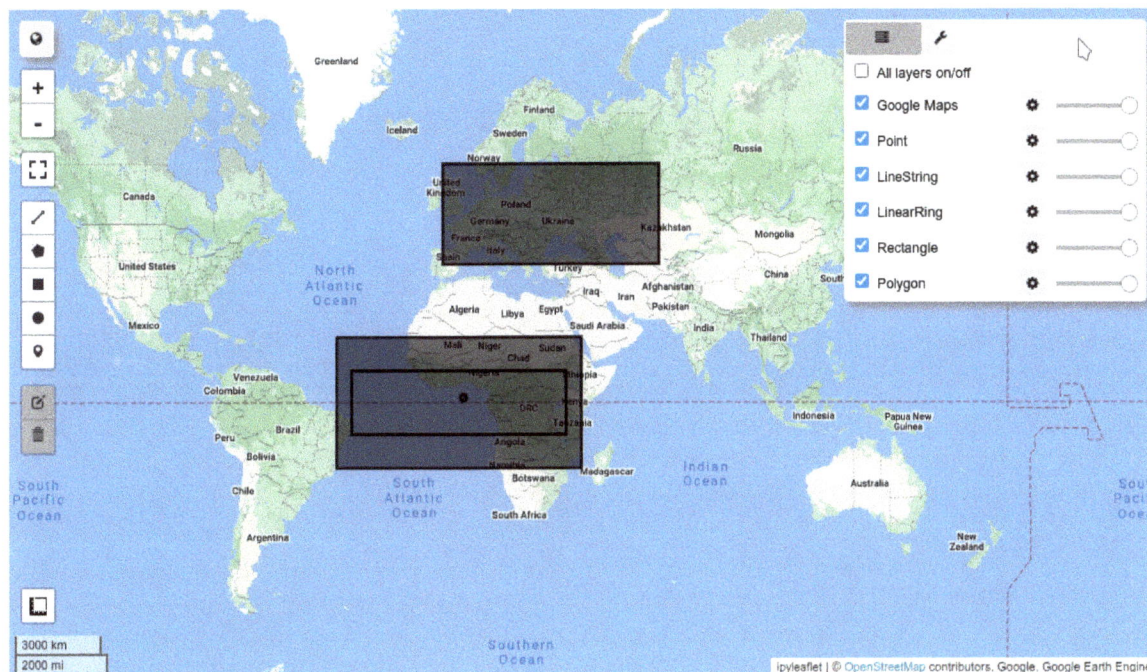

Figure 3.11: Planar geometries in Earth Engine.

Using drawing tools

To create a **Geometry** object interactively, you can use the drawing tools located on the left side of the map. The supported **Geometry** types include Point, LineString, LinearRing, Polygon, and Rectangle. Click on one of the shape icons on the drawing toolbar to create a new geometry (Fig. 3.12).

Note that the draw circle tool has not worked properly since ipyleaflet v0.14.0 (see ipyleaflet issues 876[68]). If you want to use the draw circle tool, please downgrade to ipyleaflet v0.13.3 using `pip install ipyleaflet==0.13.3`.

To retrieve the last drawn geometry object, use `Map.user_roi`, which returns an `ee.Geometry` object. Note that an `ee.Geometry` object is a server-side object. To convert it to a client-side object, use the `.getInfo()` method. For example:

```
if Map.user_roi is not None:
    print(Map.user_roi.getInfo())
```

The output should like this:

---

[68]ipyleaflet issues 876: tiny.geemap.org/iusi

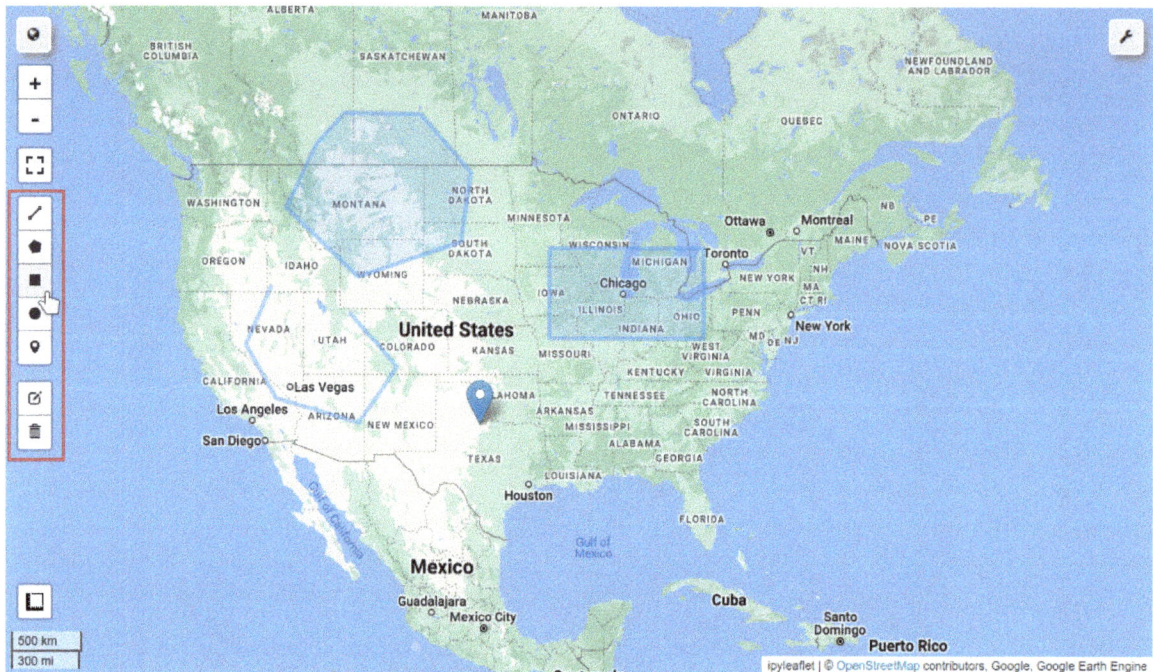

Figure 3.12: Creating geometry interactively with the drawing tools.

```
{'geodesic': False,
 'type': 'Polygon',
 'coordinates': [[[-110.603139, 35.436058],
   [-110.603139, 44.192051],
   [-89.148105, 44.192051],
   [-89.148105, 35.436058],
   [-110.603139, 35.436058]]]}
```

Alternatively, you can directly print any Earth Engine object without using the `.getInfo()` method. For example:

```
if Map.user_roi is not None:
    print(Map.user_roi)
```

The output is a collapsible JSON object, which is not very readable. You can click on the arrow to expand it.

Feature

A **Feature** is an object with a `geometry` property storing a **Geometry** object (or None) and a `properties` property storing a dictionary of other properties. An Earth Engine **Feature** object is analogous to a GeoJSON Feature object, i.e., a geometry with associated properties/attributes.

Creating Feature objects

To create a **Feature**, you can provide the constructor with a **Geometry**, and optionally a dictionary of other properties. For example:

```
polygon = ee.Geometry.Polygon(
    [[[-35, -10], [35, -10], [35, 10], [-35, 10], [-35, -10]]], None, False
```

```
    )
    polyFeature = ee.Feature(polygon, {'foo': 42, 'bar': 'tart'})
```

Similar to a **Geometry** object, a **Feature** object may be printed or added to the map for inspection and visualization. To inspect a **Feature** object, simply type its name in a code cell and run the cell. For example:

```
    polyFeature
```

The output should look like this:

```
    {'type': 'Feature',
     'geometry': {'geodesic': False,
      'type': 'Polygon',
      'coordinates': [[[-35, -10], [35, -10], [35, 10], [-35, 10], [-35, -10]]]},
     'properties': {'bar': 'tart', 'foo': 42}}
```

To visualize a **Feature** object, you can display it on an interactive map. For example:

```
    Map = geemap.Map()
    Map.addLayer(polyFeature, {}, 'feature')
    Map
```

Note that a **Geometry** is not required to create a **Feature** object. A **Feature** may simply wrap a dictionary of properties. For example:

```
    props = {'foo': ee.Number(8).add(88), 'bar': 'hello'}
    nowhereFeature = ee.Feature(None, props)
    nowhereFeature
```

The output should look like this:

```
    {'type': 'Feature',
     'geometry': None,
     'properties': {'bar': 'hello', 'foo': 96}}
```

Setting Feature properties

Besides setting properties at object creation time as shown above, you can also set properties after creation using the `feature.set()` method. Properties can be set with either a key/value pair, or with a Python dictionary. For example:

```
    feature = (
        ee.Feature(ee.Geometry.Point([-122.22599, 37.17605]))
        .set('genus', 'Sequoia')
        .set('species', 'sempervirens')
    )
    newDict = {'genus': 'Brachyramphus', 'presence': 1, 'species': 'marmoratus'}
    feature = feature.set(newDict)
    feature
```

The output should look like this:

```
    {'type': 'Feature',
     'geometry': {'type': 'Point', 'coordinates': [-122.22599, 37.17605]},
     'properties': {'genus': 'Brachyramphus', 'presence': 1, 'species': 'marmoratus'}}
```

Getting Feature properties

To retrieve a single property from a **Feature** object, use the `feature.get()` method. For example:

```
prop = feature.get('species')
prop
```

To retrieve all properties on a **Feature** all at once, use the `feature.toDictionary()` method. For example:

```
props = feature.toDictionary()
props
```

The output should look like this:

```
{'genus': 'Brachyramphus', 'presence': 1, 'species': 'marmoratus'}
```

FeatureCollection

A **FeatureCollection** is a collection of Features. A FeatureCollection is analogous to a GeoJSON FeatureCollection object which consists of a collection of geometries with associated properties/attributes. Data contained in a shapefile can be represented as a FeatureCollection.

Loading feature collections

The Earth Engine Data Catalog[55] hosts a variety of vector datasets (e.g., US Census data, country boundaries, and more) as feature collections. You can find feature collection IDs by searching the Earth Engine Data Catalog. For example, to load the TIGER roads data[69] developed by the U.S. Census Bureau:

```
Map = geemap.Map()
fc = ee.FeatureCollection('TIGER/2016/Roads')
Map.setCenter(-73.9596, 40.7688, 12)
Map.addLayer(fc, {}, 'Census roads')
Map
```

The resulting map should look like Fig. 3.13.

Similar to inspecting an **ImageCollection**, you can inspect a **FeatureCollection** by using methods such as `collection.first()` and `collection.limit()`. For example, to inspect the first three features in the **FeatureCollection**:

```
fc.limit(3)
```

The output should look like Fig. 3.14.

Creating feature collections

To create a **FeatureCollection** from scratch, you can provide the constructor with a list of **Feature** objects. The features do not need to have the same geometry type or the same properties as other features in the list. For example:

```
features = [
    ee.Feature(ee.Geometry.Rectangle(30.01, 59.80, 30.59, 60.15), {'name': 'Voronoi'}),
    ee.Feature(ee.Geometry.Point(-73.96, 40.781), {'name': 'Thiessen'}),
```

---

[69]TIGER roads data: tiny.geemap.org/qbez

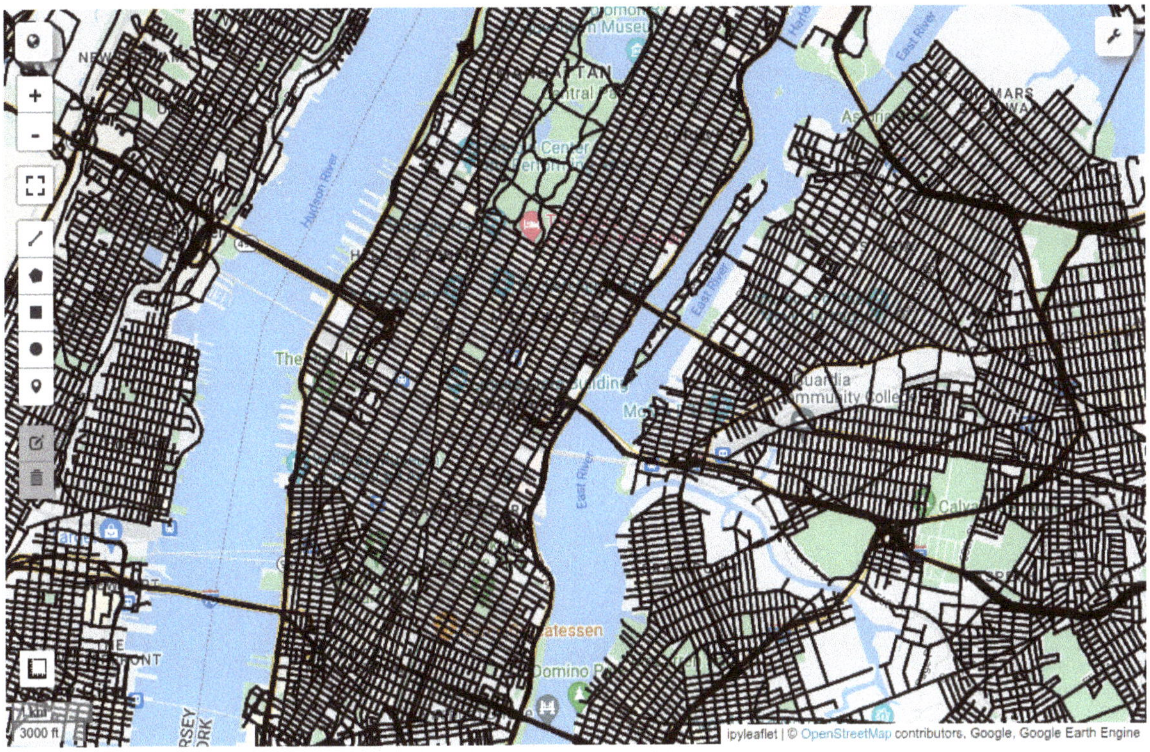

Figure 3.13: Visualizing the TIGER roads data developed by the U.S. Census Bureau.

```
▼ FeatureCollection TIGER/2016/Roads (3 elements, 5 columns)
    type: FeatureCollection
    id: TIGER/2016/Roads
    version: 1566852143799343
  ▶ columns: Object (5 properties)
  ▶ properties: Object (12 properties)
  ▼ features: List (3 elements)
    ▶ 0: Feature (LineString, 4 properties)
    ▶ 1: Feature (LineString, 4 properties)
    ▼ 2: Feature (LineString, 4 properties)
        type: Feature
        id: 00000000000000000006
      ▶ geometry: LineString (2 vertices)
      ▼ properties: Object (4 properties)
          fullname:
          linearid: 110469265916
          mtfcc: S1400
          rttyp:
```

Figure 3.14: Inspecting the first three features in an Earth Engine FeatureCollection.

```
    ee.Feature(ee.Geometry.Point(6.4806, 50.8012), {'name': 'Dirichlet'}),
]
fromList = ee.FeatureCollection(features)
```

Filtering feature collections

Filtering a **FeatureCollection** is analogous to filtering an **ImageCollection**. There are convenience methods such as `filterDate()` and `filterBounds()` as well as the `filter()` method for use with any applicable `ee.Filter`. For example, let's filter the U.S. Census states data[70] to select the state of Texas:

```
Map = geemap.Map()
states = ee.FeatureCollection('TIGER/2018/States')
feat = states.filter(ee.Filter.eq('NAME', 'Texas'))
Map.addLayer(feat, {}, 'Texas')
Map.centerObject(feat)
Map
```

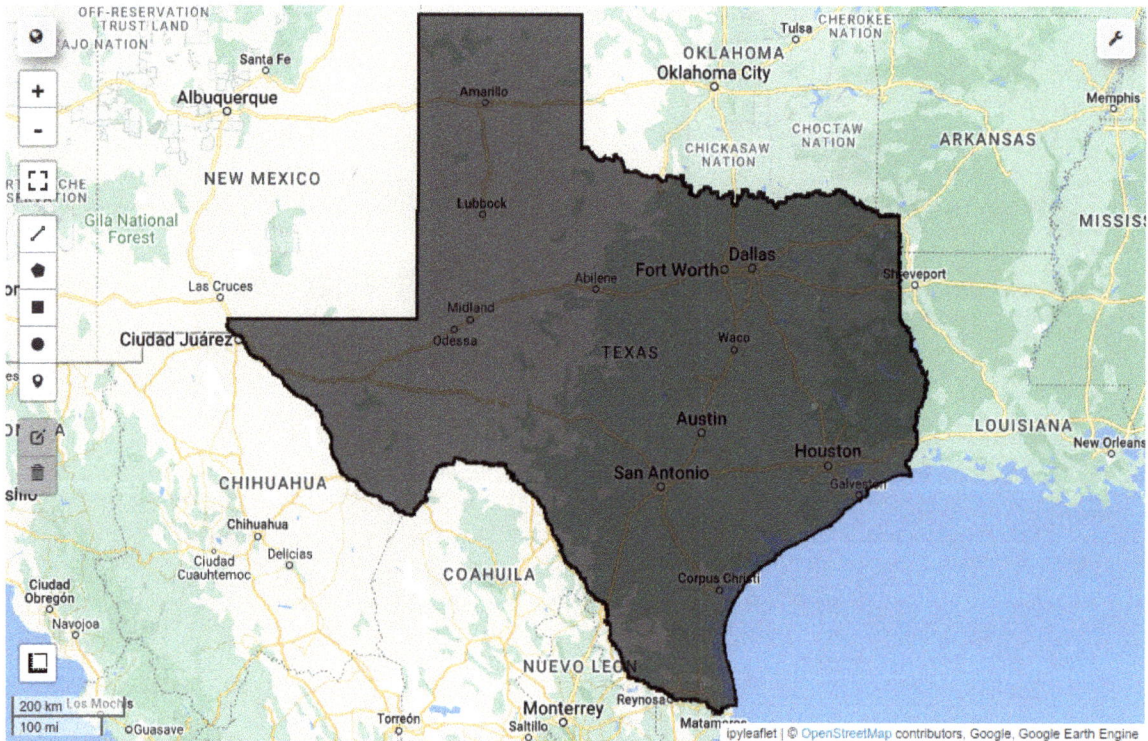

Figure 3.15: Filtering U.S. Census states to select Texas state.

Note that the result of filtering the **FeatureCollection** is returned as a **FeatureCollection**. In the example above, the filtering result is a **FeatureCollection** with only one feature, i.e., the Texas state. To get the Texas state as a **Feature**, use the `collection.first()` method and then use `feature.toDictionary()` method to inspect its contents. For example:

```
texas = feat.first()
texas.toDictionary()

{'ALAND': 676653171537,
 'AWATER': 19006305260,
```

---

[70]U.S. Census states data: tiny.geemap.org/hkoh

```
    'DIVISION': '7',
    'FUNCSTAT': 'A',
    'GEOID': '48',
    'INTPTLAT': '+31.4347032',
    'INTPTLON': '-099.2818238',
    'LSAD': '00',
    'MTFCC': 'G4000',
    'NAME': 'Texas',
    'REGION': '3',
    'STATEFP': '48',
    'STATENS': '01779801',
    'STUSPS': 'TX'}
```

To select multiple states, you can use the `ee.Filter.inList()` method. For example:

```
Map = geemap.Map()
states = ee.FeatureCollection('TIGER/2018/States')
fc = states.filter(ee.Filter.inList('NAME', ['California', 'Oregon', 'Washington']))
Map.addLayer(fc, {}, 'West Coast')
Map.centerObject(fc)
Map
```

The resulting map should look like Fig. 3.16.

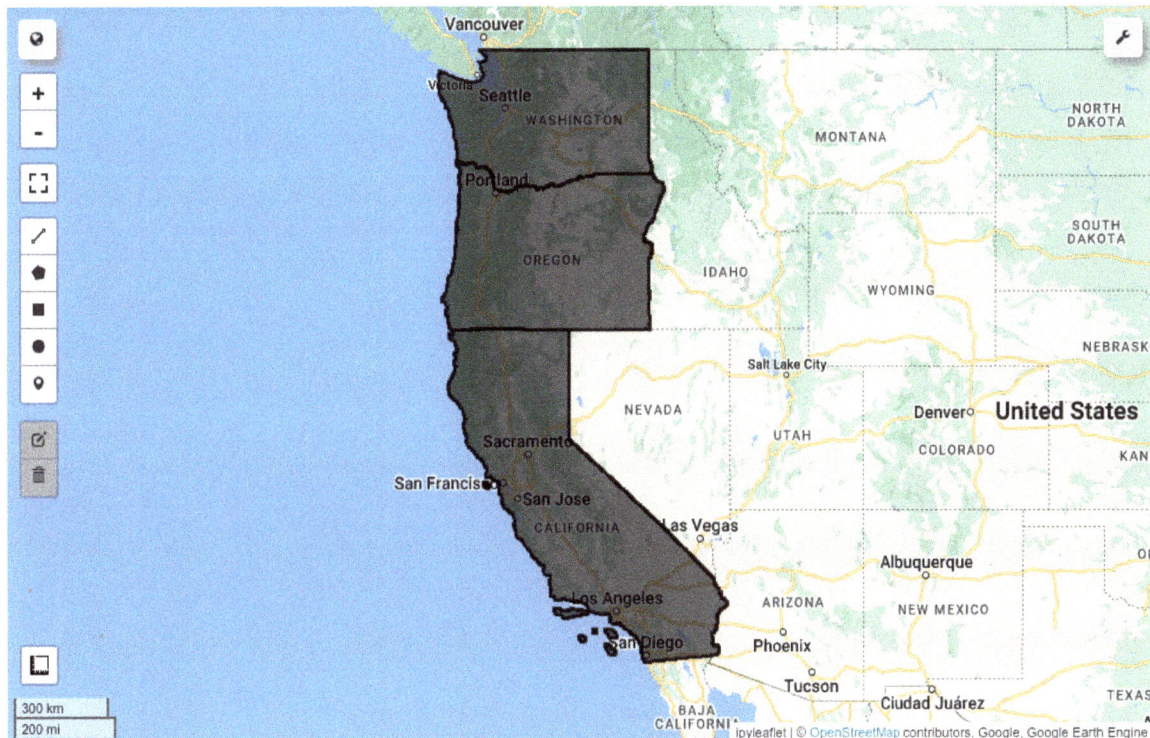

Figure 3.16: Filtering U.S. Census states to select multiple states.

Besides filtering feature collections by their attributes, you can also filter feature collections by location using the `collection.filterBounds()`. For example, if you want to select U.S. southeastern states, draw a polygon on the map and use it to filter the feature collection (see Fig. 3.17).

```
region = Map.user_roi
if region is None:
    region = ee.Geometry.BBox(-88.40, 29.88, -77.90, 35.39)
```

```
fc = ee.FeatureCollection('TIGER/2018/States').filterBounds(region)
Map.addLayer(fc, {}, 'Southeastern U.S.')
Map.centerObject(fc)
```

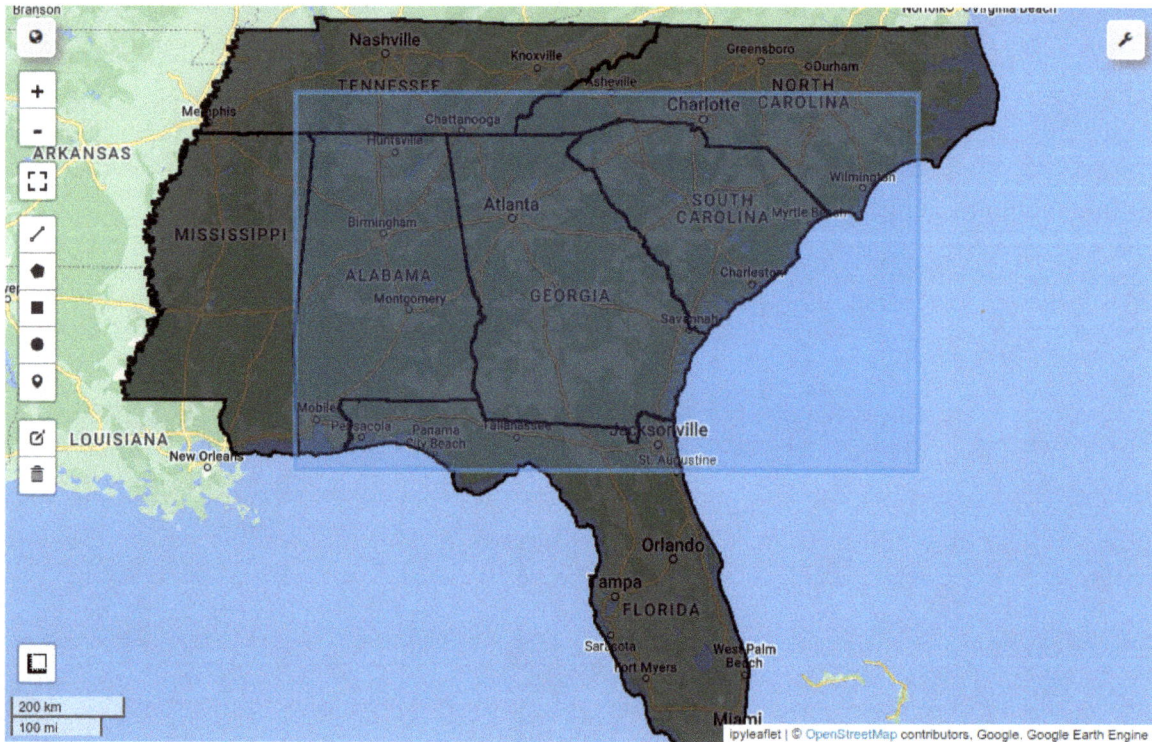

Figure 3.17: Filtering U.S. Census states to select southeastern states.

Visualizing feature collections

There are several ways to visualize a **FeatureCollection**. The simplest way is to directly pass the **FeatureCollection** to the `Map.addLayer()` method which will add it as a map layer with black outlines and slightly transparent polygons. For example:

```
Map = geemap.Map(center=[40, -100], zoom=4)
states = ee.FeatureCollection("TIGER/2018/States")
Map.addLayer(states, {}, "US States")
Map
```

The resulting map should look like Fig. 3.18.

To show only polygon outlines, use the ee.Image().paint()[71] method:

```
Map = geemap.Map(center=[40, -100], zoom=4)
states = ee.FeatureCollection("TIGER/2018/States")
image = ee.Image().paint(states, 0, 3)
Map.addLayer(image, {'palette': 'red'}, "US States")
Map
```

The resulting map should look like Fig. 3.19.

---

[71]ee.Image().paint(): tiny.geemap.org/mgbe

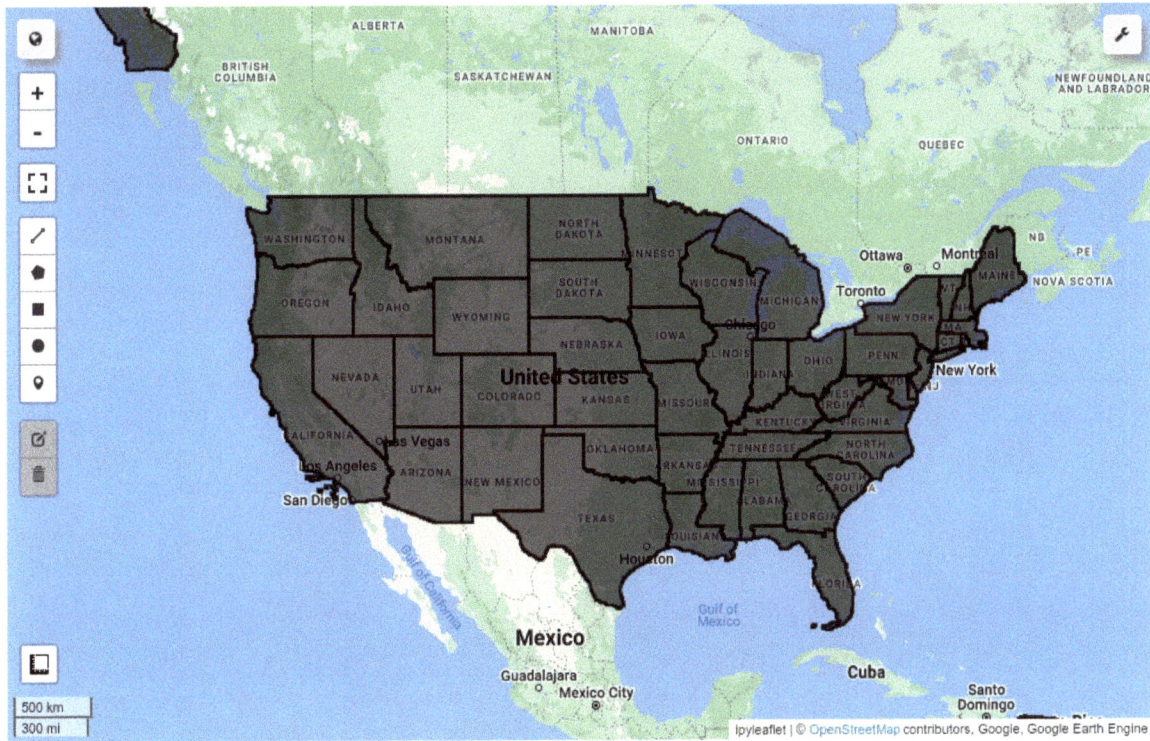

Figure 3.18: Visualizing a FeatureCollection with the default style.

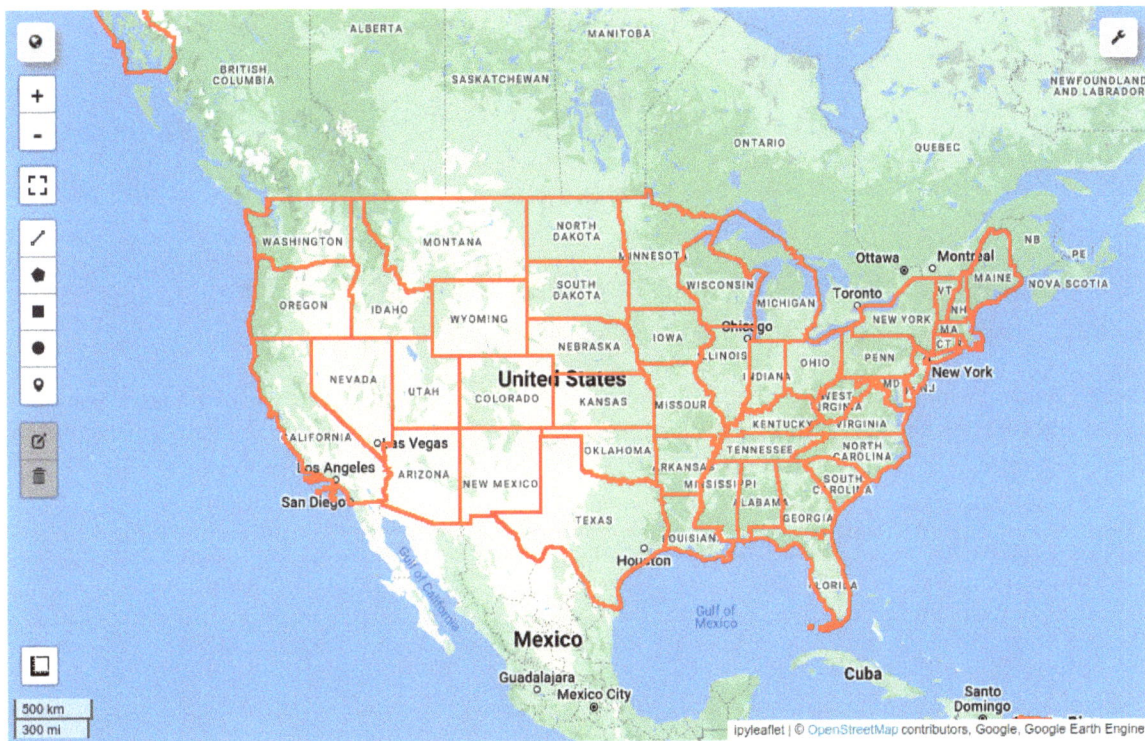

Figure 3.19: Visualizing a FeatureCollection with 'ee.Image().paint()'.

Another way to visualize a **FeatureCollection** is to use the featureCollection.style()[72] method to specify the style (e.g., color, width, lineType, fillColor) of the features. The color and fillColor parameters are specified as hexadecimal strings. The first six characters specify the red, green, blue values of the color. The last two characters specify the transparency. For example, 'FF000088' specifies a 50% transparent red.

```
Map = geemap.Map(center=[40, -100], zoom=4)
states = ee.FeatureCollection("TIGER/2018/States")
style = {'color': '0000ffff', 'width': 2, 'lineType': 'solid', 'fillColor': 'FF000080'}
Map.addLayer(states.style(**style), {}, "US States")
Map
```

The resulting map should look like Fig. 3.20.

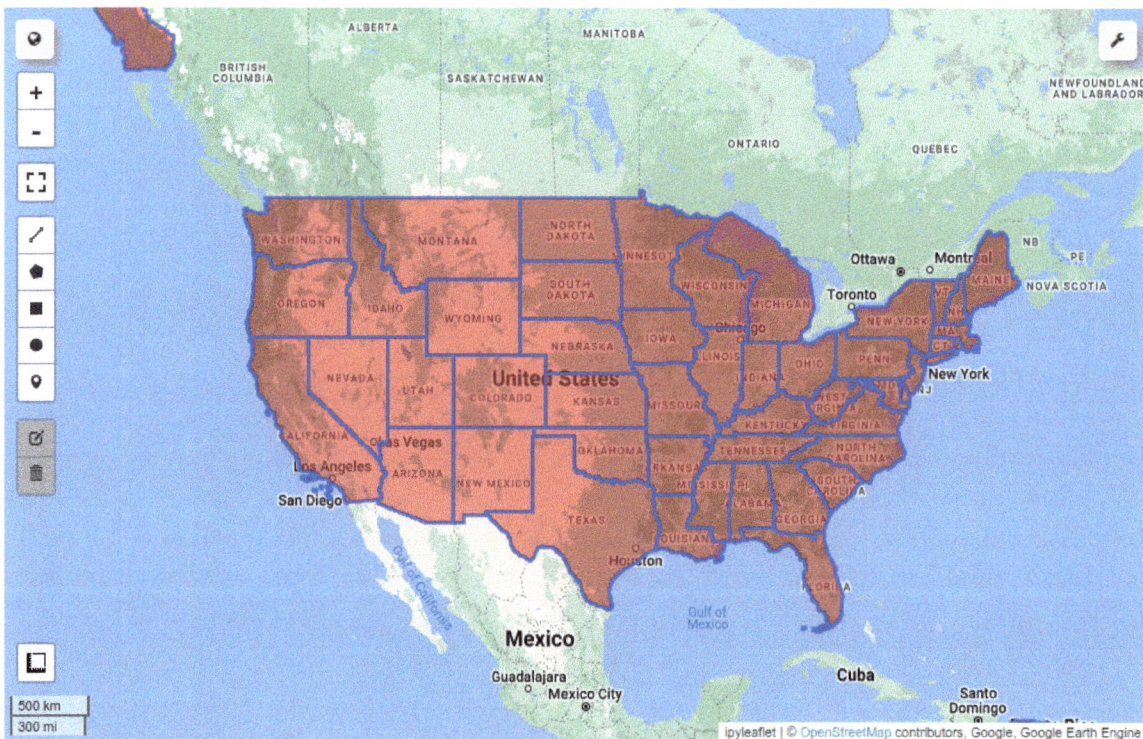

Figure 3.20: Visualizing a FeatureCollection with 'featureCollection.style()'.

Besides the built-in visualization methods provided by Earth Engine, geemap provides a Map.add_styled_vector() method for adding a vector layer with custom styles based on the values of a column. The style parameters for featureCollection.style()[72] are also supported here. For example:

```
Map = geemap.Map(center=[40, -100], zoom=4)
states = ee.FeatureCollection("TIGER/2018/States")
vis_params = {
    'color': '000000',
    'colorOpacity': 1,
    'pointSize': 3,
    'pointShape': 'circle',
    'width': 2,
    'lineType': 'solid',
    'fillColorOpacity': 0.66,
```

---

[72]featureCollection.style(): tiny.geemap.org/nbki

```
    }
    palette = ['006633', 'E5FFCC', '662A00', 'D8D8D8', 'F5F5F5']
    Map.add_styled_vector(
        states, column="NAME", palette=palette, layer_name="Styled vector", **vis_params
    )
    Map
```

Styling by attribute

Sometimes it is necessary to style a feature collection using the attributes of the features in the collection.

For example, if you have a vectorized land use and land cover map, you might want to color each land cover type with a different color. To do this, you can use the ee_vector_style() method to specify the column to be used as the attribute. You can also customize the style (e.g., color, width, lineType, fillColor) of the features. The color and fillColor parameters are specified as hexadecimal strings. The first six characters specify the red, green, blue values of the color. The last two characters specify the transparency. For example, 'FF000088' specifies a 50% transparent red. The following example illustrates how to use ee_vector_style() to color the US National Wetland Inventory (NWI[73]).

First, create an interactive map and add the Google Satellite imagery basemap:

```
    Map = geemap.Map(center=[28.00142, -81.7424], zoom=13)
    Map.add_basemap('HYBRID')
```

Specify the list of attribute values and the corresponding colors to be used for each attribute value. For example, if the attribute values are ['one', 'two', 'three', 'four', 'five'] and the corresponding colors are ['FF0000', '00FF00', '0000FF', 'FFFF00', 'FF00FF'], then the first feature will be colored red, the second green, the third blue, the fourth yellow, and the fifth magenta. Note that you can add two additional characters to the end of the color string to specify the transparency. For example, 'FF0000A8' specifies a 66% transparent red.

```
    types = [
        "Freshwater Forested/Shrub Wetland",
        "Freshwater Emergent Wetland",
        "Freshwater Pond",
        "Estuarine and Marine Wetland",
        "Riverine",
        "Lake",
        "Estuarine and Marine Deepwater",
        "Other",
    ]

    colors = [
        "#008837",
        "#7FC31C",
        "#688CC0",
        "#66C2A5",
        "#0190BF",
        "#13007C",
        "#007C88",
        "#B28653",
    ]

    fillColor = [c + "A8" for c in colors]
```

Next, use ee_vector_style() to style the NWI FeatureCollection based on the WETLAND_TY column. The fillColor parameter is used to specify the fill color and transparency of each feature, while the

---

[73]NWI: tiny.geemap.org/kcxm

color parameter is used to specify the outline color and transparency of each feature. A `color` of `'00000000'` indicates that the feature outline should be invisible.

```
fc = ee.FeatureCollection("projects/sat-io/open-datasets/NWI/wetlands/FL_Wetlands")
styled_fc = geemap.ee_vector_style(
    fc, column='WETLAND_TY', labels=types, fillColor=fillColor, color='00000000'
)
```

Now, add the styled vector layer and its corresponding legend to the map (see Fig. 3.21).

```
Map.addLayer(styled_fc, {}, 'NWI')
Map.add_legend(title='Wetland Type', labels=types, colors=colors)
Map
```

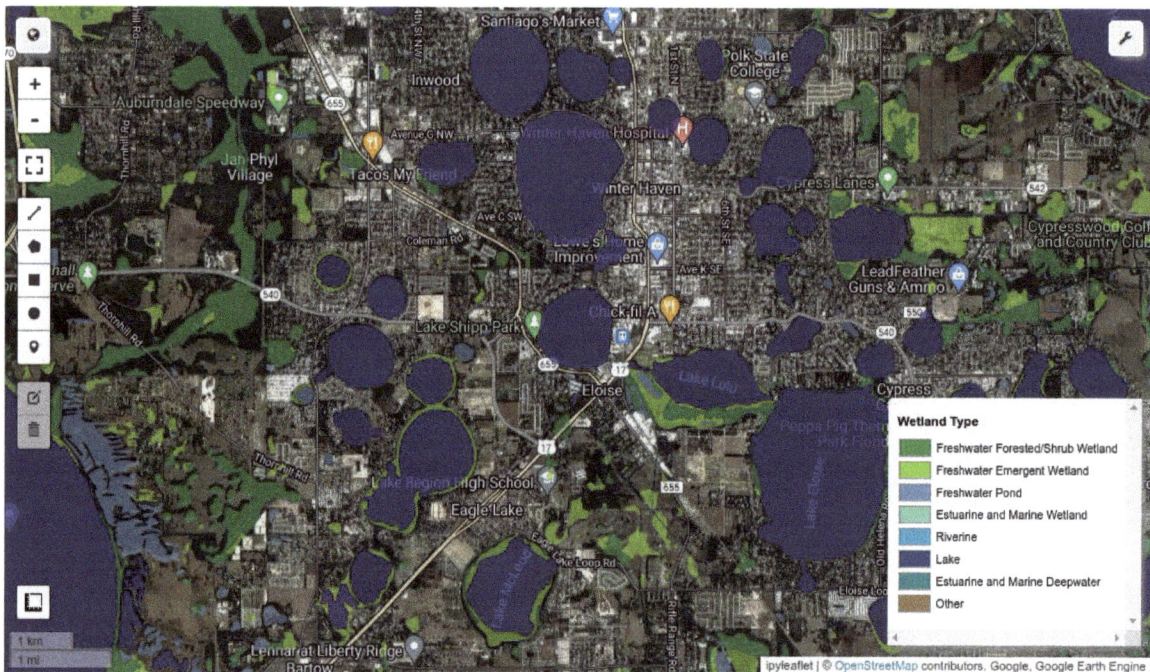

Figure 3.21: Styling a FeatureCollection of polygons by attribute.

Similarly, you can style a `FeatureCollection` of points by attribute. The following example illustrates how to use `ee_vector_style()` to color the Global Power Plant Database[74]. There are many energy sources used in electricity generation by power plants, e.g., coal, oil, hydro, natural gas, nuclear, and more. Here, we provide a list of fuel types and their corresponding colors to be used for styling each fuel type.

```
fuels = [
    'Coal',
    'Oil',
    'Gas',
    'Hydro',
    'Nuclear',
    'Solar',
    'Waste',
    'Wind',
    'Geothermal',
    'Biomass',
```

[74]Global Power Plant Database: tiny.geemap.org/vzim

```
]
colors = [
    '000000',
    '593704',
    'BC80BD',
    '0565A6',
    'E31A1C',
    'FF7F00',
    '6A3D9A',
    '5CA2D1',
    'FDBF6F',
    '229A00',
]
```

The selected fuel types can be used to filter the FeatureCollection and select only the features that match the selected fuel types. Then, use ee_vector_style() to style the FeatureCollection based on the fuel1 column:

```
fc = ee.FeatureCollection("WRI/GPPD/power_plants").filter(
    ee.Filter.inList('fuel1', fuels)
)
styled_fc = geemap.ee_vector_style(fc, column="fuel1", labels=fuels, color=colors)
```

Now, add the styled vector layer and its corresponding legend to the map (see Fig. 3.22).

```
Map = geemap.Map(center=[40, -100], zoom=4)
Map.addLayer(styled_fc, {}, 'Power Plants')
Map.add_legend(title="Power Plant Fuel Type", labels=fuels, colors=colors)
Map
```

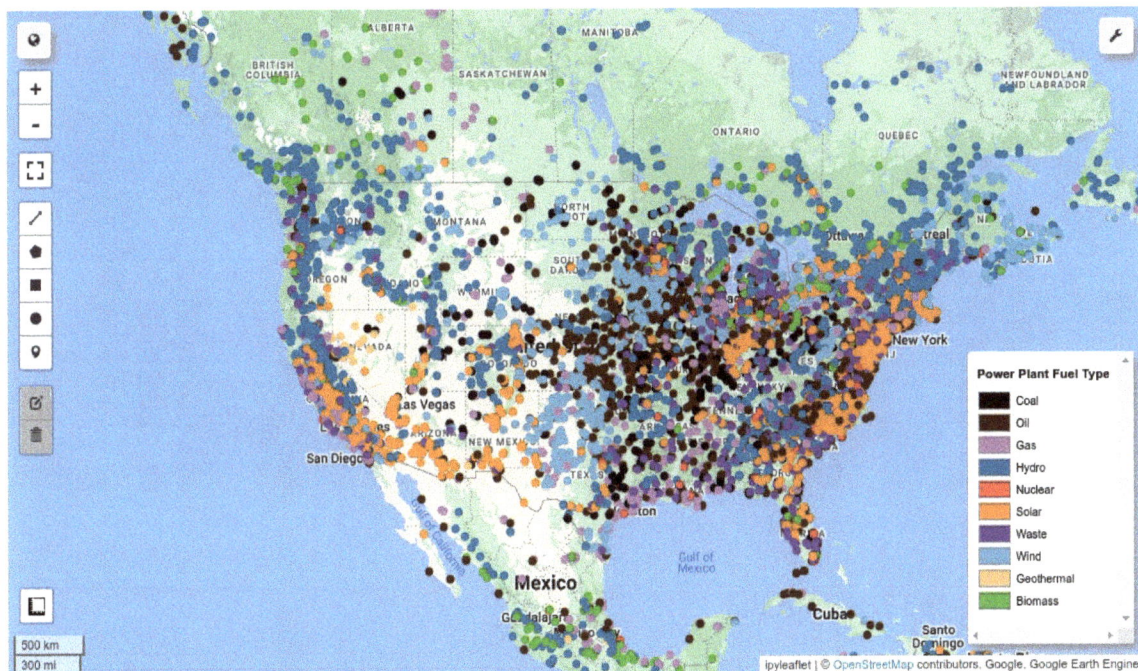

Figure 3.22: Styling a FeatureCollection of points by attribute.

Unlike polygons and points, polylines can only be styled with the color, width, and lineType styling options. The following example shows how to style the TIGER: US Census Roads[75] based on route type. The available route types are Interstate (I), U.S. (U), State recognized (S), Common Name (M),

County (C), and Other (O). For example, you can specify the colors and the widths to be used for each route type:

```
types = ['I', 'U', 'S', 'M', 'C', 'O']
labels = ['Interstate', 'U.S.', 'State recognized', 'Common Name', 'County', 'Other']
colors = ['E31A1C', 'FF7F00', '6A3D9A', '000000', 'FDBF6F', '229A00']
width = [8, 5, 4, 2, 1, 1]
```

Next, use ee_vector_style() to style the FeatureCollection based on the rttyp column containing the route type:

```
fc = ee.FeatureCollection('TIGER/2016/Roads')
styled_fc = geemap.ee_vector_style(
    fc, column='rttyp', labels=types, color=colors, width=width
)
```

Now, add the styled vector layer and its corresponding legend to the map (see Fig. 3.23).

```
Map = geemap.Map(center=[40.7424, -73.9724], zoom=13)
Map.addLayer(styled_fc, {}, 'Census Roads')
Map.add_legend(title='Route Type', labels=labels, colors=colors)
Map
```

Figure 3.23: Styling a FeatureCollection of polylines by attribute.

3.4   Earth Engine Data Catalog

The Earth Engine Data Catalog[55] hosts a variety of geospatial datasets. As of March 2023, the catalog contains over 1000 datasets[76] with a total size of over 80 petabytes. Some notable datasets include: Landsat, Sentinel, MODIS, NAIP, etc. For a complete list of datasets in CSV or JSON formats, see the

---

[75]TIGER: US Census Roads: tiny.geemap.org/slht

Earth Engine Datasets List[77].

Searching for datasets

The Earth Engine Data Catalog is searchable. You can search datasets by name, keyword, or tag. For example, entering "elevation" in the search box will filter the catalog to show only datasets containing "elevation" in their name, description, or tags. At the time of writing, 63 datasets are returned for this search query. Scroll down the list to find the NASA SRTM Digital Elevation 30m[78] dataset as an example of the information we can find in the catalog. On each dataset page, you can find the information, including Dataset Availability, Dataset Provider, Earth Engine Snippet, Tags, Description, Code Example, and more (see Fig. 3.24). One important piece of information is the **Image/ImageCollection/FeatureCollection** ID of each dataset, which is essential for accessing the dataset through the Earth Engine JavaScript or Python APIs.

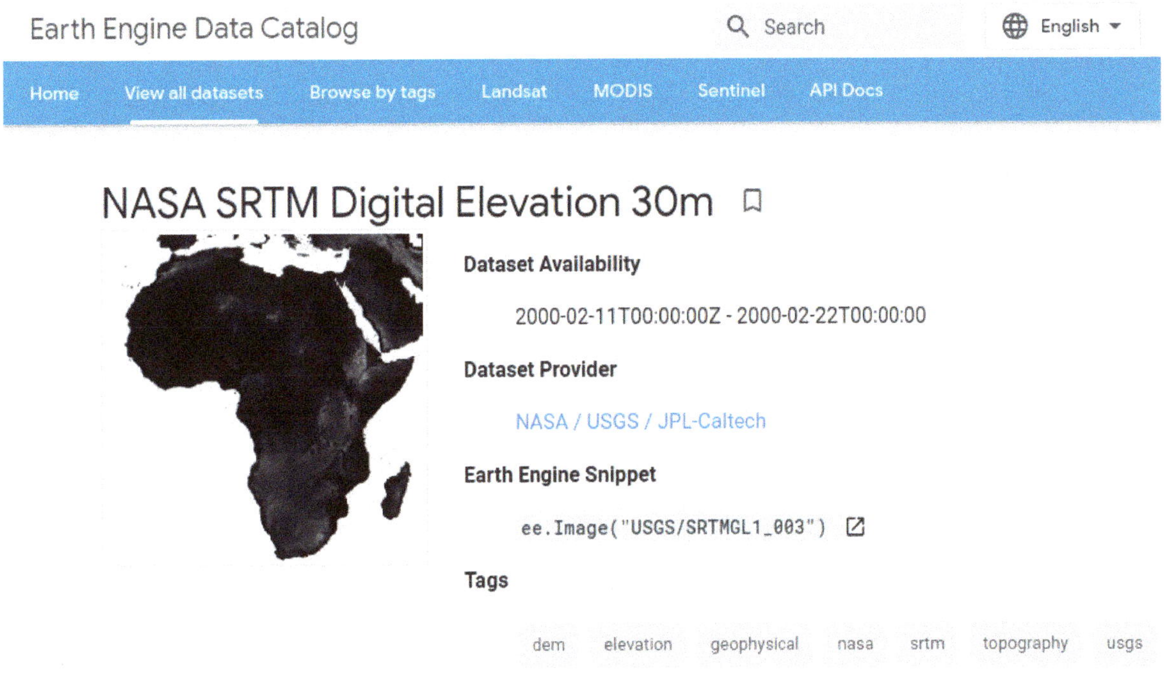

Figure 3.24: The NASA SRTM Digital Elevation dataset in the Earth Engine Data Catalog.

Besides searching the Earth Engine Data Catalog using the website, you can also search the datalog using geemap in a Jupyter environment. Click on the globe icon in the upper-left corner of the map, then click the data tab on the popup dialog. Enter a keyword (e.g., "elevation") in the search box and press enter. The search results will populate the dropdown menu (see Fig. 3.25). Select a dataset from the dropdown list. The information about the dataset will be displayed in the panel below, including Data Availability, Earth Engine Snippet, Earth Engine Dataset ID, and Dataset Thumbnail (see Fig. 3.26).

If you are running Jupyter notebook, you can click the **import** button on the search result dialog to import the dataset into the notebook. A code snippet will be automatically generated and inserted into a code cell below the map.

---

[76]1000 datasets: tiny.geemap.org/xxqj

[77]Earth Engine Datasets List: tiny.geemap.org/zqbc

[78]NASA SRTM Digital Elevation 30m: tiny.geemap.org/tbak

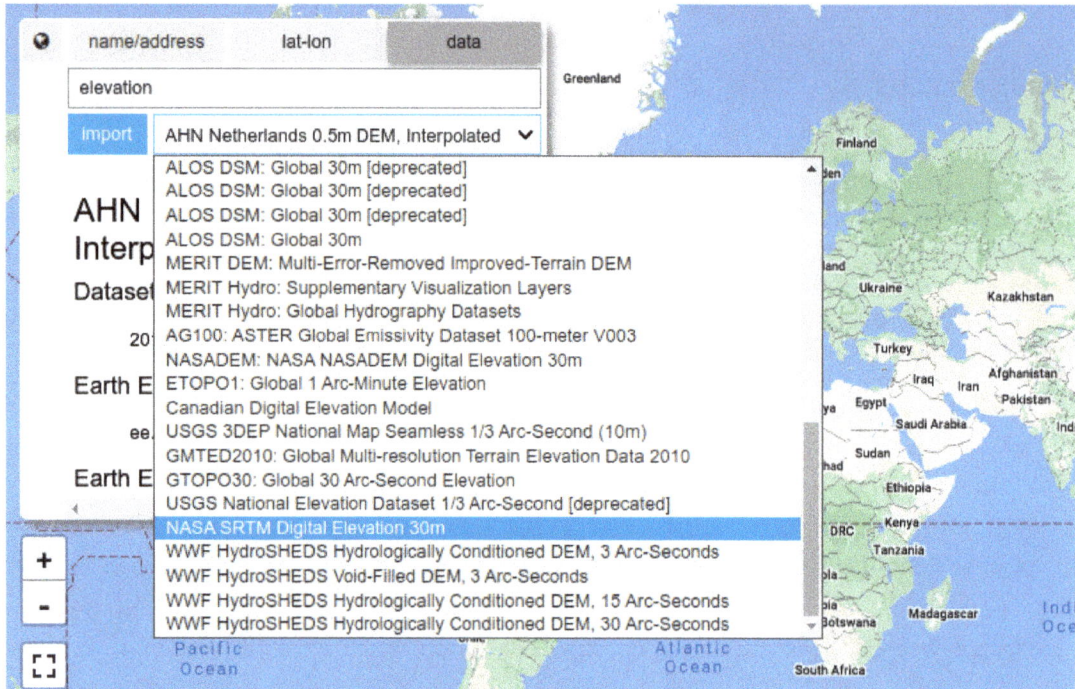

Figure 3.25: Searching data from the Earth Engine Data Catalog within geemap.

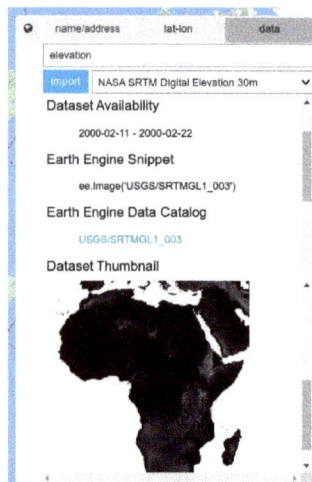

Figure 3.26: Displaying search results from the Earth Engine Data Catalog.

```
dataset_xyz = ee.Image('USGS/SRTMGL1_003')
Map.addLayer(dataset_xyz, {}, "USGS/SRTMGL1_003")
```

Note that the **import** button only works for Jupyter Notebook. JupyterLab and Google Colab do not support creating a new code cell programmatically (see jupyterlab issue 8415[79]).

Alternatively, you can copy the Earth Engine Snippet (e.g., ee.Image('USGS/SRTMGL1_003')) from the search result dialog (see Fig. 3.26) and paste it into an existing code cell.

```
Map = geemap.Map()
dem = ee.Image('USGS/SRTMGL1_003')
vis_params = {
    'min': 0,
    'max': 4000,
    'palette': ['006633', 'E5FFCC', '662A00', 'D8D8D8', 'F5F5F5'],
}
Map.addLayer(dem, vis_params, 'SRTM DEM')
Map
```

The resulting map should look like Fig. 3.27.

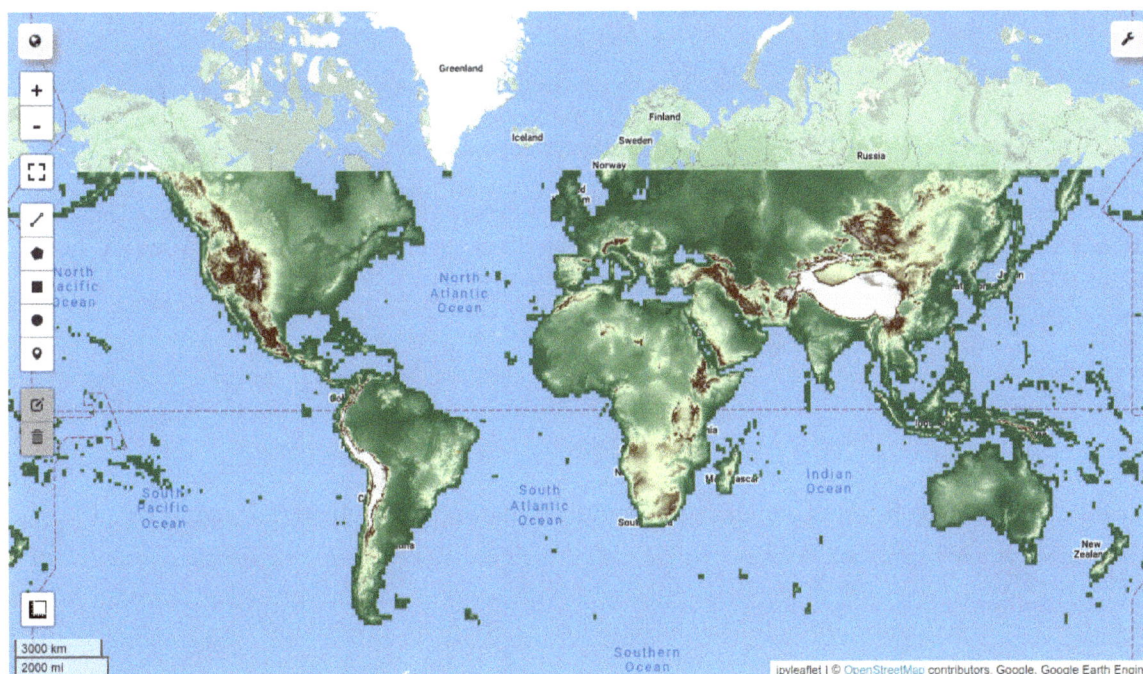

Figure 3.27: The NASA SRTM Digital Elevation data.

Using the datasets module

As mentioned earlier, the Earth Engine Data Catalog hosts over 1000 datasets[76] amounting to more than 80 petabytes of data. It could be challenging to memorize all the dataset IDs and access them through the Earth Engine Python API. Although we can search the data catalog through the webpage or use the geemap user interface as introduced in the previous section, it would be useful if we can

---

[79]jupyterlab issue 8415: tiny.geemap.org/nlmp

access the dataset IDs programmatically without leaving a code cell. The geemap datasets module[80] provides an easy way to access the dataset IDs through dot notation.

First, let's import the `geemap.datasets` module:

```
from geemap.datasets import DATA
```

Once the datasets module is imported the entire Earth Engine Data Catalog can be accessed through dot notation. Simply type `DATA.` into a code cell and press **TAB** to see the list of available datasets (see Fig. 3.28).

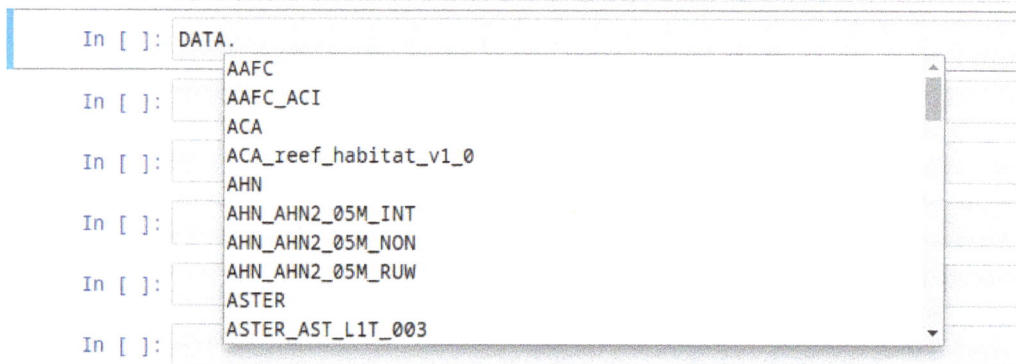

Figure 3.28: Accessing the Earth Engine Data Catalog through dot notation.

Scroll down the list to find your desired dataset, for example, `DATA.USGS_GAP_CONUS_2011`. Then press **Enter**, and the dataset ID `USGS/GAP/CONUS/2011` will be displayed in the output cell. You can directly use the dataset ID `DATA.USGS_GAP_CONUS_2011` in your code, for example:

```
Map = geemap.Map(center=[40, -100], zoom=4)
dataset = ee.Image(DATA.USGS_GAP_CONUS_2011)
Map.addLayer(dataset, {}, 'GAP CONUS')
Map
```

The resulting map should look like Fig. 3.29.

To view the metadata of the selected dataset, use the `get_metadata()` method. For example:

```
from geemap.datasets import get_metadata

get_metadata(DATA.USGS_GAP_CONUS_2011)
```

The output should look like Fig. 3.30.

Click the hyperlink USGS/GAP/CONUS/2011[81] on the JupyterLab cell output to see the dataset's metadata in a new tab in your browser.

3.5 Getting image metadata

Before using satellite images for geospatial analysis, it is important to explore the image metadata, such as band names, projection information, properties, and other metadata. The example below shows how to get the metadata of a Landsat 9 image.

---

[80]datasets module: geemap.org/datasets
[81]USGS/GAP/CONUS/2011: tiny.geemap.org/gulj

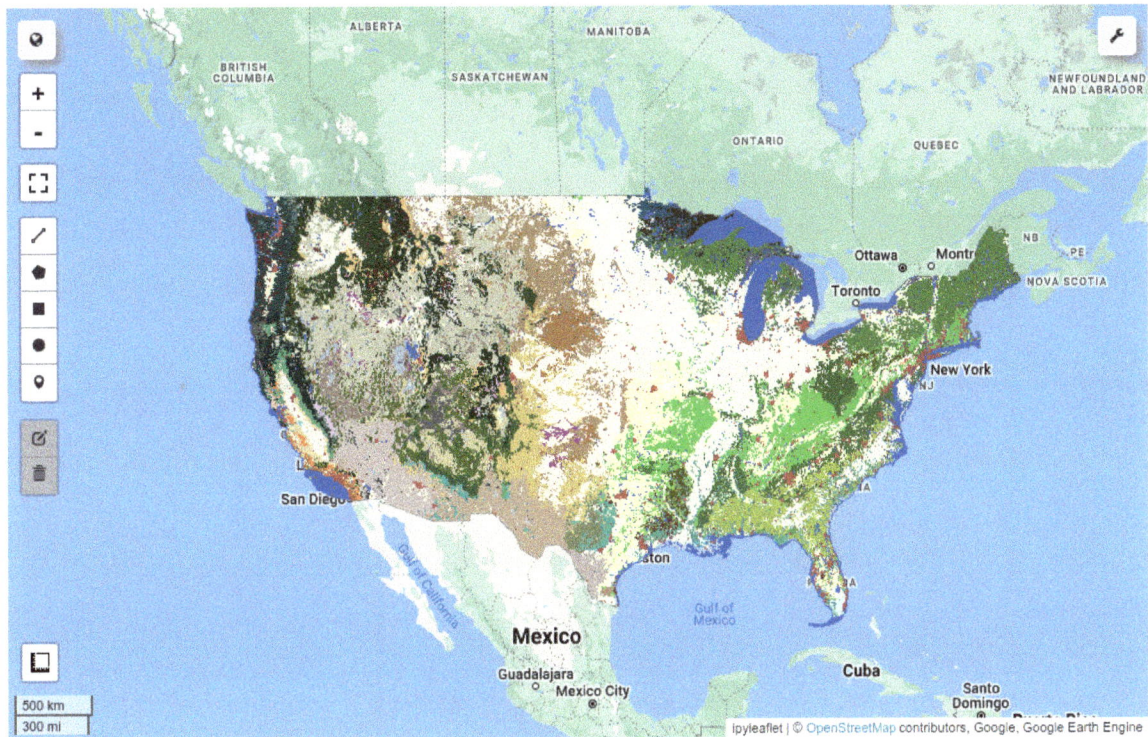

Figure 3.29: The National Gap Analysis Program Land Cover Data.

USGS GAP CONUS 2011

Dataset Availability

2011-01-01 - 2012-01-01

Earth Engine Snippet

ee.Image('USGS/GAP/CONUS/2011')

Earth Engine Data Catalog

USGS/GAP/CONUS/2011

Dataset Thumbnail

Figure 3.30: The metadata of the National Gap Analysis Program Land Cover Data.

```
image = ee.Image('LANDSAT/LC09/C02/T1_L2/LC09_044034_20220503')
```

First, let's look at the band names of the image:

```
image.bandNames()
```

```
['SR_B1',
 'SR_B2',
 'SR_B3',
 'SR_B4',
 'SR_B5',
 'SR_B6',
 'SR_B7',
 'SR_QA_AEROSOL',
 'ST_B10',
 'ST_ATRAN',
 'ST_CDIST',
 'ST_DRAD',
 'ST_EMIS',
 'ST_EMSD',
 'ST_QA',
 'ST_TRAD',
 'ST_URAD',
 'QA_PIXEL',
 'QA_RADSAT']
```

Each band has its own name, data type, scale, mask and projection.

To check the projection of a band, use `image.projection()`:

```
image.select('SR_B1').projection()
```

```
{'type': 'Projection',
 'crs': 'EPSG:32610',
 'transform': [30, 0, 458985, 0, -30, 4264215]}
```

To check the spatial resolution of a band, use `image.projection().nominalScale()`:

```
image.select('SR_B1').projection().nominalScale()
```

```
30
```

To retrieve all property names of an image, use `image.propertyNames()`:

```
image.propertyNames()
```

```
['DATA_SOURCE_ELEVATION',
 'WRS_TYPE',
 'system:id',
 'REFLECTANCE_ADD_BAND_1',
 'REFLECTANCE_ADD_BAND_2',
 'DATUM',
 'REFLECTANCE_ADD_BAND_3',
 'REFLECTANCE_ADD_BAND_4',
 'REFLECTANCE_ADD_BAND_5',
 'REFLECTANCE_ADD_BAND_6',
 'REFLECTANCE_ADD_BAND_7',
 'system:footprint',
 'REFLECTIVE_SAMPLES',
 'system:version',
 'GROUND_CONTROL_POINTS_VERSION',
 'SUN_AZIMUTH',
 'UTM_ZONE',
```

```
  ...
  ]
```

To get the value of a specific property (e.g., CLOUD_COVER), use the image.get() function:

```
image.get('CLOUD_COVER')
```

```
0.06
```

To get the acquisition time of a Landsat image:

```
image.get('DATE_ACQUIRED')
```

```
'2022-05-03'
```

Keep in mind that not all Earth Engine images have the DATE_ACQUIRED property. However, most images will have the system:time_start property, which stores the image acquisition time as the number of milliseconds since 1970-01-01T00:00:00Z.

```
image.get('system:time_start')
```

```
1651603548097
```

To convert the acquisition time in milliseconds to a human-readable date string, use the .format() function:

```
date = ee.Date(image.get('system:time_start'))
date.format('YYYY-MM-dd')
```

```
'2022-05-03'
```

To get all properties of an image, use the image.toDictionary() method:

```
image.toDictionary()
```

```
{'ALGORITHM_SOURCE_SURFACE_REFLECTANCE': 'LaSRC_1.5.0',
 'ALGORITHM_SOURCE_SURFACE_TEMPERATURE': 'st_1.3.0',
 'CLOUD_COVER': 0.07,
 'CLOUD_COVER_LAND': 0.09,
 'COLLECTION_CATEGORY': 'T1',
 'COLLECTION_NUMBER': 2,
 'DATA_SOURCE_AIR_TEMPERATURE': 'MODIS',
 'DATA_SOURCE_ELEVATION': 'GLS2000',
 'DATA_SOURCE_OZONE': 'MODIS',
 'DATA_SOURCE_PRESSURE': 'Calculated',
 'DATA_SOURCE_REANALYSIS': 'GEOS-5 FP-IT',
 'DATA_SOURCE_WATER_VAPOR': 'MODIS',
 'DATE_ACQUIRED': '2022-05-03',
 'DATE_PRODUCT_GENERATED': 1651724371000,
 'DATUM': 'WGS84',
 'EARTH_SUN_DISTANCE': 1.0081313,
 'ELLIPSOID': 'WGS84',
 ...
 }
```

Similarly, geemap has an image_props() function for getting the metadata of an image. The difference is that it automatically converts the image acquisition time in milliseconds to a human-readable date string. For example, check the system:time_start and system:time_end properties in the output:

```
props = geemap.image_props(image)
props
```

```
{'ALGORITHM_SOURCE_SURFACE_REFLECTANCE': 'LaSRC_1.5.0',
 'ALGORITHM_SOURCE_SURFACE_TEMPERATURE': 'st_1.3.0',
 'CLOUD_COVER': 0.06,
 'CLOUD_COVER_LAND': 0.09,
 'COLLECTION_CATEGORY': 'T1',
 'COLLECTION_NUMBER': 2,
 'DATA_SOURCE_AIR_TEMPERATURE': 'MODIS',
 'DATA_SOURCE_ELEVATION': 'GLS2000',
 ...
 'system:id': 'LANDSAT/LC09/C02/T1_L2/LC09_044034_20220503',
 'system:index': 'LC09_044034_20220503',
 'system:time_end': '2022-05-03 18:45:48',
 'system:time_start': '2022-05-03 18:45:48',
 'system:version': 1651763920643564
}
```

3.6   Calculating descriptive statistics

To visualize a multi-spectral image, it is often desirable to set visualization parameters, such as min, max, and bands. Geemap provides several functions for calculating descriptive statistics of an image, such as image_min_value(), image_max_value(), image_mean_value(), and image_stats(). For example, to get the minimum value of each band of an image, use image_min_value():

```
image = ee.Image('LANDSAT/LC09/C02/T1_L2/LC09_044034_20220503')
geemap.image_min_value(image)
```

```
{'QA_PIXEL': 21762,
 'QA_RADSAT': 0,
 'SR_B1': 1,
 'SR_B2': 8,
 'SR_B3': 288,
 'SR_B4': 481,
 'SR_B5': 3410,
 'SR_B6': 6302,
 'SR_B7': 7095,
 'SR_QA_AEROSOL': 1,
 'ST_ATRAN': 8821,
 'ST_B10': 36413,
 'ST_CDIST': 0,
 'ST_DRAD': 170,
 'ST_EMIS': 8373,
 'ST_EMSD': 0,
 'ST_QA': 134,
 'ST_TRAD': 5991,
 'ST_URAD': 294}
```

To get the maximum value of each band of an image, use image_max_value():

```
geemap.image_max_value(image)
```

```
{'QA_PIXEL': 54724,
 'QA_RADSAT': 127,
 'SR_B1': 53671,
 'SR_B2': 54313,
 'SR_B3': 57598,
 'SR_B4': 57531,
 'SR_B5': 56620,
 'SR_B6': 59018,
 'SR_B7': 60538,
 'SR_QA_AEROSOL': 228,
 'ST_ATRAN': 9516,
 'ST_B10': 52088,
```

```
'ST_CDIST': 2706,
'ST_DRAD': 423,
'ST_EMIS': 9915,
'ST_EMSD': 1386,
'ST_QA': 1080,
'ST_TRAD': 12909,
'ST_URAD': 816}
```

To get the mean value of each band of an image, use `image_mean_value()`:

```
geemap.image_mean_value(image)
```

```
{'QA_PIXEL': 21879.45624974622,
 'QA_RADSAT': 0.000970456318786452,
 'SR_B1': 8104.119979979327,
 'SR_B2': 8461.536303658795,
 'SR_B3': 9240.731292093615,
 'SR_B4': 9298.173920971509,
 'SR_B5': 12888.685823419664,
 'SR_B6': 12140.18577036555,
 'SR_B7': 10736.956490617888,
 'SR_QA_AEROSOL': 142.6421931965974,
 'ST_ATRAN': 9076.834480797119,
 'ST_B10': 43469.766724609784,
 'ST_CDIST': 620.0690476796947,
 'ST_DRAD': 337.6147726655202,
 'ST_EMIS': 9783.119355764633,
 'ST_EMSD': 68.50123066692646,
 'ST_QA': 245.14052551309015,
 'ST_TRAD': 8969.705392231179,
 'ST_URAD': 635.921206837569}
```

To get the descriptive statistics (e.g., `min`, `max`, `mean`, `std`, and `sum`) of each band of an image, use `image_stats()`:

```
geemap.image_stats(image)
```

3.7 Using the inspector tool

Geemap has an inspector tool for querying Earth Engine data interactively, such as retrieving the pixel values of an image or getting the attribute of a geometry at the location of a mouse click. You can find the inspector tool in the upper-left corner of the toolbar (see Fig. 3.31).

Figure 3.31: The inspector tool for querying Earth Engine data interactively.

First, let's add some Earth Engine datasets to the map. In this example, we will use two raster datasets

(SRTM[82] and Landsat[83]) and one vector dataset (US Census States[84]).

```python
Map = geemap.Map(center=(40, -100), zoom=4)
dem = ee.Image('USGS/SRTMGL1_003')
landsat7 = ee.Image('LANDSAT/LE7_TOA_5YEAR/1999_2003').select(
    ['B1', 'B2', 'B3', 'B4', 'B5', 'B7']
)
states = ee.FeatureCollection("TIGER/2018/States")
vis_params = {
    'min': 0,
    'max': 4000,
    'palette': ['006633', 'E5FFCC', '662A00', 'D8D8D8', 'F5F5F5'],
}
Map.addLayer(dem, vis_params, 'SRTM DEM')
Map.addLayer(
    landsat7,
    {'bands': ['B4', 'B3', 'B2'], 'min': 20, 'max': 200, 'gamma': 2.0},
    'Landsat 7',
)
Map.addLayer(states, {}, "US States")
Map
```

Next, activate the inspector tool by clicking on the identity icon on the toolbar (see Fig. 3.31). Then, move the mouse cursor over the map and click on the map. The pixel values of each image and the attributes of the intersecting polygon at the mouse click location will be displayed in a panel on the right side of the map (see Fig. 3.32). By default, the inspector tool will query all visible layers. To query only a specific layer, toggle other layers off by unchecking the checkboxes next to the layer names.

3.8 Converting JavaScript to Python

In the past, most code examples in the Earth Engine Developer Guide[85] were written exclusively in JavaScript. Since 2021, more and more Python examples have been added to the Developer Guide. As of March 2023, there are still a significant number of JavaScript examples in the Developer Guide that do not have corresponding Python examples. Geemap has functionality for converting Earth Engine JavaScript to Python and Jupyter notebooks automatically, including interactive conversion and batch conversion.

Interactive conversion

You can convert most JavaScript examples in the Earth Engine Developer Guide to Python interactively without coding using geemap. For example, let's navigate to the Image Visualization example[86] and copy the JavaScript code snippet using the copy button in the upper-right corner of the code section (see Fig. 3.33).

Next, create and display an interactive map.

```python
Map = geemap.Map()
Map
```

Click the convert icon on the toolbar. Paste the copied JavaScript code snippet above into the text area and click the convert button (Fig. 3.34). The JavaScript code snippet will be converted to Python and

---

[82]SRTM: tiny.geemap.org/bwsx

[83]Landsat: tiny.geemap.org/lrdk

[84]US Census States: tiny.geemap.org/fbyf

[85]Earth Engine Developer Guide: tiny.geemap.org/dfjo

[86]Image Visualization example: tiny.geemap.org/dzzd

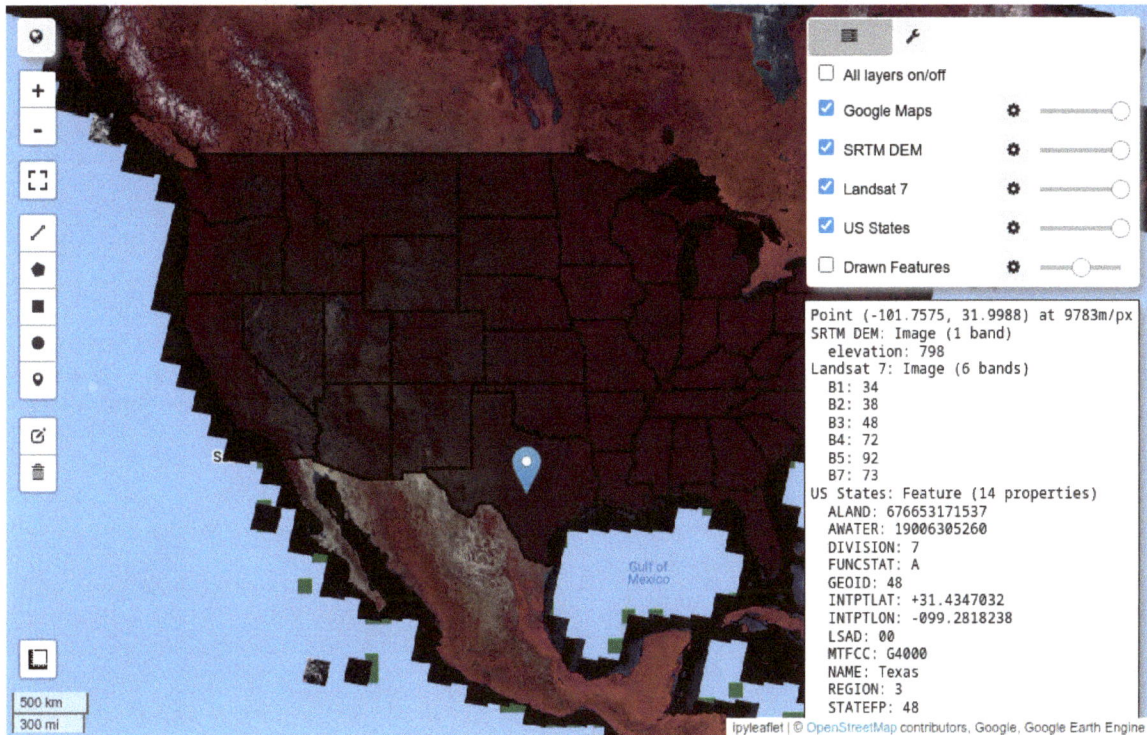

Figure 3.32: Using the geemap inspector tool.

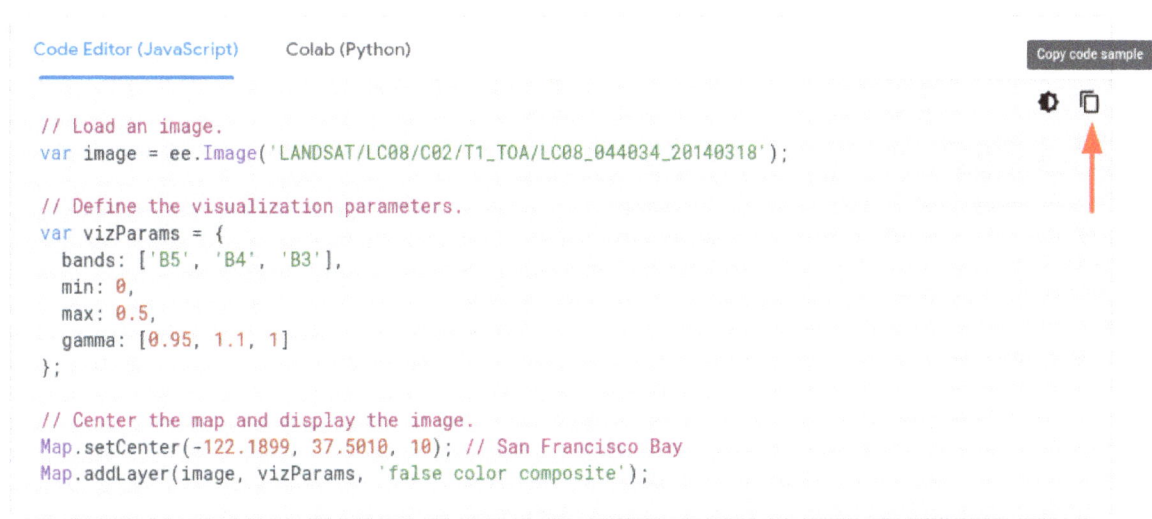

Figure 3.33: An Earth Engine JavaScript code snippet from the Earth Engine Developer Guide.

displayed in the text area.

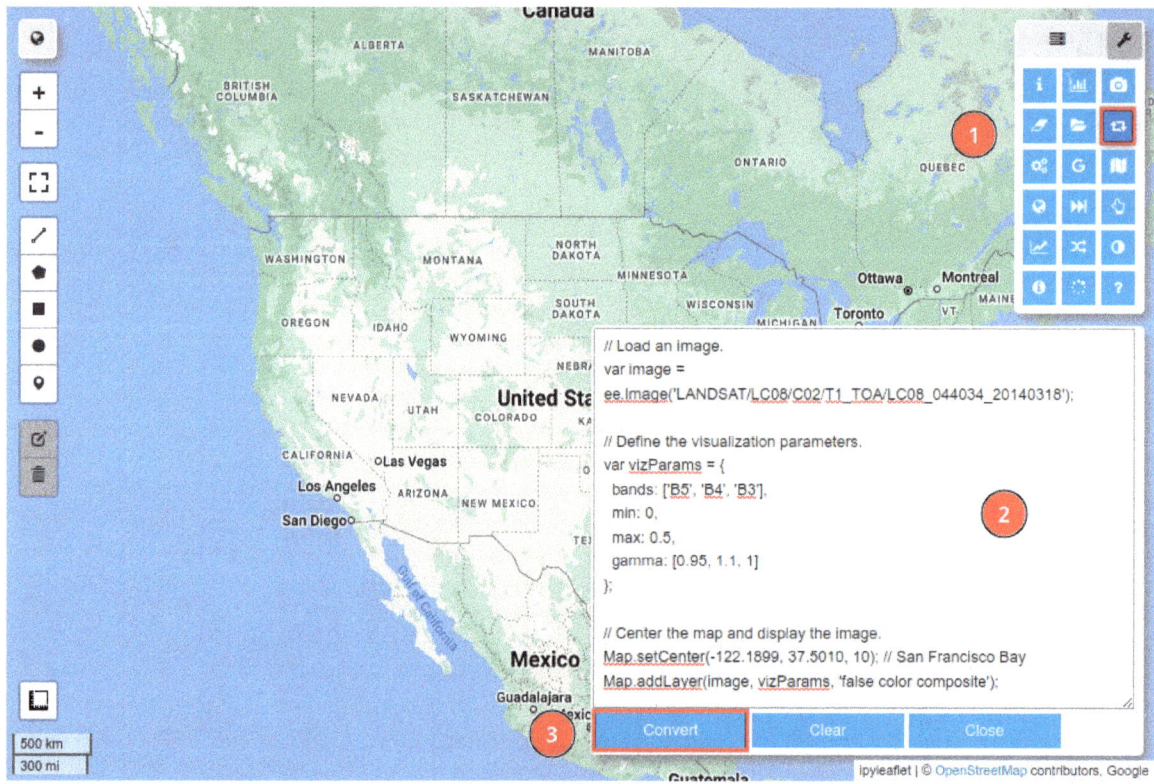

Figure 3.34: Converting an Earth Engine JavaScript to Python.

If you are running Jupyter Notebook, a new code cell will be created below the map and populated with the converted Python code. If you are running JupyterLab or Google Colab, you will need to copy the converted Python code and paste it into a new cell manually as JupyterLab and Google Colab does not support create a new cell programmatically (see jupyterlab issues 8415[87]). The converted Python code should look like this:

```python
# Load an image.
image = ee.Image('LANDSAT/LC08/C02/T1_TOA/LC08_044034_20140318')

# Define the visualization parameters.
vizParams = {'bands': ['B5', 'B4', 'B3'], 'min': 0, 'max': 0.5, 'gamma': [0.95, 1.1, 1]}

# Center the map and display the image.
Map.setCenter(-122.1899, 37.5010, 10)
# San Francisco Bay
Map.addLayer(image, vizParams, 'False color composite')
```

Executing the code cell above will generate a map that looks like Fig. 3.35.

Batch conversion

To convert an Earth Engine JavaScript snippet to Python programmatically, wrap the JavaScript snippet with triple quotes and assign it to a variable. Then, use the `js_snippet_to_py()` function to convert the

---

[87]jupyterlab issues 8415: tiny.geemap.org/krwm

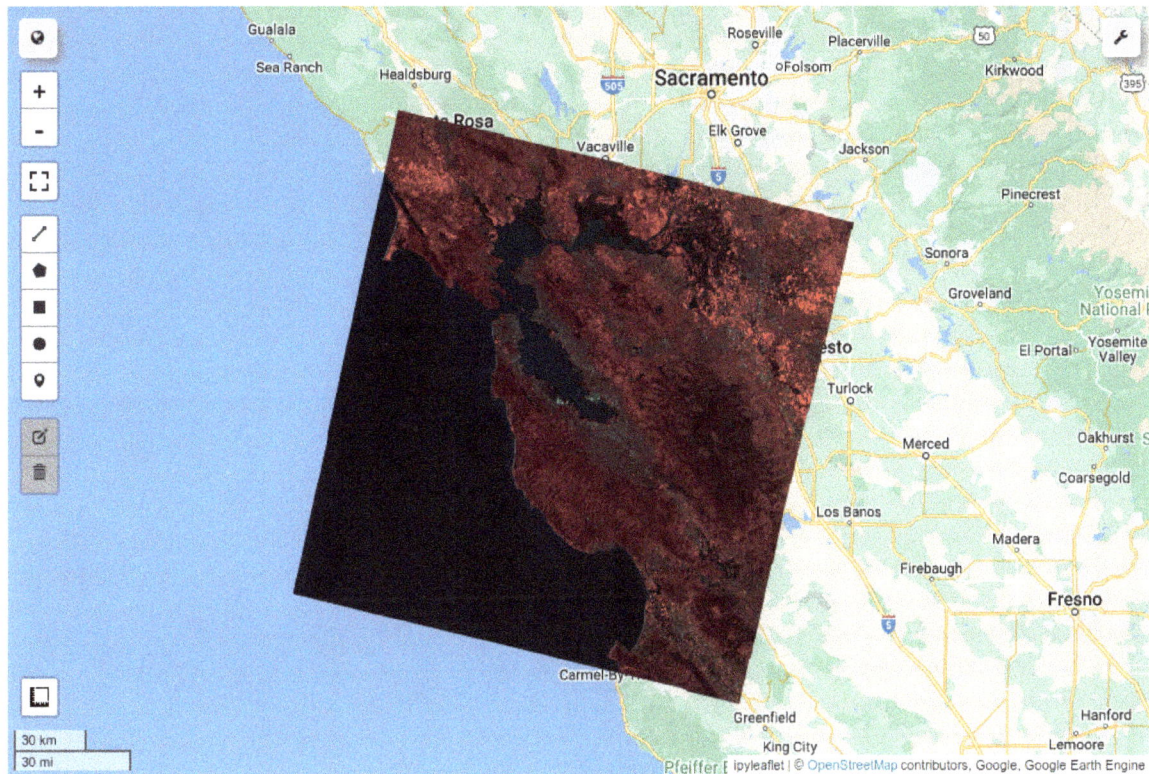

Figure 3.35: A map generated from a Python snippet converted from an Earth Engine JavaScript snippet.

JavaScript snippet to Python. For example, let's use the Normalized Difference Water Index (NDWI) example[88] from the Earth Engine Developer Guide:

```
snippet = """
// Load an image.
var image = ee.Image('LANDSAT/LC08/C02/T1_TOA/LC08_044034_20140318');

// Create an NDWI image, define visualization parameters and display.
var ndwi = image.normalizedDifference(['B3', 'B5']);
var ndwiViz = {min: 0.5, max: 1, palette: ['00FFFF', '0000FF']};
Map.addLayer(ndwi, ndwiViz, 'NDWI');
Map.centerObject(image)
"""

geemap.js_snippet_to_py(snippet, add_new_cell=True, import_ee=False)
```

If you are running Jupyter Notebook or JupyterLab, a new code cell will be created and populated with the converted Python code as follows. If you are running Google Colab, you can use the interactive conversion method described in the previous section to convert the JavaScript snippet to Python. The js_snippet_to_py function will not work in Google Colab as it does not support creating a new code cell programmatically.

```
# Load an image.
image = ee.Image('LANDSAT/LC08/C02/T1_TOA/LC08_044034_20140318')

# Create an NDWI image, define visualization parameters and display.
ndwi = image.normalizedDifference(['B3', 'B5'])
```

[88]example: tiny.geemap.org/ximm

```
ndwiViz = {'min': 0.5, 'max': 1, 'palette': ['00FFFF', '0000FF']}
Map.addLayer(ndwi, ndwiViz, 'NDWI')
Map.centerObject(image)
Map
```

Executing the cell above will generate a map that looks like Fig. 3.36.

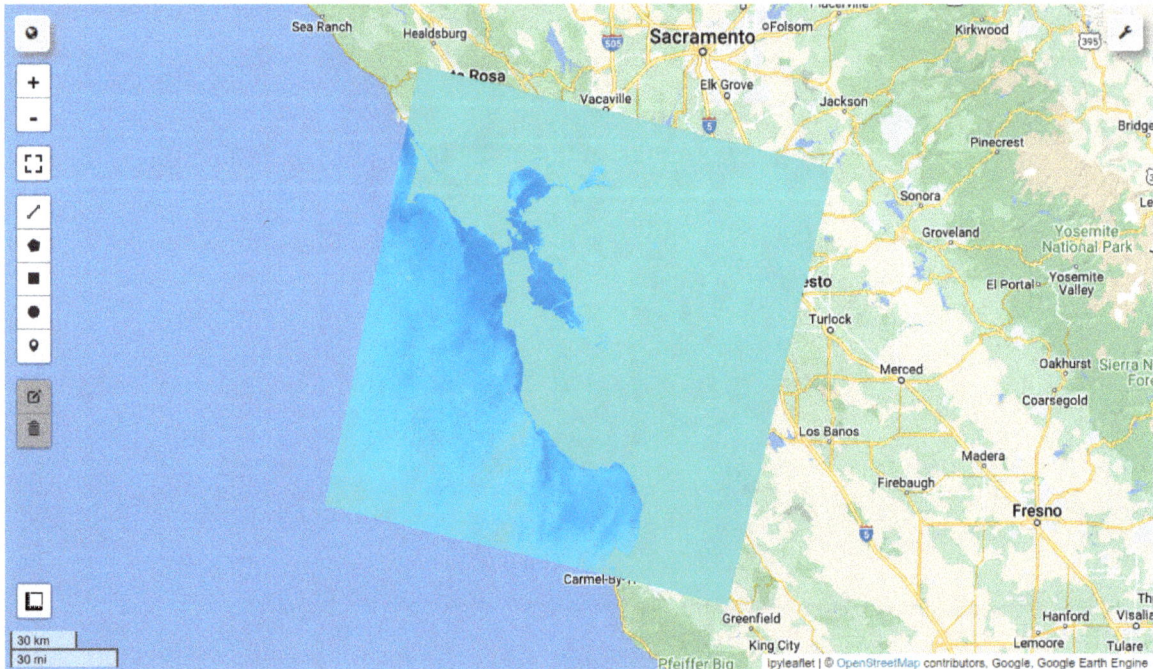

Figure 3.36: Landsat 8 NDWI, San Francisco bay area, USA. Cyan are low values, blue are high values.

In addition to converting Earth Engine JavaScript snippets to Python one by one, geemap provides a convenient way to perform batch conversions of multiple Earth Engine JavaScript code to Python scripts. By specifying the input directory containing the JavaScript files and the output directory, you can use the js_to_python_dir() function to recursively convert all the JavaScript files within the input directory to Python files within the output directory.

Furthermore, you can also use the py_to_ipynb_dir() function to recursively convert all Python files within the input directory to Jupyter Notebook files within the output directory. This feature makes it simple and efficient to convert multiple Earth Engine JavaScript snippets to Python.

```
import os
from geemap.conversion import *

out_dir = os.getcwd()
js_dir = get_js_examples(out_dir)
js_to_python_dir(in_dir=js_dir, out_dir=out_dir, use_qgis=False)
py_to_ipynb_dir(js_dir)
```

Note that the automated conversion functionality in geemap is designed to work effectively for most of the JavaScript examples found in the Earth Engine Developer Guide. Nevertheless, if your Earth Engine scripts have a complex structure (such as numerous nested functions or for loops), the automated conversion may not function correctly. In these instances, manual editing may be necessary after the conversion process is complete.

3.9 Calling JavaScript functions from Python

In the preceding section, we demonstrated how to convert Earth Engine JavaScript snippets to Python. However, for Earth Engine JavaScript snippets with complex structures (such as nested map functions or for loops), conversion to Python may prove challenging. Fortunately, it's possible to directly call Earth Engine JavaScript functions from Python using the Open Earth Engine Library (OEEL[89]). Geemap streamlines this process by providing a wrapper for OEEL, enabling you to call Earth Engine JavaScript functions from Python with just one line of code.

This section requires the installation of several software packages, including Node.js, Git, and OEEL. **Note that the section will not work in Google Colab**. It could be challenging to install these software packages on some operating systems. If you are unable to install these software packages, you can skip this section and move on to the next section.

This section presents four examples of how to call Earth Engine JavaScript functions from Python using OEEL. To utilize OEEL, you will need to install Node.js and Git. You can download Node.js from the Node.js website[90] and Git from the git-scm website[91]. Follow the instructions on the respective websites to complete the installation of Node.js and Git. Once Node.js and Git have been installed, you can install the OEEL Python package by executing the following command in the **Terminal** or **Anaconda Prompt**:

```
# %pip install oeel
```

Once the OEEL Python package has been installed successfully, you can import oeel and begin the initialization process by running the following code:

```
import oeel
```

The Open Earth Engine Library (OEEL)[92] contains a collection Earth Engine JavaScript functions, which can be called from Python. Once oeel is imported successfully, call the requiredJS() function to load the OEEL JavaScript library into Jupyter Notebook:

```
oeel = geemap.requireJS()
```

The OEEL JavaScript library is now accessible from Python. Simply type oeel. in a code cell and press Tab to see the list of available functions (see Fig. 3.37).

Figure 3.37: Accessing the OEEL JavaScript library through dot notation with autocompletion.

---

[89]OEEL: tiny.geemap.org/rfpf

[90]Node.js website: nodejs.org/en/download

[91]git-scm website: git-scm.com/downloads

[92]Open Earth Engine Library (OEEL): tiny.geemap.org/vpsu

The following example shows how to call the `oeel.Algorithms.Sentinel2.cloudfree()` function from the OEEL JavaScript library, specifying the `maxCloud` and `S2Collection` parameters for the function. The result is an `ImageCollection` containing Sentinel-2 imagery with a cloud pixel percentage less than the specified `maxCloud` threshold. The resulting `ImageCollection` can be further processed using other Earth Engine Python API functions, such as `filterDate()` and `size()`:

```python
ic = ee.ImageCollection("COPERNICUS/S2_SR")
icSize = (
    oeel.Algorithms.Sentinel2.cloudfree(maxCloud=20, S2Collection=ic)
    .filterDate('2020-01-01', '2020-01-02')
    .size()
)
print('Cloud free imagery: ', icSize.getInfo())
```

In addition to calling JavaScript functions in the OEEL JavaScript library, you can also call JavaScript functions in custom JavaScript modules provided that they are formatted as follows:

```javascript
var generateRasterGrid = function(origin, dx, dy, proj) {
    var coords = origin.transform(proj).coordinates();
    origin = ee.Image.constant(coords.get(0)).addBands(ee.Image.constant(coords.get(1)));

    var pixelCoords = ee.Image.pixelCoordinates(proj);

    var grid = pixelCoords
      .subtract(origin)
      .divide([dx, dy]).floor()
      .toInt().reduce(ee.Reducer.sum()).bitwiseAnd(1).rename('grid');

    var xy = pixelCoords.reproject(proj.translate(coords.get(0), coords.get(1)).scale(dx, dy));

    var id = xy.multiply(ee.Image.constant([1, 1000000])).reduce(ee.Reducer.sum()).rename('id');

    return grid
      .addBands(id)
      .addBands(xy);
  }

exports.generateRasterGrid = generateRasterGrid;
```

It is important to note that any custom JavaScript module you wish to call from Python must contain functions (e.g., var `generateRasterGrid` = `function()`) and at least one export statement (e.g., `exports.generateRasterGrid` = `generateRasterGrid`) in order to make JavaScript functions accessible from Python. An example of a custom JavaScript module can be found here[93].

Geemap can load a custom JavaScript module from an HTTP URL, a local file, or an Earth Engine user repository. To start, try loading a custom JavaScript module from an HTTP URL using the following code:

```python
url = 'https://tinyurl.com/27xy4oh9'
lib = geemap.requireJS(lib_path=url)
```

Note that the HTTP URL should be passed to the `lib_path` parameter. Once the custom JavaScript module has been loaded successfully, you can call `lib.availability` to view the list of available functions:

```python
lib.availability
```

There are three functions available in this custom JavaScript module:

---

[93]here: tinyurl.com/27xy4oh9

```
{'generateGrid': 'function',
 'generateRasterGrid': 'function',
 'grid_test': 'function'}
```

You can call the functions in the custom JavaScript module by typing `lib.function_name()` and provide the required parameters for the function. For example, to call the `generateGrid()` function, type the following code:

```
grid = lib.generateGrid(-180, -70, 180, 70, 10, 10, 0, 0)
grid.first()
```

The `generateGrid()` function generates a coordinate grid based on specified parameters and returns a `FeatureCollection`. Use `grid.first()` to print out the first feature in the collection. The output should look like this:

```
{'type': 'Feature',
 'geometry': {'geodesic': False,
  'type': 'Polygon',
  'coordinates': [[[-180, -50],
    [-170, -50],
    [-170, -40],
    [-180, -40],
    [-180, -50]]]},
 'id': '0',
 'properties': {'nx': '0', 'ny': '0'}}
```

The `FeatureCollection` resulting from the JavaScript function is like any other `FeatureCollection` generated with the Earth Engine Python API. You can add it to the map using the `Map.addLayer()` method:

```
Map = geemap.Map()
style = {'fillColor': '00000000'}
Map.addLayer(grid.style(**style), {}, 'Grid')
Map
```

The resulting map should look like Fig. 3.38.

Similarly, you can load a custom JavaScript module from a local file. After running the example above, a file named `grid.js` should have been created in the current working directory. Open the file in a text editor, and you should see the `grid_test()` function in the file that looks like this:

```
var grid_test = function() {

    var gridRaster = generateRasterGrid(ee.Geometry.Point(0, 0), 10, 10, ee.Projection('EPSG:4326'))
    Map.addLayer(gridRaster.select('id').randomVisualizer(), {}, 'Grid raster')

    var gridVector = generateGrid(-180, -70, 180, 70, 10, 10, 0, 0)
    Map.addLayer(gridVector, {}, 'Grid vector')
}
```

Note that the function uses `Map`, which is a global variable in the Earth Engine JavaScript environment referring to the interactive map in the Earth Engine JavaScript Code Editor. To use the `Map` object in Python, you first need to create a `Map` object using `geemap.Map()` and pass it to the `Map` parameter of the `requireJS()` function. Try **restarting the kernel** and run the following code:

```
Map = geemap.Map()
lib = geemap.requireJS(lib_path='grid.js', Map=Map)

grid = lib.generateGrid(-180, -70, 180, 70, 10, 10, 0, 0)
grid.first()
```

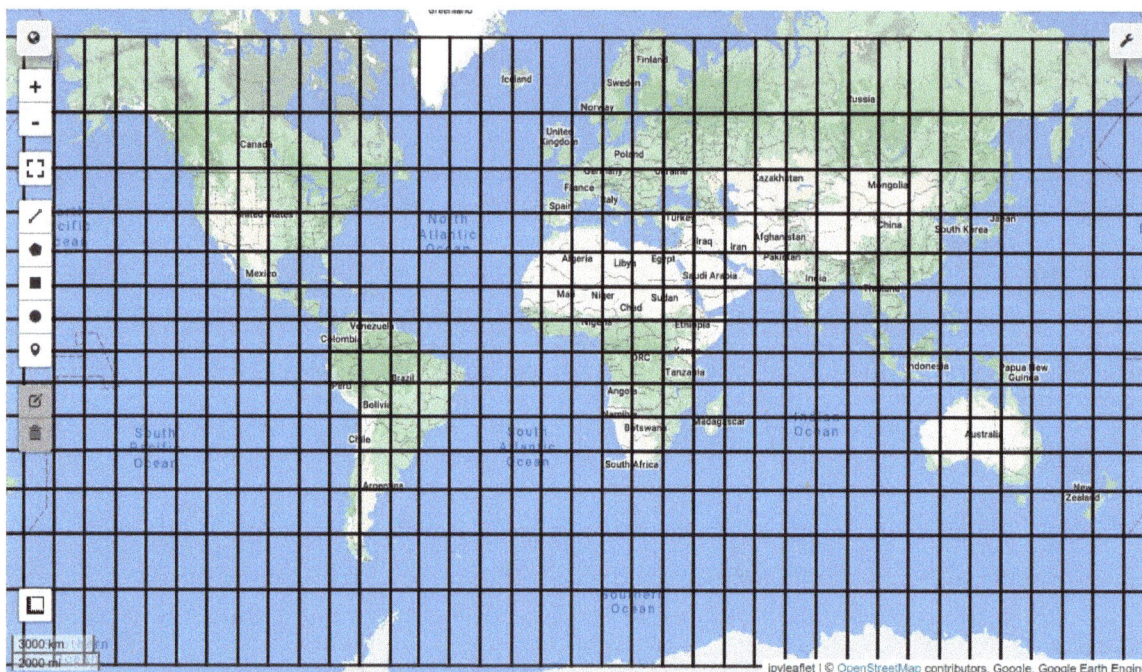

Figure 3.38: A coordinate grid generated by a JavaScript function.

Then, call the `grid_test()` function to generate coordinate grids and add them to the map (see Fig. 3.39).

```
lib.grid_test()
Map
```

Lastly, you can load a custom JavaScript module from an Earth Engine user repository. To do this, you need to specify the user repository name and the path to the JavaScript module, such as users/gena/-packages:grid:

```
lib = geemap.requireJS('users/gena/packages:grid')
```

When loading a custom JavaScript module from an Earth Engine user repository, the entire repository will be cloned to the current working directory. If this is the first time you are cloning an Earth Engine user repository, you will be prompted to authenticate with Google. Simply follow the on-screen instructions to authenticate and clone the user repository to your computer.

Once the custom JavaScript module has been loaded successfully, call `lib.availability` to view the list of available functions. Then, select one of the available functions and provide the necessary parameters. For instance, you can call the `generateGrid()` function to generate a coordinate grid:

```
grid = lib.generateGrid(-180, -70, 180, 70, 10, 10, 0, 0)

Map = geemap.Map()
style = {'fillColor': '00000000'}
Map.addLayer(grid.style(**style), {}, 'Grid')
Map
```

The map should look like Fig. 3.38.

Figure 3.39: A raster grid generated by a JavaScript function.

3.10 Summary

Throughout this chapter, we have covered a variety of GEE data types, such as `Image`, `ImageCollection`, `Geometry`, `Feature`, and `FeatureCollection`. Additionally, we delved into the Earth Engine Data Catalog and accessed a variety of geospatial datasets. By the end of this chapter, you should feel confident in your ability to search and load datasets from the Earth Engine Data Catalog as well as convert Earth Engine JavaScript snippets to Python.

In the following chapter, we will focus on how to access and visualize geospatial datasets that are stored locally.

4. Using Local Geospatial Data

4.1 Introduction

Geospatial data is available in various formats, with vector data and raster data being the two primary types. Vector data is typically represented as a collection of points, lines, or polygons, while raster data is represented as a two-dimensional grid of values. In this chapter, we will explore how to read and visualize both vector data and raster data using geemap.

We will specifically focus on raster data formats such as GeoTIFF, Cloud Optimized GeoTIFF (COG[59]), and SpatioTemporal Asset Catalog (STAC[94]), as well as vector data formats such as Shapefile, GeoJSON, KML, and GeoDataFrame. Additionally, we will cover how to convert local vector and raster data into Earth Engine data formats.

Finally, we will introduce several useful functions for downloading vector data from OpenStreetMap. By the end of this chapter, you should be able to access and visualize local vector and raster data with confidence.

4.2 Technical requirements

To follow along with this chapter, you will need to have geemap and several optional dependencies installed. If you have already followed Section 1.5 - *Installing geemap*, then you should already have a conda environment with all the necessary packages installed. Otherwise, you can create a new conda environment and install pygis[30] with the following commands, which will automatically install geemap and all the required dependencies:

```
conda create -n gee python
conda activate gee
conda install -c conda-forge mamba
mamba install -c conda-forge pygis
```

Next, launch JupyterLab by typing the following commands in your terminal or Anaconda prompt:

```
jupyter lab
```

Alternatively, you can use geemap with a Google Colab cloud environment without installing anything on your local computer. Click 04_local_data.ipynb[95] to launch the notebook in Google Colab.

Once in Colab, you can uncomment the following line and run the cell to install pygis, which includes geemap and all the necessary dependencies:

```
# %pip install pygis
```

The installation process may take 2-3 minutes. Once pygis has been installed successfully, click the **RESTART RUNTIME** button that appears at the end of the installation log or go to the **Runtime** menu and select **Restart runtime**. After that, you can start coding.

---

[94]STAC: stacspec.org

[95]04_local_data.ipynb: tiny.geemap.org/ch04

To begin, import the necessary libraries that will be used in this chapter:

```
import ee
import geemap
```

Initialize the Earth Engine Python API:

```
geemap.ee_initialize()
```

If this is your first time running the code above, you will need to authenticate Earth Engine first. Follow the instructions in Section 1.7 - *Earth Engine authentication* to authenticate Earth Engine.

4.3   Local raster datasets

In this section, we will explore how to work with local raster datasets. Built upon localtileserver[20] , geemap simplifies the process of loading local raster datasets with one line of code. The following examples demonstrate how to visualize single-band and multi-band raster datasets.

Single-band imagery

Let's start by downloading a sample single-band imagery from GitHub. We will use a small subset of the Shuttle Radar Topography Mission (SRTM) dataset. The download_file() function can be used to download a file from a URL, including Google Drive. The following code downloads a GeoTIFF file from GitHub:

```
url = 'https://github.com/giswqs/data/raw/main/raster/srtm90.tif'
filename = 'dem.tif'
geemap.download_file(url, filename)
```

Next, we can use the Map.add_raster() method to add the local raster to the map. We can specify parameters such as cmap and layer_name. Since this is an elevation dataset, it would be useful to add a corresponding colorbar to the map. The following code shows how to add the raster dataset and a colorbar to the map:

```
Map = geemap.Map()
Map.add_raster(filename, cmap='terrain', layer_name="DEM")
vis_params = {'min': 0, 'max': 4000, 'palette': 'terrain'}
Map.add_colorbar(vis_params, label='Elevation (m)')
Map
```

Open the layer control widget and drag the slider for the layer to change the layer opacity interactively (see Fig. 4.1).

The example above uses the terrain palette. To check the list of available palettes, use the list_colormaps() function in the geemap.colormaps module:

```
import geemap.colormaps as cm
cm.list_colormaps()
```

To preview the available palettes, use the plot_colormaps() function:

```
cm.plot_colormaps(width=12, height=0.4)
```

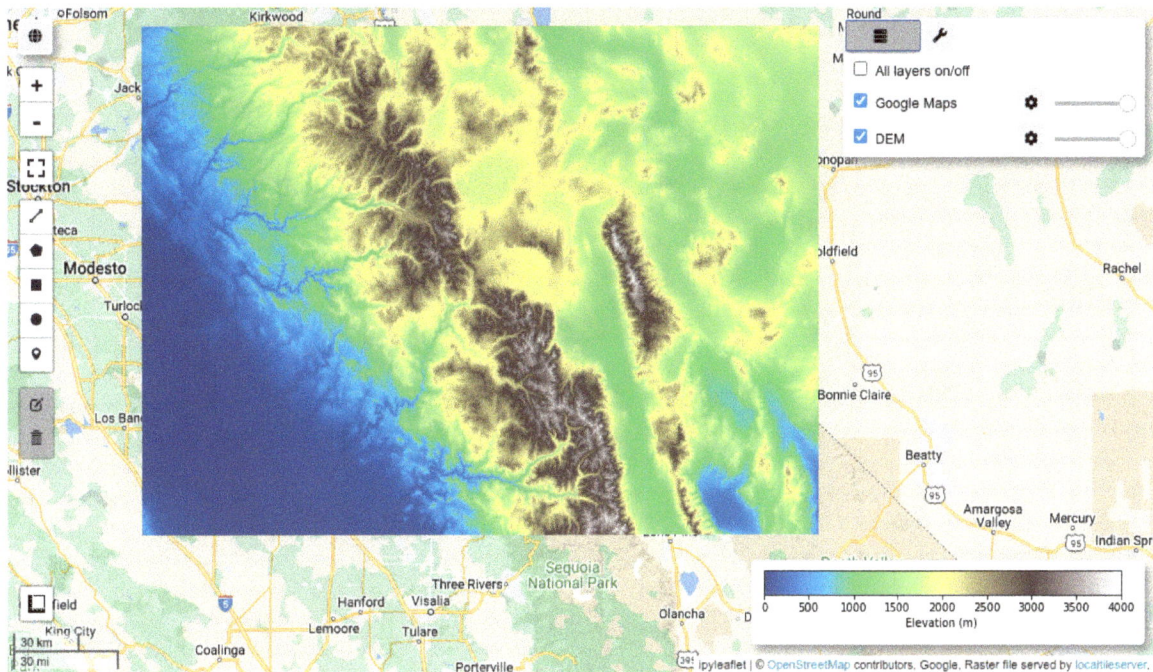

Figure 4.1: Visualizing a single-band imagery.

Multi-band imagery

Similar to the single-band imagery introduced earlier, you can also visualize multi-band local raster datasets using `Map.add_raster()`. However, you cannot specify the `palette` parameter for multi-band raster datasets. Instead, you can use the `bands` parameter to indicate which bands to visualize.

Let's download a sample multi-band imagery from GitHub. This is a small subset of a Landsat imagery with four spectral bands, including the red, green, blue, and NIR bands:

```
url = 'https://github.com/giswqs/leafmap/raw/master/examples/data/cog.tif'
filename = 'cog.tif'
geemap.download_file(url, filename)
```

To load a multi-band raster dataset using `Map.add_raster()`, we can specify the `bands` parameter. It is important to note that band indices start from 1.

For example, let's say we have a satellite imagery with four spectral bands: red, green, blue, and near-infrared (NIR). We can load this imagery and visualize the NIR, red, and green bands using the following code:

```
Map = geemap.Map()
Map.add_raster(filename, band=[4, 1, 2], layer_name="Color infrared")
Map
```

In this example, the bands parameter is set to [4, 1, 2], indicating that the NIR band is the first band, followed by the red band, and the green band. The map should look like Fig. 4.2.

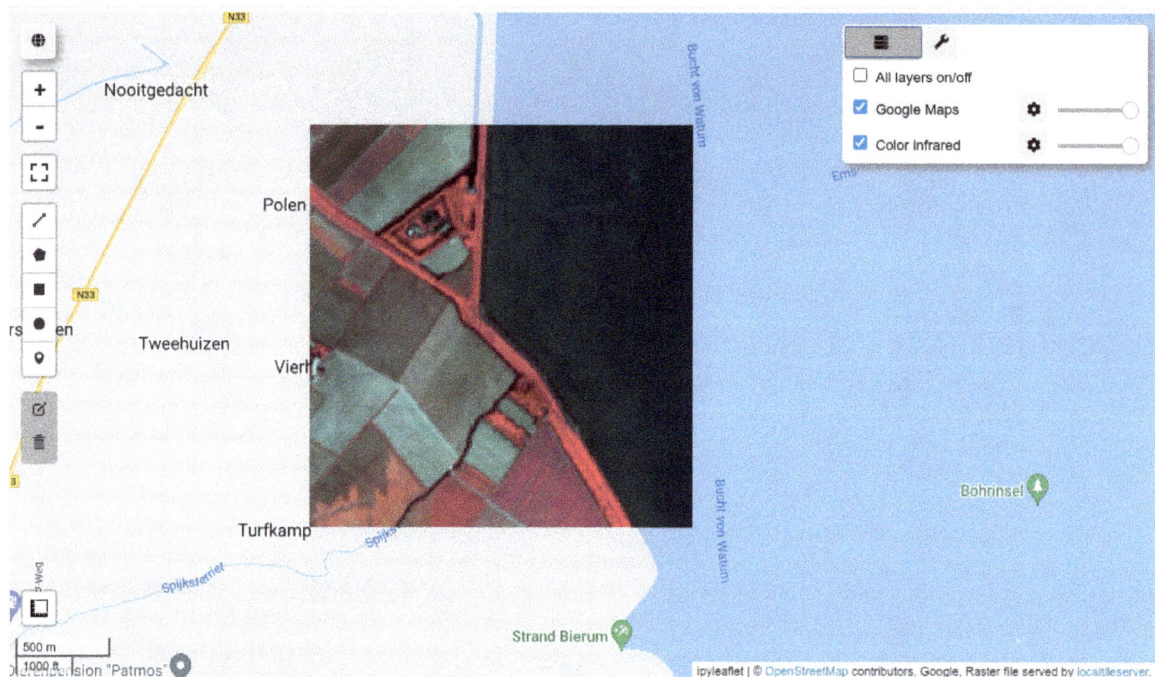

Figure 4.2: Visualizing a multi-band raster image.

Interactive raster GUI

Besides loading raster datasets programmatically with `Map.add_raster()`, you can also load raster datasets interactively. Click the **Open** button in the toolbar control to open the raster dataset selection dialog. Click the **Raster** tab on the popup dialog and then click the **Select** button to select a raster dataset form the local disk. Once the raster dataset is selected, you can customize the visualization parameters such as band combination, minimum value, maximum value, nodata value, and color palette. For example, you can open the DEM dataset that we downloaded earlier and set the color palette to `terrain` using the following steps:

1. Click the **Open** button in the toolbar control.

2. Click the **Raster** tab and choose the DEM dataset.

3. Enter a layer name, such as "Raster".

4. Adjust visualization parameters as desired, such as band, min, max, and nodata.

5. Set the color palette to `terrain`.

6. Click **Apply** to add the raster to the map.

By using this interactive method, you can quickly explore and visualize different raster datasets without having to write any code (see Fig. 4.3).

For loading a multi-band raster dataset interactively, specify the band parameter, such as [4, 1, 2] as shown in Fig. 4.4.

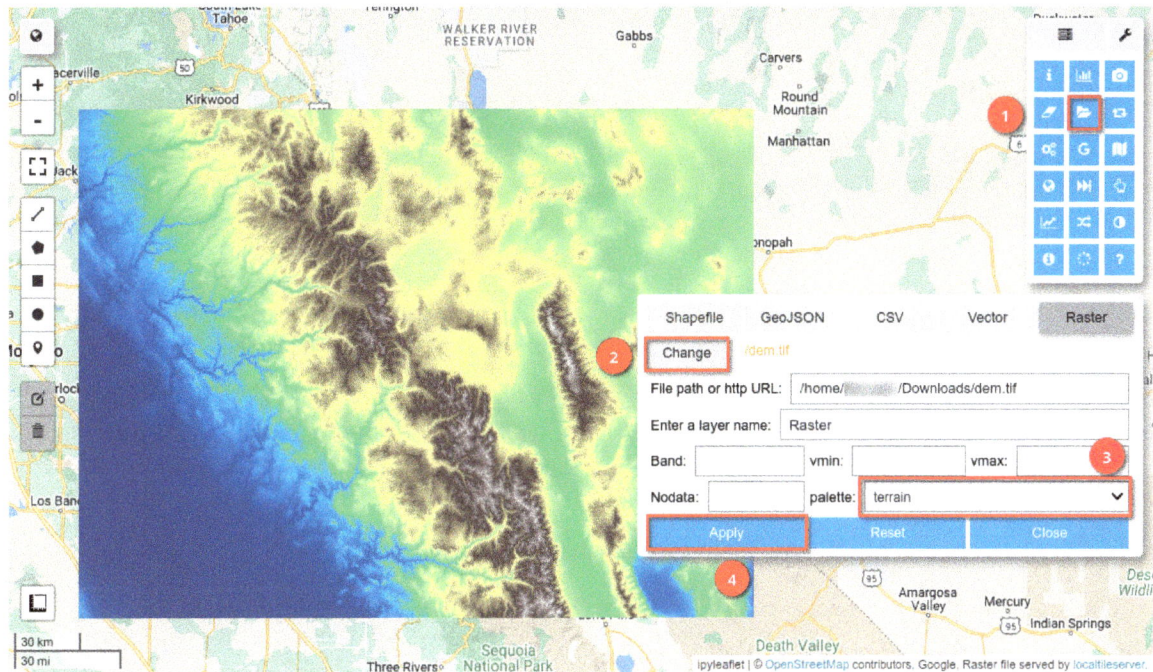

Figure 4.3: An interactive GUI for loading and visualizing single-band raster datasets.

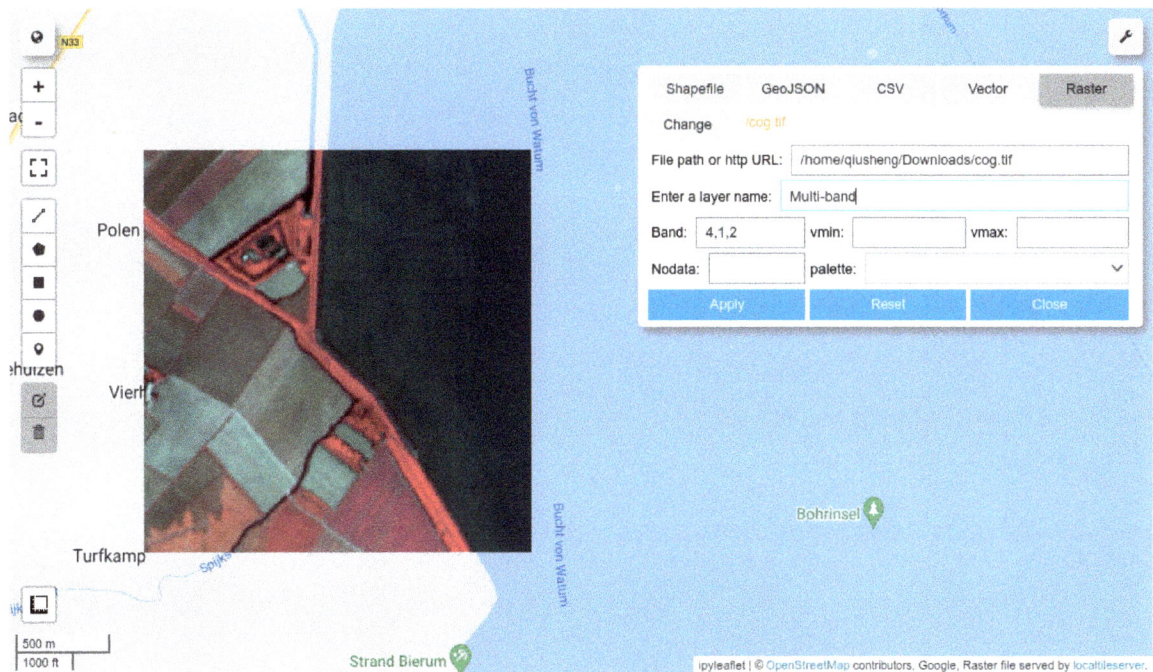

Figure 4.4: An interactive GUI for loading and visualizing multi-band raster datasets.

4.4   Cloud Optimized GeoTIFF (COG)

Visualizing COG

A Cloud Optimized GeoTIFF (COG) is a regular GeoTIFF file that is optimized for serving on an HTTP file server, with an internal organization that enables more efficient workflows on a cloud environment. This is achieved by allowing clients to issue HTTP GET requests to ask for only the parts of a file that they need. For more information about COG, please visit `https://www.cogeo.org`.

The Maxar Open Data Program[96] hosts numerous COG datasets that are freely available to the public. With the help of TiTiler[97] , geemap provides functionality to load COG datasets onto the map with just a few lines of code.

As an example, let's take a look at how to visualize COG datasets for mapping the California and Colorado fires[98] from the Maxar Open Data Program. We will begin by inspecting the pre-event imagery acquired on February 16, 2018. First we need to find the URL of the COG dataset:

```
url = 'https://tinyurl.com/24bo8umr'
```

To get the bounding box the dataset covers, you can use the `cog_bounds` function:

```
geemap.cog_bounds(url)
```

The output is a list of four numbers, representing the bounding box of the dataset in the order of [minX, minY, maxX, maxY]:

```
[-108.63447456563128,
 38.963980238226654,
 -108.38008268561431,
 40.025815049929754]
```

To retrieve the center coordinates of the dataset you can use the `cog_center` function:

```
geemap.cog_center(url)
```

The output is a tuple of two numbers representing the center coordinates of the dataset in the order of [longitude, latitude]:

```
(-108.5072786256228, 39.49489764407821)
```

To retrieve band names from the file, you can use the `cog_bands` function:

```
geemap.cog_bands(url)
```

The output is a list of band names:

```
['1', '2', '3']
```

To convert the COG dataset to a tile layer that can be added to the map you can use the `cog_tile` function:

```
geemap.cog_tile(url)
```

The output is a tile layer URL generated by TiTiler:

---

[96]Maxar Open Data Program: www.maxar.com/open-data

[97]TiTiler: developmentseed.org/titiler

[98]California and Colorado fires: tiny.geemap.org/ztxf

```
'https://titiler.xyz/cog/tiles/WebMercatorQuad/{z}/{x}/{y}@1x?url=https%3A%2F%2Fopendata.digitalglobe.com%2
    Fevents%2Fcalifornia-fire-2020%2Fpre-event%2F2018-02-16%2Fpine-gulch-fire20%2F1030010076004E00.tif&bidx=1&
    bidx=2&bidx=3&rescale=1.0%2C251.0'
```

Next, create an interactive map and add the pre-event COG dataset to the map using the `Map.add_cog_layer()` method:

```
Map = geemap.Map()
Map.add_cog_layer(url, name="Fire (pre-event)")
Map
```

You can also add a post-event COG dataset to the map for comparison.

```
url2 = 'https://tinyurl.com/2awjl66w'
Map.add_cog_layer(url2, name="Fire (post-event)")
Map
```

Pan and zoom the map to a specific area and toggle the layers on and off to see the difference between the pre-event and post-event imagery (see Fig. 4.5).

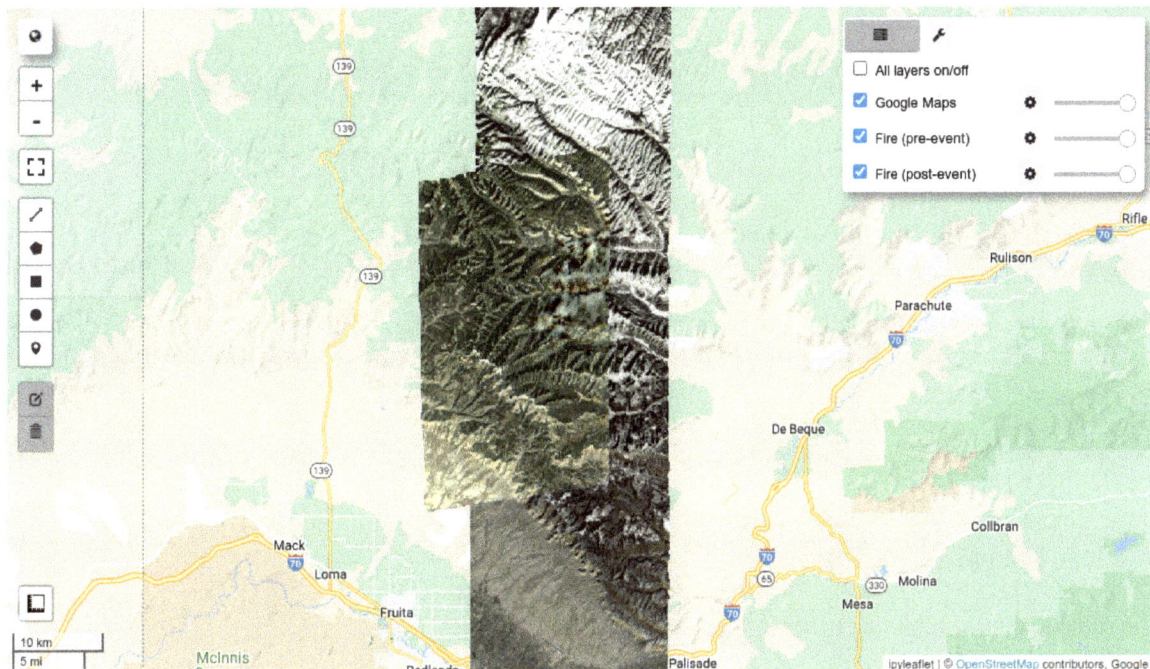

Figure 4.5: Visualizing Cloud Optimized GeoTIFFs from the Maxar Open Data Program.

Creating COG

There are several ways to create Cloud Optimized GeoTIFFs (COGs). One common method is to convert a regular GeoTIFF to COG, which can be done using geemap with just one line of code.

To determine whether a dataset is already in COG format, you can use the `cog_validate()` function. This function returns `True` if the dataset is in COG format, and `False` if it is not.

Let's verify if the DEM dataset we used earlier is in COG format. You can pass an HTTP URL directly to the `cog_validate()` function as follows:

```
url = "https://github.com/giswqs/data/raw/main/raster/srtm90.tif"
geemap.cog_validate(url)
```

The output should look like this:

```
(True,
 [],
 ['The file is greater than 512xH or 512xW, it is recommended to include internal overviews'])
```

As can be seen from the output above, the dataset is a COG, but it does not include internal overviews, which does not fully utilize the COG capabilities. To get more information about the dataset, set verbose=True:

```
geemap.cog_validate(url, verbose=True)
```

The output gives more information about the dataset:

```
Info(Path='https://github.com/giswqs/data/raw/main/raster/srtm90.tif', Driver='GTiff', COG=True, Compression='
    LZW', ColorSpace=None, COG_errors=None, COG_warnings=['The file is greater than 512xH or 512xW, it is
    recommended to include internal overviews'], Profile=Profile(Bands=1, Width=4269, Height=2465, Tiled=True,
    Dtype='int16', Interleave='BAND', AlphaBand=False, InternalMask=False, Nodata=None, ColorInterp=('gray',)
    , ColorMap=False, Scales=(1.0,), Offsets=(0.0,)), GEO=Geo(CRS='EPSG:4326', BoundingBox
    =(-120.75592734926073, 36.63401097629554, -117.30451019614512, 38.6269234341147), Origin
    =(-120.75592734926073, 38.6269234341147), Resolution=(0.0008084837557075694, -0.0008084837557075694),
    MinZoom=7, MaxZoom=11), Tags={'Image Metadata': {'AREA_OR_POINT': 'Area'}, 'Image Structure': {'
    COMPRESSION': 'LZW', 'INTERLEAVE': 'BAND'}}, Band_Metadata={'Band 1': BandMetadata(Description='elevation
    ', ColorInterp='gray', Offset=0.0, Scale=1.0, Metadata={})}, IFD=[IFD(Level=0, Width=4269, Height=2465,
    Blocksize=(256, 256), Decimation=0)])
```

To convert a regular GeoTIFF dataset to a full-fledged COG dataset, you can use the image_to_cog() function.

The image_to_cog() function takes a file path or URL to the input GeoTIFF file, and an output file path to save the COG file. For example, to convert a local GeoTIFF file to COG, you can use the following code:

```
out_cog = "cog.tif"
geemap.image_to_cog(url, out_cog)
```

Now, let's validate the new COG dataset:

```
geemap.cog_validate(out_cog)
```

The output should look like this:

```
(True, [], [])
```

As can be seen above, the new dataset is a full-fledged COG. Now, let's load the two raster datasets onto the map. The local raster dataset can be loaded using Map.add_raster() while the remote COG file can be loaded using Map.add_cog_layer():

```
Map = geemap.Map()
Map.add_raster(out_cog, cmap="terrain", layer_name="Local COG")
Map.add_cog_layer(url, cmap="gist_earth", name="Remote COG")
vis_params = {'min': 0, 'max': 4000, 'palette': 'gist_earth'}
Map.add_colorbar(vis_params, label='Elevation (m)')
Map
```

The resulting map should look like Fig. 4.6.

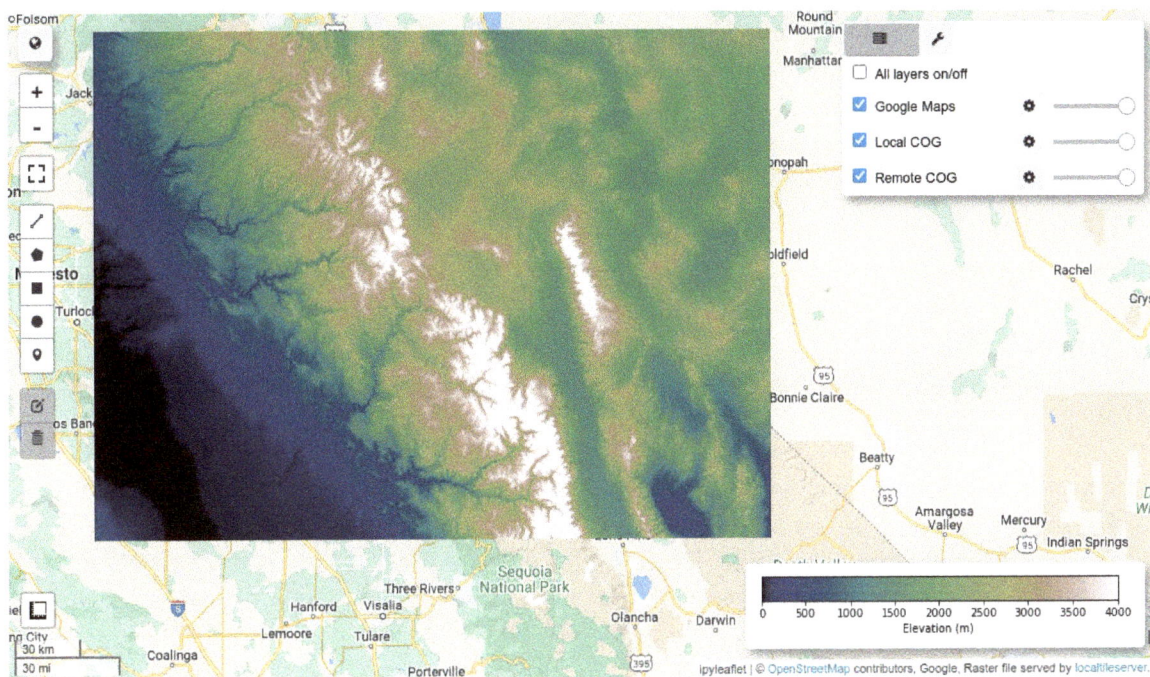

Figure 4.6: Creating a Cloud Optimized GeoTIFF (COG) and visualizing it.

Converting NumPy arrays to COG

The previous section demonstrated how to convert a regular GeoTIFF to a COG. When performing geospatial analysis with Python, it is common to store intermediate results in the form of NumPy arrays in memory. To save a NumPy array to a COG file, you can use the `numpy_to_cog()` function.

The following example shows how to compute the Normalized Difference Vegetation Index (NDVI) from a Landsat imagery and save the result as a COG. First, let's download the sample dataset using the `download_file()` function:

```
url = 'https://github.com/giswqs/leafmap/raw/master/examples/data/cog.tif'
in_cog = 'cog.tif'
out_cog = "ndvi.tif"
geemap.download_file(url, in_cog, overwrite=True)
```

Next, you can read the COG as a NumPy array using the `image_to_numpy` function:

```
arr = geemap.image_to_numpy(in_cog)
```

You can check the shape of the NumPy array by displaying the `shape` property:

```
arr.shape
```

The output is a tuple containing the number of bands, height, and width of the COG:

```
(4, 206, 343)
```

Compute the NDVI using the equation `(nir - red)/(nir + red)`:

```
ndvi = (arr[3] - arr[0]) / (arr[3] + arr[0])
```

Similarly, check the shape of the NDVI NumPy array, which should have the same height and width as the original COG file:

```
ndvi.shape
```

The output should look like this:

```
(206, 343)
```

To convert the NumPy array containing the NDVI values to a COG file, we can use the `numpy_to_cog()` function. The `profile` parameter of this function can be a file path to an existing COG file or a dictionary containing the profile information such as `driver`, `dtype`, `compress`, and so on.

In this example, we can use the profile information from the original COG file used to compute the NDVI values. Note that using the profile information from the original COG file ensures that the output COG file has the same properties as the input COG file, such as data type, compression, and so on. For a complete list of properties, see the GDAL GeoTIFF creation options[99].

```
geemap.numpy_to_cog(ndvi, out_cog, profile=in_cog)
```

Next, you can add the resulting COG onto the map and visualize it with a color palette (Fig. 4.7):

```
Map = geemap.Map()
Map.add_raster(in_cog, band=[4, 1, 2], layer_name="Color infrared")
Map.add_raster(out_cog, cmap="Greens", layer_name="NDVI")
Map
```

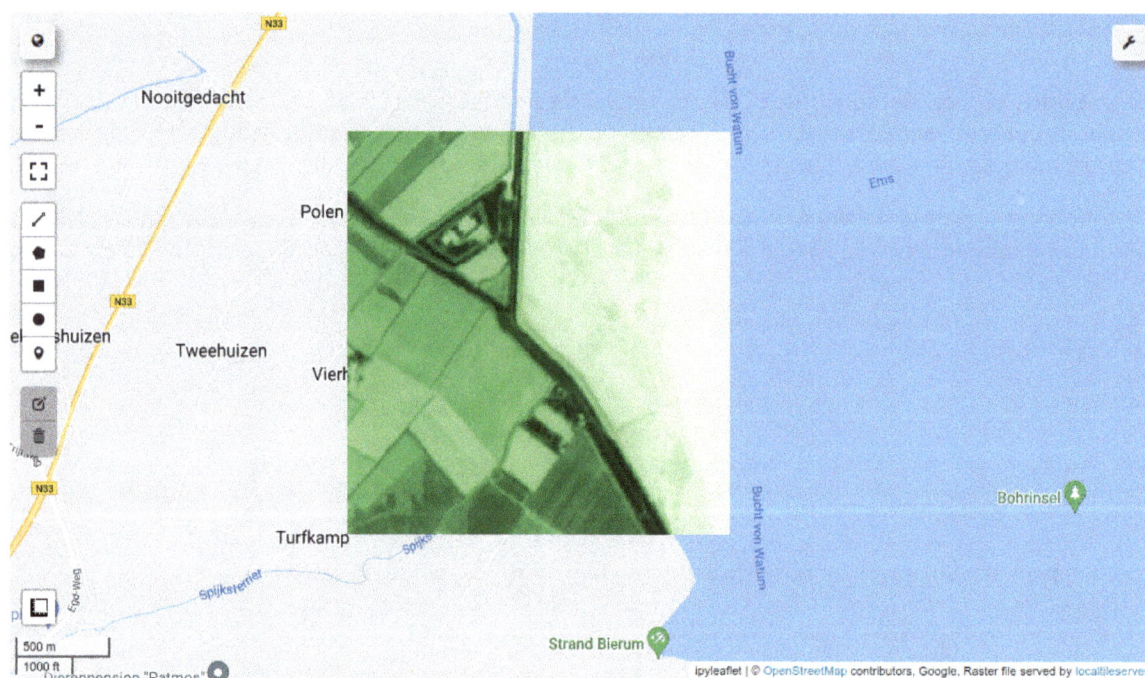

Figure 4.7: Visualizing an NDVI image on the map.

---

[99]GDAL GeoTIFF creation options: tiny.geemap.org/npvm

Clipping image by mask

When performing remote sensing analysis, we often need to extract a smaller area of interest (AOI) from a larger raster image. There are several reasons for doing so, such as reducing the size of the dataset, reducing the processing time, focusing on a specific feature, removing unwanted features, and so on. In this section, we will demonstrate how to extract a smaller AOI from a larger raster image using the `clip_image()` function. First, let's download a raster dataset from the web:

```
url = 'https://github.com/giswqs/data/raw/main/raster/srtm90.tif'
dem = 'dem.tif'
geemap.download_file(url, dem)
```

Next, create an interactive map and add the raster dataset to the map:

```
Map = geemap.Map()
Map.add_raster(dem, cmap='terrain', layer_name="DEM")
Map
```

Next, define a mask to extract the image. The mask can be a string representing a file path to a vector dataset (e.g., geojson, shp), or a list of coordinates (e.g., [[lon,lat], [lon,lat]]), or a dictionary representing a GeoJSON feature (e.g., `Map.user_roi`). For example, the mask can be a filepath to a vector dataset:

```
mask = (
    'https://raw.githubusercontent.com/giswqs/leafmap/master/examples/data/mask.geojson'
)
```

Or you can draw a polygon on the map, then use `Map.user_roi` as the mask:

```
mask = Map.user_roi
```

Or specify a list of coordinates [lon, lat] as the mask:

```
mask = [
    [-119.679565, 37.256566],
    [-119.679565, 38.061067],
    [-118.24585, 38.061067],
    [-118.24585, 37.256566],
    [-119.679565, 37.256566],
]
```

Finally, clip the raster image by the mask and add the clipped image to the map (Fig. 4.8):

```
output = 'clip.tif'
geemap.clip_image(dem, mask, output)
Map.add_raster(output, cmap='coolwarm', layer_name="Clip Image")
Map
```

4.5 SpatioTemporal Asset Catalog (STAC)

The SpatioTemporal Asset Catalog (STAC) specification provides a common language to describe a range of geospatial information so that it can more easily be indexed and discovered. A **SpatioTemporal Asset** is any file that represents information about the earth captured in a certain space and time. STAC aims to enable that next generation of geospatial search engines, while also supporting web best practices so geospatial information is more easily surfaced in traditional search engines. More information about STAC can be found at the STAC website[100]. The STAC Index website[101] is a one-stop-shop for discovering STAC catalogs, collections, APIs, software and tools. In this example, we will use a

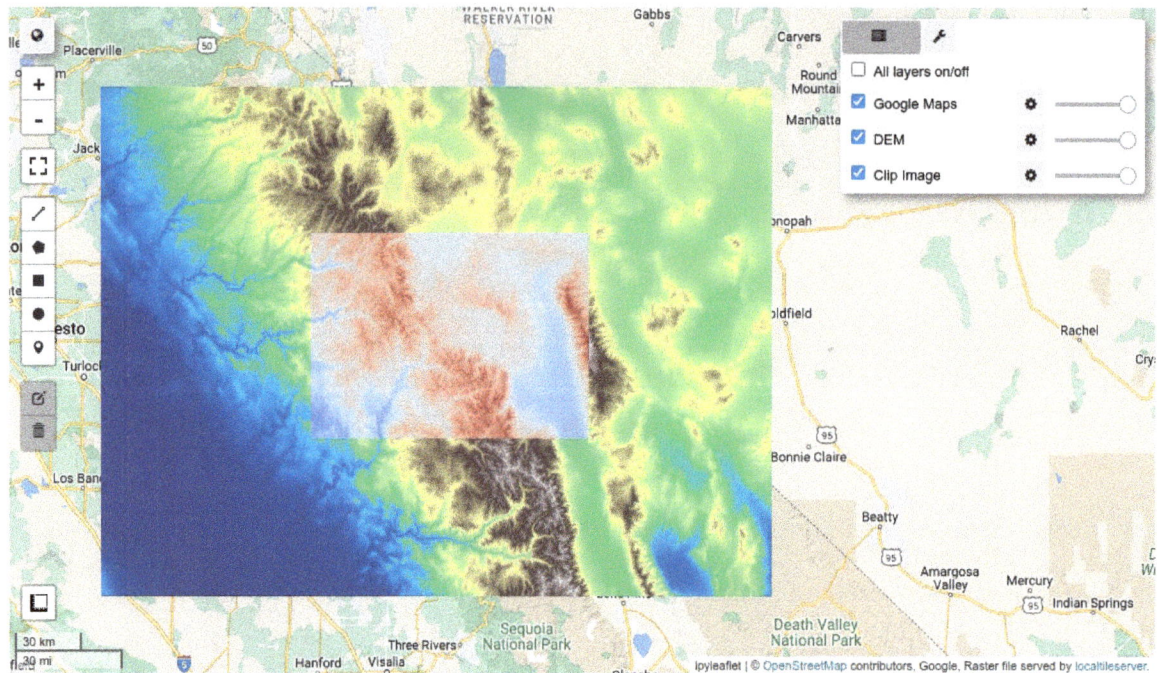

Figure 4.8: Clipping a raster image by mask.

STAC item from the SPOT Orthoimages of Canada[102] available through the link below:

```
url = 'https://tinyurl.com/22vptbws'
```

To get the bounding box the dataset covers you can call the `stac_bounds` function:

```
geemap.stac_bounds(url)
```

The output should look like this:

```
[-111.6453245, 60.59892389999882, -110.1583693, 61.30928879999903]
```

To retrieve the center coordinates of the dataset you can call the `stac_center` function:

```
geemap.stac_center(url)
```

The output should look like this:

```
(-110.90184690000001, 60.95410634999892)
```

To retrieve band names you can use the `stac_bands` function:

```
geemap.stac_bands(url)
```

The output should look like this:

```
['pan', 'B1', 'B2', 'B3', 'B4']
```

---

[100]STAC website: stacspec.org

[101]STAC Index website: stacindex.org

[102]SPOT Orthoimages of Canada: tiny.geemap.org/ngbw

To convert the STAC asset to a tile layer that can be added to the map you can call the `stac_tile` function:

```
geemap.stac_tile(url, bands=['B3', 'B2', 'B1'])
```

The output should look like this:

```
'https://titiler.xyz/stac/tiles/WebMercatorQuad/{z}/{x}/{y}@1x?url=https%3A%2F%2Fcanada-spot-ortho.s3.amazonaws
    .com%2Fcanada_spot_orthoimages%2Fcanada_spot5_orthoimages%2FS5_2007%2FS5_11055_6057_20070622%2
    FS5_11055_6057_20070622.json&assets=B3&assets=B2&assets=B1'
```

Lastly, to create an interactive map and add the STAC asset to the map you can use the `add_stac_layer()` function. You can also specify the bands to use for visualization, for example, to visualize the panchromatic band, set `bands=['pan']`. To visualize the asset with a false-color composite, set `bands=['B3', 'B2', 'B1']`.

```
Map = geemap.Map()
Map.add_stac_layer(url, bands=['pan'], name='Panchromatic')
Map.add_stac_layer(url, bands=['B3', 'B2', 'B1'], name='False color')
Map
```

The resulting map should look like Fig. 4.9.

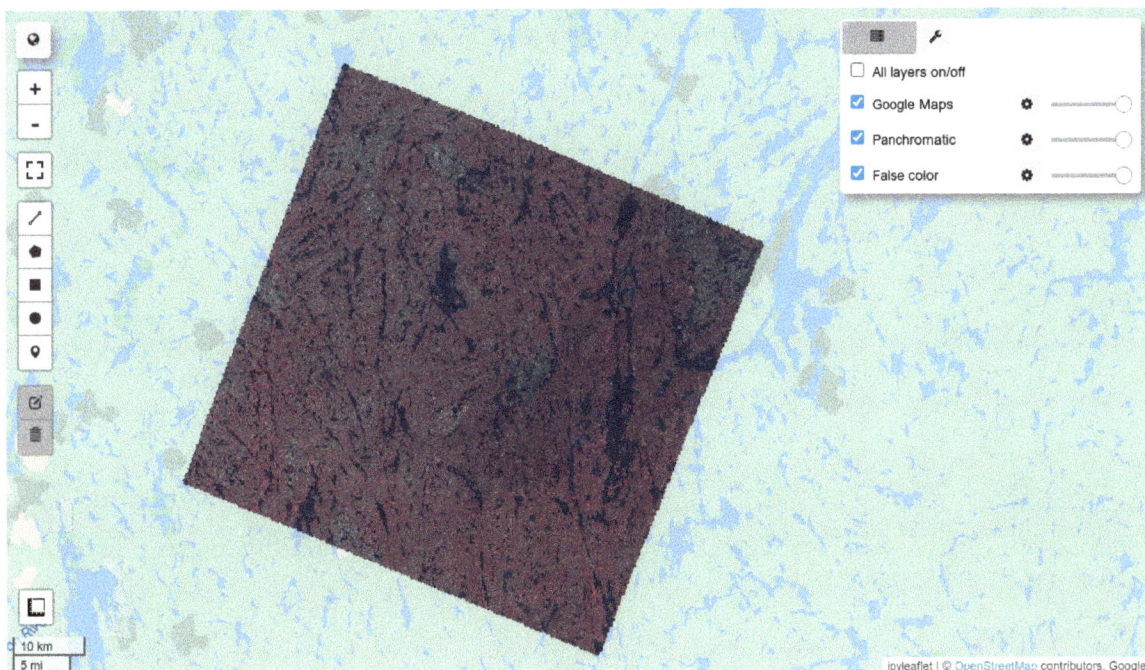

Figure 4.9: Visualizing a STAC asset on the map.

4.6 Vector datasets

Built upon fiona[103] and GeoPandas[19] , geemap can read almost any vector-based spatial data format, including GeoJSON, ESRI shapefiles, GeoPackage, KML, and many others. In this section, you will learn how to read and visualize local vector datasets using geemap with a few lines of code. Under the hood, geemap reads any vector dataset and convert it to GeoJSON to be added to the map.

---

[103]fiona: tiny.geemap.org/yhkv

GeoJSON

GeoJSON is an open standard based on the JSON format that's designed for representing simple geographical features, along with their non-spatial attributes. To add a GeoJSON dataset to the map, use the `add_geojson()` function. The dataset can be a local file path or an HTTP URL. The following example shows how to add a GeoJSON dataset of undersea Internet cables[104] to the map (Fig. 4.10):

```
in_geojson = (
    'https://github.com/gee-community/geemap/blob/master/examples/data/cable_geo.geojson'
)
Map = geemap.Map()
Map.add_geojson(in_geojson, layer_name="Cable lines", info_mode="on_hover")
Map
```

Figure 4.10: Visualizing undersea Internet cables.

By default, polylines are rendered as solid black lines. The `info_mode` parameter can be `on_hover`, `on_click`, or `None`.

- When `info_mode="on_hover"`, a popup dialog will display the feature attributes when the user hovers over a feature.
- When `info_mode="on_click"`, a popup dialog will display the feature attributes when the user clicks on the feature.
- When `info_mode=None`, no popup dialog will be displayed.

The feature style can be customized by providing a style callback function. The style callback function takes a feature and returns a dictionary of style properties for displaying that feature. The following example shows how to change the line color of the undersea Internet cable lines:

```
Map = geemap.Map()
Map.add_basemap("CartoDB.DarkMatter")
```

[104]undersea Internet cables: tiny.geemap.org/bqdh

```
callback = lambda feat: {"color": "#" + feat["properties"]["color"], "weight": 2}
Map.add_geojson(in_geojson, layer_name="Cable lines", style_callback=callback)
Map
```

The resulting map should look like Fig. 4.11.

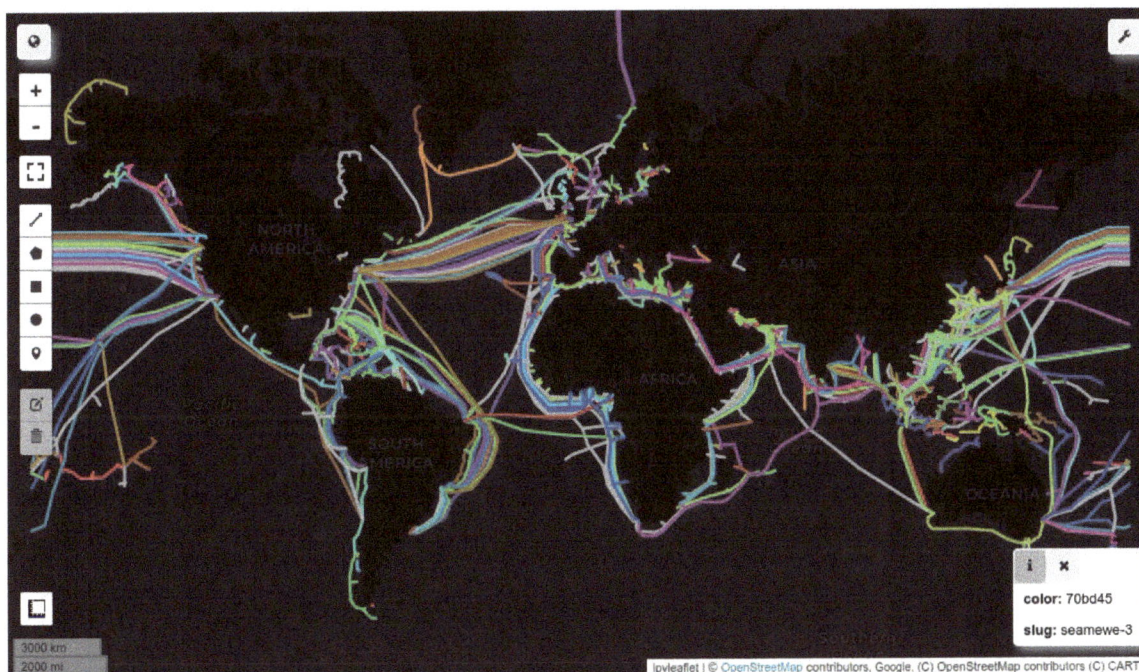

Figure 4.11: Visualizing undersea Internet cables with colors.

The anonymous lambda function in the example retrieves a hex color code from the feature's properties and returns a dictionary of style properties for the feature. The style callback function is called for each feature in the dataset to render it with color based on the feature's properties.

Polygon datasets can also be added to the map using `add_geojson()`. The following example shows how to add a GeoJSON dataset of countries to the map with random fill colors (Fig. 4.12):

```
url = "https://github.com/gee-community/geemap/blob/master/examples/data/countries.geojson"
Map = geemap.Map()
Map.add_geojson(
    url, layer_name="Countries", fill_colors=['red', 'yellow', 'green', 'orange']
)
Map
```

Specify the `fill_colors` parameter to provide a list of colors for rendering polygons. If you want more control over the style, then you can provide a callback function to the `style_callback` argument. The style callback function takes a feature and returns a dictionary of style properties for the feature, such as `color`, `weight`, and `fillColor`. The following example shows how to change the fill color of the country polygons using a style callback function (Fig. 4.13):

```
import random

Map = geemap.Map()

def random_color(feature):
    return {
        'color': 'black',
```

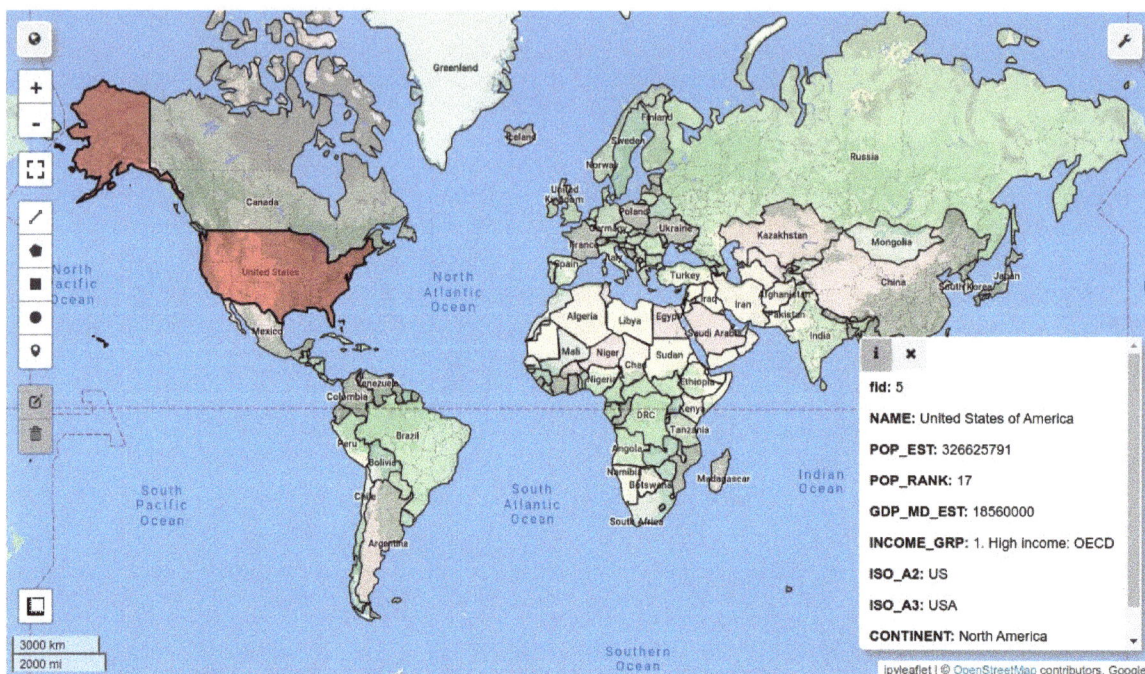

Figure 4.12: Visualizing a GeoJSON dataset of countries with random colors.

```
        'weight': 3,
        'fillColor': random.choice(['red', 'yellow', 'green', 'orange']),
    }

Map.add_geojson(url, layer_name="Countries", style_callback=random_color)
Map
```

Besides changing the base style of the polygons, you can also change the hover style of the polygons. Simply provide a dictionary of style properties to the hover_style parameter. The following example shows how to change the opacity of the polygons from 0.1 to 0.7 when the mouse hovers over them (Fig. 4.14):

```
Map = geemap.Map()

style = {
    "stroke": True,
    "color": "#0000ff",
    "weight": 2,
    "opacity": 1,
    "fill": True,
    "fillColor": "#0000ff",
    "fillOpacity": 0.1,
}

hover_style = {"fillOpacity": 0.7}

Map.add_geojson(url, layer_name="Countries", style=style, hover_style=hover_style)
Map
```

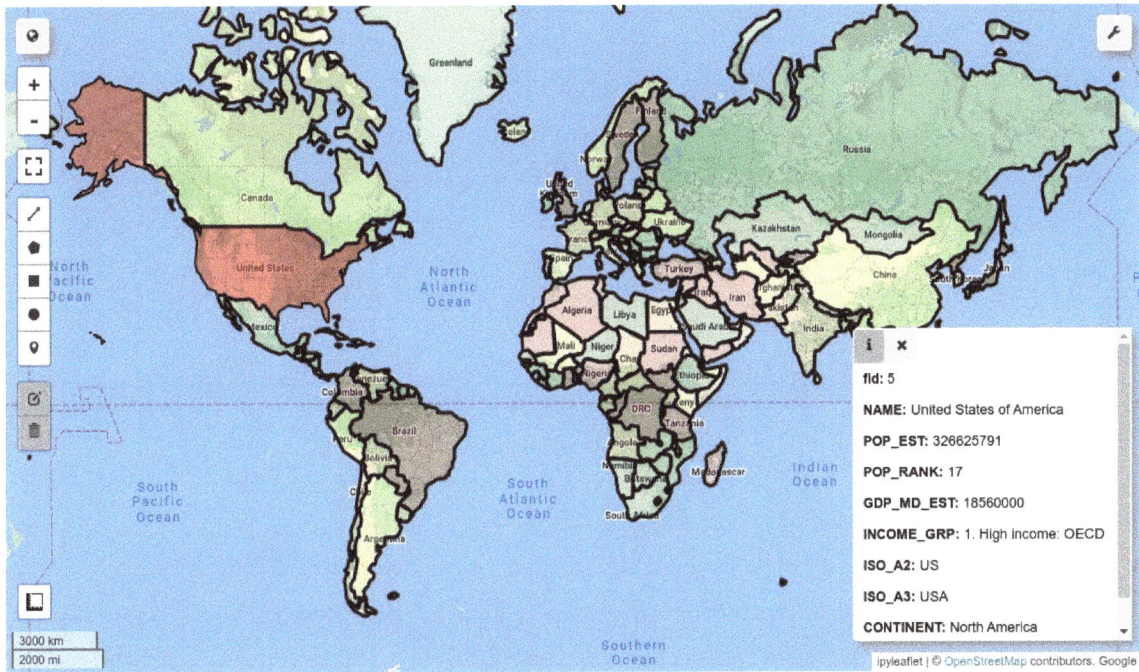

Figure 4.13: Visualizing a GeoJSON dataset of countries with a style callback.

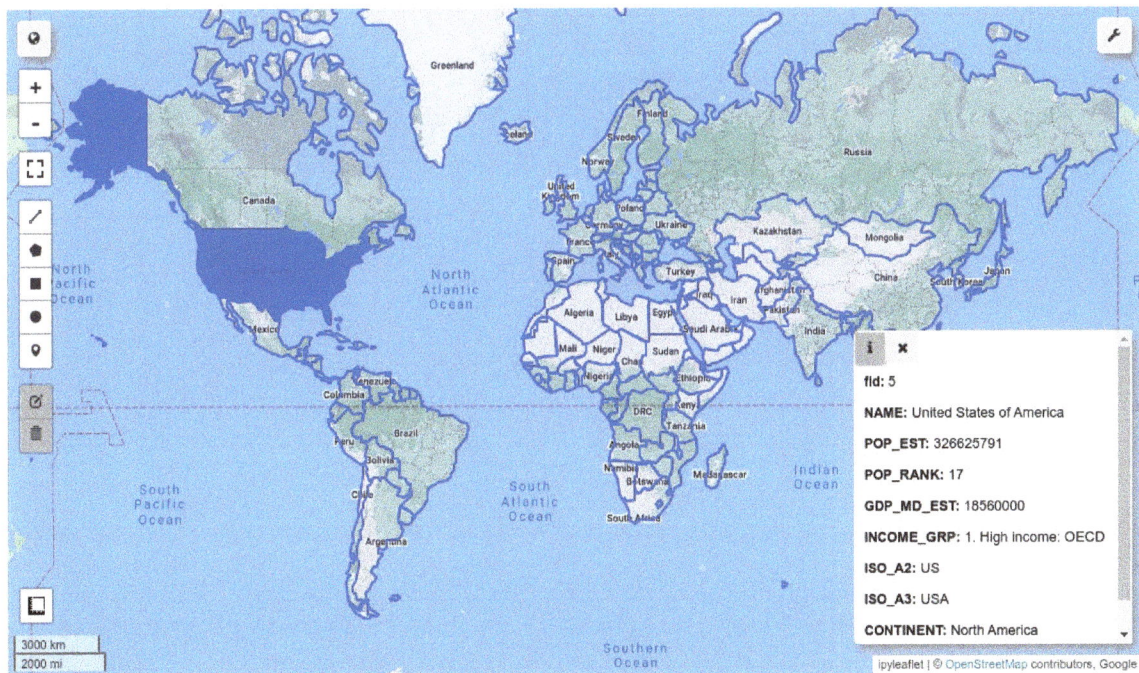

Figure 4.14: Visualizing a GeoJSON dataset of countries with a hover style callback.

Shapefile

To visualize an ESRI shapefile on the map, you can use the `add_shp()` method.

First, download a sample shapefile from the geemap GitHub repository:

```
url = "https://github.com/gee-community/geemap/blob/master/examples/data/countries.zip"
geemap.download_file(url)
```

The downloaded file is a ZIP archive which will be automatically unzipped to the current working directory. You should be able to find the shapefile named `countries.shp` in the current working directory. Then use the `add_shp()` method to add the shapefile to the map (Fig. 4.15). The style of the shapefile can be further customized by providing optional parameters, such as `style`, `hover_style`, `style_callback`, and `fill_colors`. These optional parameters are the same as those of the `add_geojson()` function introduced above.

```
Map = geemap.Map()
in_shp = "countries.shp"
Map.add_shp(in_shp, layer_name="Countries")
Map
```

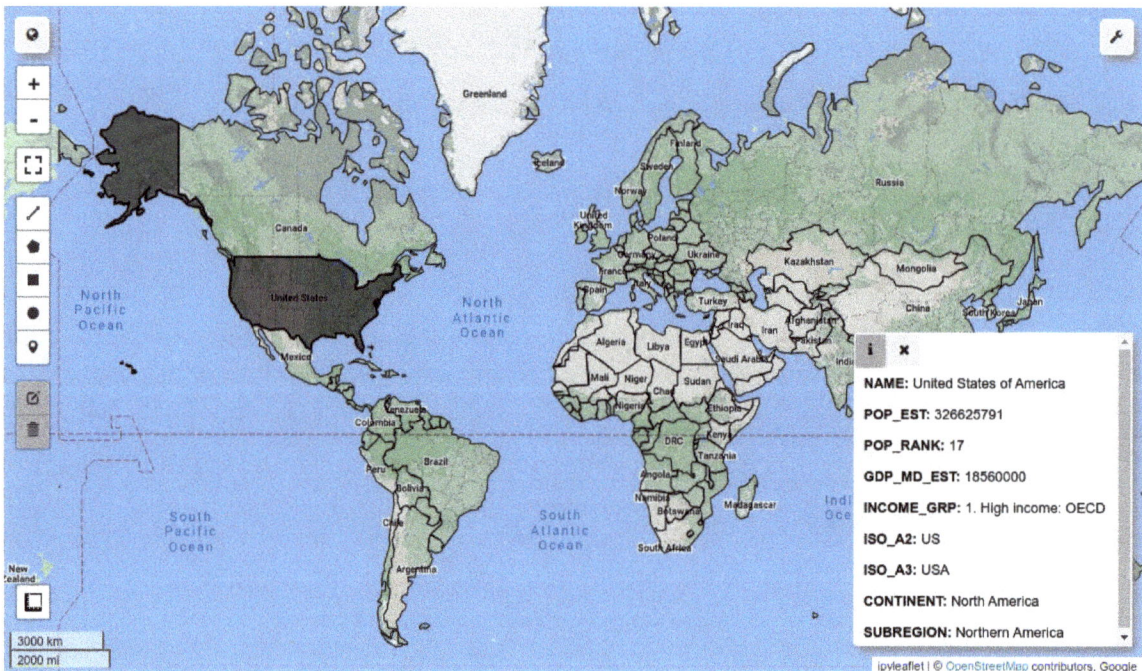

Figure 4.15: Visualizing a ESRI shapefile on the map.

KML

To visualize a KML/KMZ file on the map, use the `add_kml()` method. The input KML/KMZ file can be a local file or an HTTP URL. Let's use the sample KML file from the geemap GitHub repository:

```
in_kml = "https://github.com/gee-community/geemap/blob/master/examples/data/us_states.kml"
```

Use the `add_kml()` method to add the KML file to the map (Fig. 4.16). The style of the KML file can be customized by specifying optional parameters, such as `style`, `hover_style`, `style_callback`, and `fill_` colors. These optional parameters are the same as those of the `add_geojson()` method introduced above.

```
Map = geemap.Map(center=[40, -100], zoom=4)
Map.add_kml(in_kml, layer_name="US States")
Map
```

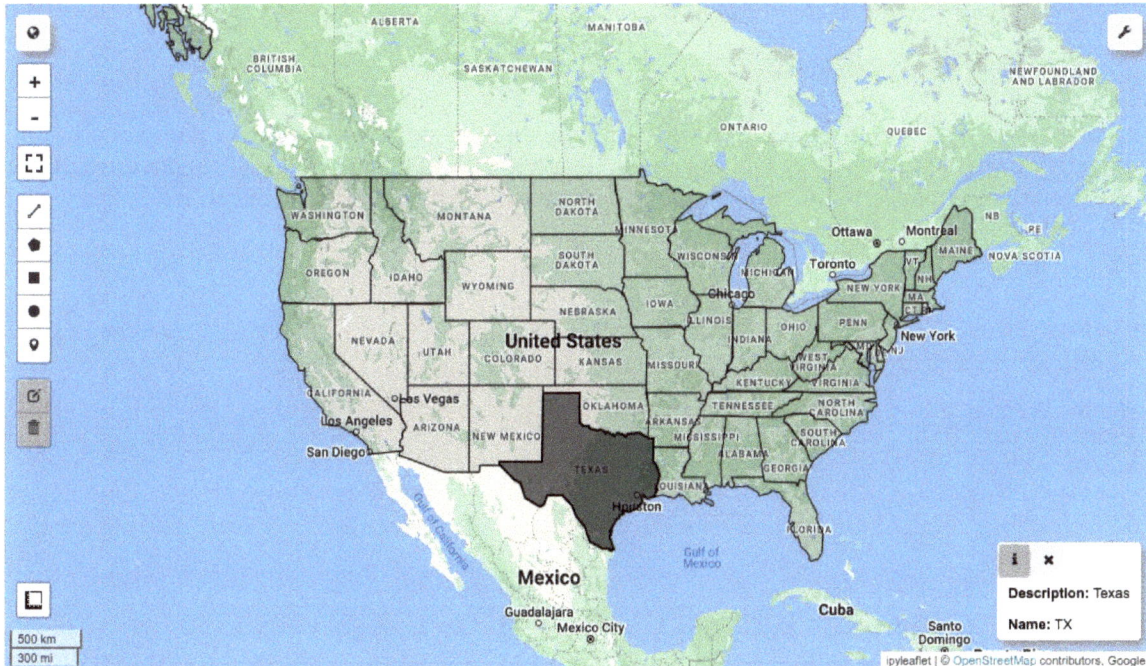

Figure 4.16: Visualizing a KML file on the map.

GeoDataFrame

Geemap also supports adding GeoPandas `GeoDataFrame` objects to the map. Under the hood, a **GeoDataFrame** is converted to a GeoJSON dataset and then added to the map using the `add_geojson()` method. First, import the geopandas package:

```
import geopandas as gpd
```

Then, you can use `read_file()` to read any vector file formats supported by GeoPandas, such as GeoPackage, Shapefile, GeoJSON, KML, etc. The following example shows how to create a **GeoDataFrame** from a Shapefile and add it to the map:

```
Map = geemap.Map(center=[40, -100], zoom=4)
gdf = gpd.read_file('countries.shp')
Map.add_gdf(gdf, layer_name="US States")
Map
```

The map should look like Fig. 4.15.

Other vector formats

For other vector formats, use the `add_vector()` method to add vector datasets to the map. The function supports any vector formats supported by GeoPandas, such as GeoPackage, Shapefile, GeoJSON, KML, etc. The following example loads a GeoPackage file and adds it to the map:

```
Map = geemap.Map()
data = 'https://github.com/gee-community/geemap/blob/master/examples/data/countries.gpkg'
Map.add_vector(data, layer_name="Countries")
Map
```

The map should look like Fig. 4.15.

4.7 Creating points from XY

You can convert a CSV file containing latitude and longitude coordinates to a vector dataset with commonly used formats, such as GeoJSON, shapefile, GeoPandas GeoDataFrame, etc. The converted vector dataset can then be added to the map using the add_vector() function introduced above.

CSV to vector

Let's use the sample dataset us_cities.csv from the geemap GitHub repository. First, use the csv_to_df() function to convert the CSV file to a Pandas DataFrame (Fig. 4.17):

```
data = 'https://github.com/gee-community/geemap/blob/master/examples/data/us_cities.csv'
geemap.csv_to_df(data)
```

	name	sov_a3	latitude	longitude	pop_max	region
0	San Bernardino	USA	34.12038	-117.30003	1745000	West
1	Bridgeport	USA	41.17998	-73.19996	1018000	Northeast
2	Rochester	USA	43.17043	-77.61995	755000	Northeast
3	St. Paul	USA	44.94399	-93.08497	734854	Midwest
4	Billings	USA	45.78830	-108.54000	104552	West
...	...	...	...	...	...	...

Figure 4.17: Converting a CSV file to a Pandas DataFrame.

Note the latitude and longitude columns in the DataFrame. To convert the CSV file to a GeoJSON object, use the csv_to_geojson() function. Specify the output filename and latitude and longitude parameters:

```
geemap.csv_to_geojson(
    data, 'cities.geojson', latitude="latitude", longitude='longitude'
)
```

To convert the CSV file to an ESRI Shapefile, use the csv_to_shp() function:

```
geemap.csv_to_shp(data, 'cities.shp', latitude="latitude", longitude='longitude')
```

To convert the CSV file to a GeoPandas GeoDataFrame, use the csv_to_gdf() function (Fig. 4.18):

```
geemap.csv_to_gdf(data, latitude="latitude", longitude='longitude')
```

	name	sov_a3	latitude	longitude	pop_max	region	geometry
0	San Bernardino	USA	34.12038	-117.30003	1745000	West	POINT (-117.30003 34.12038)
1	Bridgeport	USA	41.17998	-73.19996	1018000	Northeast	POINT (-73.19996 41.17998)
2	Rochester	USA	43.17043	-77.61995	755000	Northeast	POINT (-77.61995 43.17043)
3	St. Paul	USA	44.94399	-93.08497	734854	Midwest	POINT (-93.08497 44.94399)
4	Billings	USA	45.78830	-108.54000	104552	West	POINT (-108.54000 45.78830)
...	...	...	...	...	...	...	...

Figure 4.18: Converting a CSV file to a GeoPandas GeoDataFrame.

Note the `geometry` column in the GeoDataFrame compared to the Pandas DataFrame shown in Fig. 4.17. It is also possible to convert a CSV file to a GeoPackage file using the `csv_to_vector()` function:

```
geemap.csv_to_vector(data, 'cities.gpkg', latitude="latitude", longitude='longitude')
```

Adding points from XY

Besides converting a CSV file containing latitude and longitude coordinates to vector dataset, you can also use the `add_points_from_xy()` method to add data to the map. Let's use the same `us_cities.csv` dataset as above and the `us_regions.csv` dataset from the geemap GitHub repository:

```
cities = 'https://github.com/gee-community/geemap/blob/master/examples/data/us_cities.csv'
regions = (
    'https://github.com/gee-community/geemap/blob/master/examples/data/us_regions.geojson'
)
```

To add points from a CSV file, use the `add_points_from_xy()` method:

```
Map = geemap.Map(center=[40, -100], zoom=4)
Map.add_points_from_xy(cities, x="longitude", y="latitude")
Map
```

The result should look like Fig. 4.19.

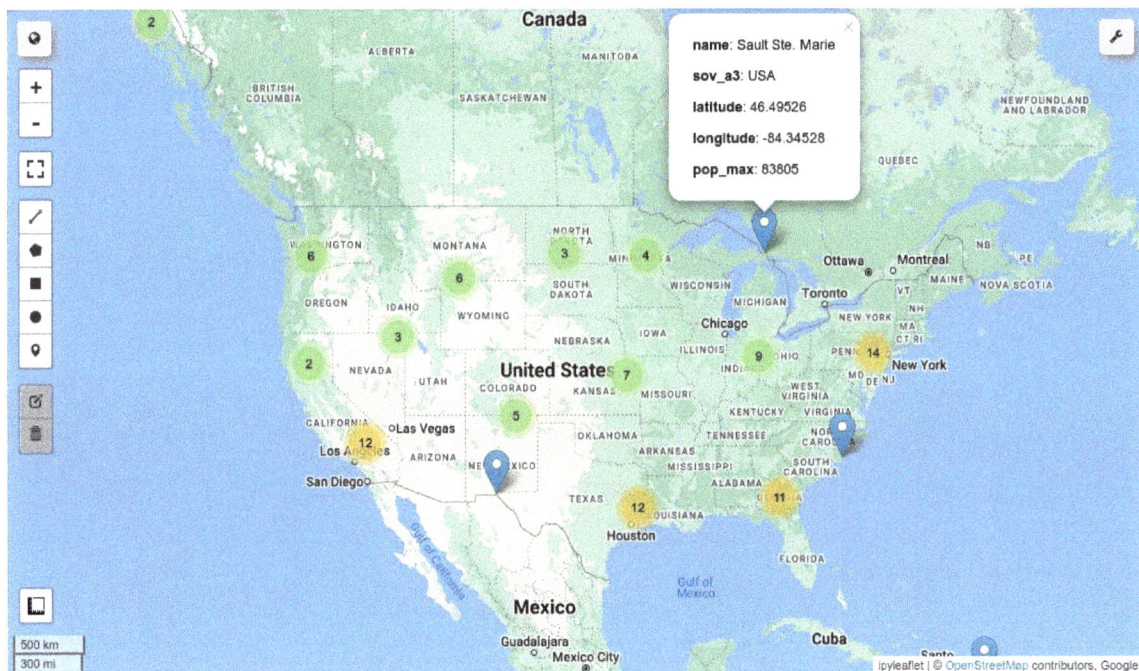

Figure 4.19: Adding points from a CSV file.

If the CSV file has a categorical column, you can specify the column name to use as the marker colors on the map. You can also specify the `icon_names` parameter to provide a list of icon names for the markers. The available icons can be found at the Font Awesome v4 website[105]. If `spin=True`, the icons will spin. Note that the `spin` parameter is only available for the ipyleaflet plotting backend and won't work for other plotting backends, such as folium. Set the `add_legend=True` parameter to add a legend to the map (Fig. 4.20).

```python
Map = geemap.Map(center=[40, -100], zoom=4)

Map.add_geojson(regions, layer_name='US Regions')

Map.add_points_from_xy(
    cities,
    x='longitude',
    y='latitude',
    layer_name='US Cities',
    color_column='region',
    icon_names=['gear', 'map', 'leaf', 'globe'],
    spin=True,
    add_legend=True,
)
Map
```

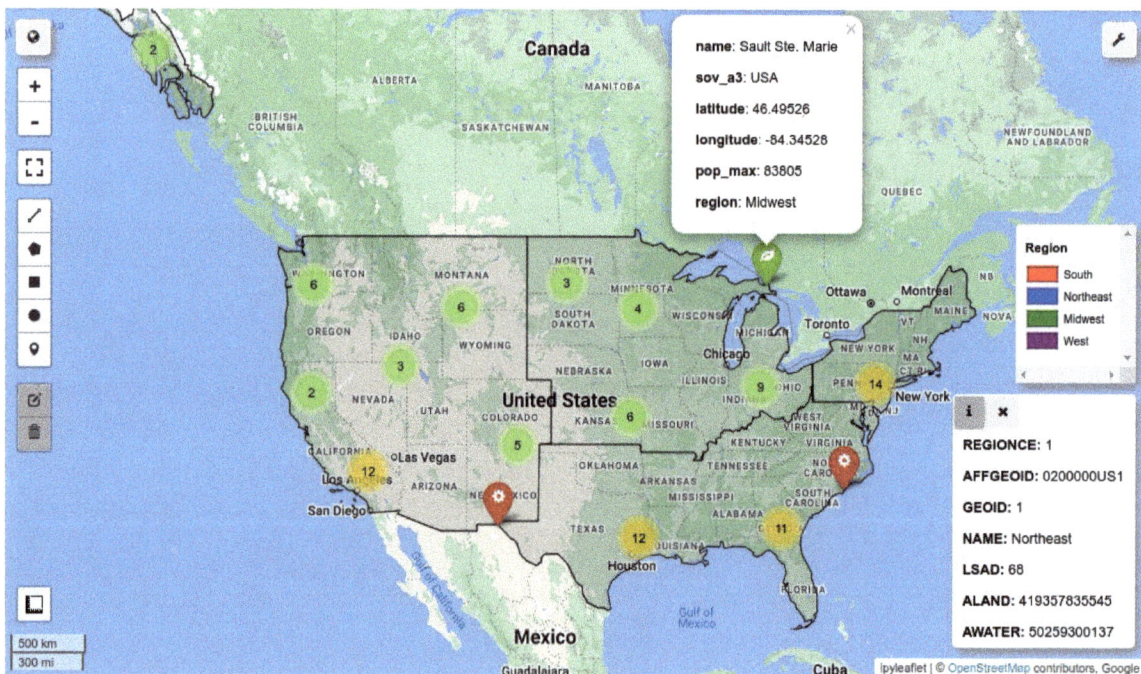

Figure 4.20: Adding points from a CSV file with custom markers.

Circle markers from points

Besides creating a marker cluster from a CSV file, you can also create circle markers from a CSV file. Let's use the same `us_cities.csv` dataset as above:

```python
data = 'https://github.com/gee-community/geemap/blob/master/examples/data/us_cities.csv'
```

Then, use the `add_circle_markers_from_xy()` method to create circle markers from the CSV file. Specify optional parameters for the circle markers, such as `radius`, `color`, `fill_color`, `fill_opacity`, etc. as desired.

```python
Map = geemap.Map(center=[40, -100], zoom=4)
Map.add_circle_markers_from_xy(
```

[105]Font Awesome v4 website: fontawesome.com/v4/icons

```
        data,
        x="longitude",
        y="latitude",
        radius=10,
        color="blue",
        fill_color="black",
        fill_opacity=0.5,
    )
Map
```

The resulting map should look like Fig. 4.21.

Figure 4.21: Creating circle markers from a CSV file with latitude and longitude coordinates.

4.8   Vector data to Earth Engine

Geemap provides functions for converting small vector datasets (< 32 MB) to an Earth Engine feature collection. The supported vector formats include GeoJSON, ESRI Shapefile, GeoPandas GeoDataFrame, and any other GeoPandas supported vector formats. For example, to convert a GeoJSON file to an Earth Engine `FeatureCollection`, use the `geojson_to_ee()` function:

```
in_geojson = (
    'https://github.com/gee-community/geemap/blob/master/examples/data/countries.geojson'
)
Map = geemap.Map()
fc = geemap.geojson_to_ee(in_geojson)
Map.addLayer(fc.style(**{'color': 'ff0000', 'fillColor': '00000000'}), {}, 'Countries')
Map
```

The resulting map should look like Fig. 4.22.

To convert a shapefile to an Earth Engine `FeatureCollection`, use the `shp_to_ee()` function:

```
url = "https://github.com/gee-community/geemap/blob/master/examples/data/countries.zip"
```

Figure 4.22: Converting a GeoJSON file to an Earth Engine 'FeatureCollection'.

```
geemap.download_file(url, overwrite=True)

in_shp = "countries.shp"
fc = geemap.shp_to_ee(in_shp)
```

To convert a GeoPandas GeoDataFrame to an Earth Engine `FeatureCollection`, use the `gdf_to_ee()` function:

```
import geopandas as gpd

gdf = gpd.read_file(in_shp)
fc = geemap.gdf_to_ee(gdf)
```

To convert any other GeoPandas-supported vector format to an Earth Engine `FeatureCollection`, use the `vector_to_ee()` function:

```
fc = geemap.vector_to_ee(url)
```

4.9   Joining attribute tables

You can join the attribute table of a local vector dataset to an Earth Engine `FeatureCollection` using the `ee_join_table()` function. The values of a column in the attribute table of the local vector dataset must match the values of a column in the `FeatureCollection`. The following example shows to how to join the latitude and longitude columns of a CSV file to an Earth Engine `FeatureCollection` containing a collection of countries:

```
Map = geemap.Map()
countries = ee.FeatureCollection(geemap.examples.get_ee_path('countries'))
Map.addLayer(countries, {}, 'Countries')
Map
```

Display the attribute table of the `FeatureCollection` to see what columns are available (Fig. 4.23):

```
geemap.ee_to_df(countries)
```

	GDP_MD_EST	ISO_A2	POP_RANK	ISO_A3	CONTINENT	POP_EST	INCOME_GRP	SUBREGION	NAME
0	25810.0	TJ	13	TJK	Asia	8468555	5. Low income	Central Asia	Tajikistan
1	21010.0	KG	13	KGZ	Asia	5789122	5. Low income	Central Asia	Kyrgyzstan
2	40000.0	KP	15	PRK	Asia	25248140	5. Low income	Eastern Asia	North Korea
3	628400.0	BD	17	BGD	Asia	157826578	5. Low income	Southern Asia	Bangladesh
4	71520.0	NP	15	NPL	Asia	29384297	5. Low income	Southern Asia	Nepal
...	...	...	...	...	...	...	...	...	...

Figure 4.23: The attribute table of an ee.FeatureCollection.

Read the CSV file to a Pandas DataFrame (Fig. 4.24):

```
data = (
    'https://github.com/gee-community/geemap/blob/master/examples/data/country_centroids.csv'
)
df = geemap.csv_to_df(data)
df
```

	country	latitude	longitude
0	AD	42.54625	1.60155
1	AE	23.42408	53.84782
2	AF	33.93911	67.70995
3	AG	17.06082	-61.79643
4	AI	18.22055	-63.06862
...	...	...	...

Figure 4.24: Reading a CSV file to a Pandas DataFrame.

Join the Pandas DataFrame to the `FeatureCollection`:

```
fc = geemap.ee_join_table(countries, data, src_key='ISO_A2', dst_key='country')
```

Display the attribute table of the new `FeatureCollection` (Fig. 4.25):

```
geemap.ee_to_df(fc)
```

	latitude	longitude	GDP_MD_EST	ISO_A2	POP_RANK	ISO_A3	CONTINENT	POP_EST	INCOME_GRP	SUBREGION	NAME
0	38.86103	71.27609	25810.0	TJ	13	TJK	Asia	8468555	5. Low income	Central Asia	Tajikistan
1	41.20438	74.76610	21010.0	KG	13	KGZ	Asia	5789122	5. Low income	Central Asia	Kyrgyzstan
2	40.33985	127.51009	40000.0	KP	15	PRK	Asia	25248140	5. Low income	Eastern Asia	North Korea
3	23.68499	90.35633	628400.0	BD	17	BGD	Asia	157826578	5. Low income	Southern Asia	Bangladesh
4	28.39486	84.12401	71520.0	NP	15	NPL	Asia	29384297	5. Low income	Southern Asia	Nepal
...	...	...	...	...	...	...	...	...	...	...	...

Figure 4.25: The attribute table of an 'ee.FeatureCollection'.

Note the `latitude` and `longitude` columns are now joined to the `FeatureCollection`. You can also add the new `FeatureCollection` to the map and use the Inspector tool to see the properties of each feature (Fig. 4.26):

```
Map.addLayer(fc, {}, 'Countries with attr')
Map
```

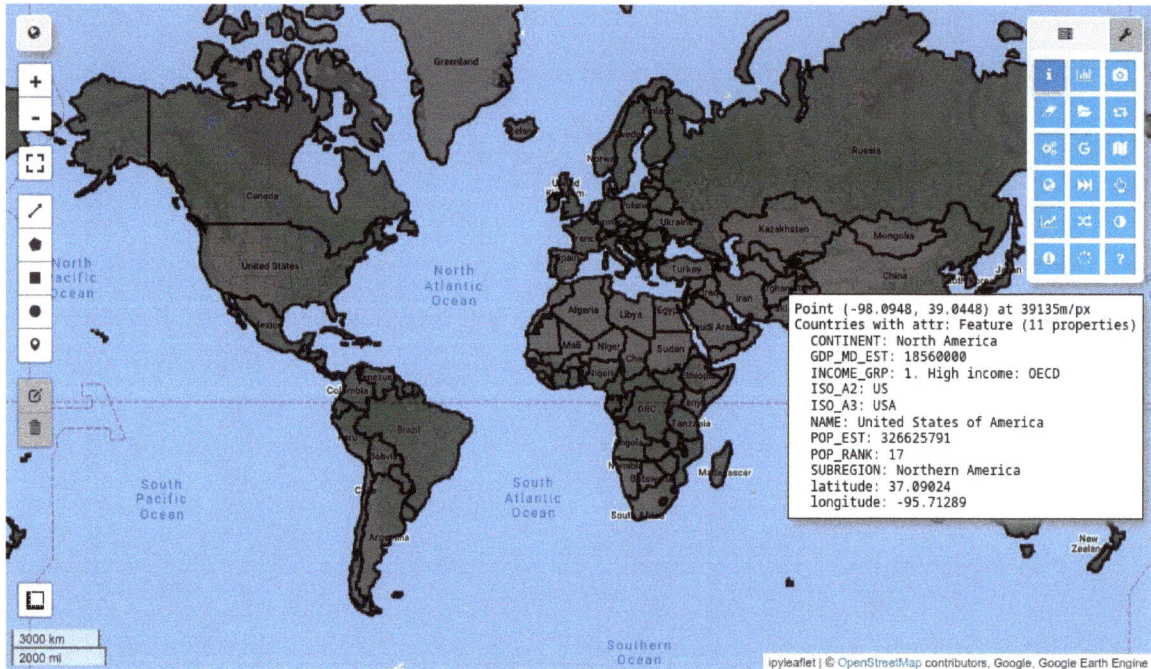

Figure 4.26: The attribute table of an 'ee.FeatureCollection' joined with an attribute table from a local dataset.

4.10   Converting NetCDF to ee.Image

NetCDF (network Common Data Form) is a file format for storing multidimensional scientific data (variables) such as temperature, humidity, pressure, wind speed, and direction. Each variable in a netCDF file has a data type, such as integer, float, or double, and an associated set of dimensions. Dimensions define the size of each dimension of the variable. In this section, you will learn how to convert a NetCDF file to an Earth Engine `Image` object using the `netcdf_to_ee()` function. First, download a sample NetCDF file of global wind speed from GitHub:

```python
import os
url = 'https://github.com/gee-community/geemap/blob/master/examples/data/wind_global.nc'
nc_file = 'wind_global.nc'
if not os.path.exists(nc_file):
    geemap.download_file(url)
```

The downloaded NetCDF file contains three variables: `u_wind`, `v_wind`, and `wind_global`. The `u_wind` and `v_wind` variables represent the eastward and northward components of wind, respectively. The `wind_global` variable is the magnitude of the wind speed. To convert the NetCDF file of global wind direction to an `ee.Image` object, use the `netcdf_to_ee()` function. Specify the `variable` parameter to select the variable to convert to an `ee.Image`:

```python
Map = geemap.Map()
img = geemap.netcdf_to_ee(nc_file=nc_file, var_names='u_wind')
vis_params = {'min': -20, 'max': 25, 'palette': 'YlOrRd', 'opacity': 0.6}
Map.addLayer(img, vis_params, "u_wind")
Map
```

The resulting map should look like Fig. 4.27:

The example above only converts the `u_wind` variable to an `ee.Image`. To convert more variables in the

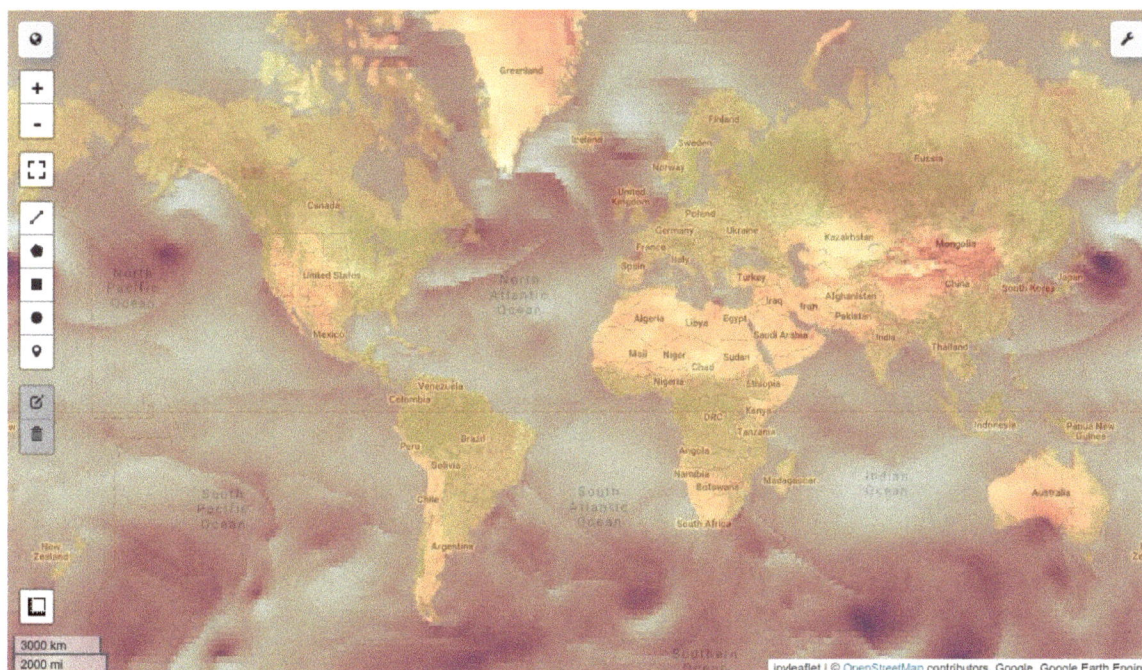

Figure 4.27: Converting a NetCDF file of global wind speed to an Earth Engine Image.

NetCDF file to an ee.Image, provide the var_names parameter as a list of variable names (Fig. 4.28):

```
Map = geemap.Map()
img = geemap.netcdf_to_ee(nc_file=nc_file, var_names=['u_wind', 'v_wind'])
Map.addLayer(
    img,
    {'bands': ['v_wind'], 'min': -20, 'max': 25, 'palette': 'coolwarm', 'opacity': 0.8},
    "v_wind",
)
Map
```

4.11 OpenStreetMap data

Geemap provides various functions for retrieving data from OpenStreetMap (OSM). The data can be saved as vector formats such as GeoJSON, ESRI Shapefile, and a GeoPandas GeoDataFrame.

OSM to GeoDataFrame

For example, to download OpenStreetMap data by place name as a GeoDataFrame through the Nominatim API[106] , use the osm_to_gdf() function (Fig. 4.29):

```
gdf = geemap.osm_to_gdf("Knoxville, Tennessee")
gdf
```

The osm_to_gdf() function can also be handy for exporting administrative boundary data from OpenStreetMap.

---

[106]Nominatim API: tiny.geemap.org/mwsr

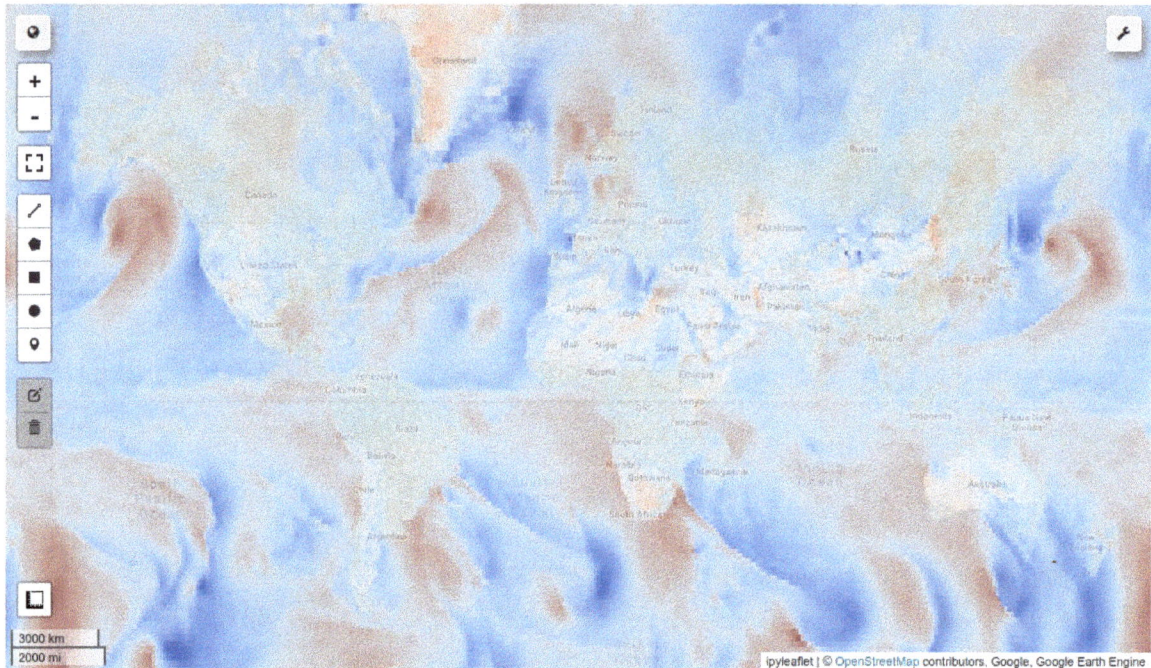

Figure 4.28: Converting a NetCDF file of global wind speed to an Earth Engine Image.

	geometry	bbox_north	bbox_south	bbox_east	bbox_west	place_id	osm_type	osm_id	lat	lon	display_name	class	type	importance
0	MULTIPOLYGON (((-84.01236 35.90634, -84.01291 ...	36.067428	35.849497	-83.688543	-84.161625	282327163	relation	197353	35.960395	-83.921026	Knoxville, Knox County, Tennessee, United States	boundary	administrative	0.795774

Figure 4.29: Downloading OpenStreetMap data as a GeoPandas GeoDataFrame.

OSM to ee.FeatureCollection

Using the osm_to_gdf() function above with the gdf_to_ee() function, you should be able to download an OSM dataset as a GeoDataFrame and convert it to a feature collection. Alternatively, you can use the osm_to_ee() function to directly convert OSM data to an Earth Engine feature collection (Fig. 4.30):

```
Map = geemap.Map()
fc = geemap.osm_to_ee("Knoxville, Tennessee")
Map.addLayer(fc, {}, "Knoxville")
Map.centerObject(fc, 11)
Map
```

Downloading OSM data

Besides downloading administrative boundary data from OpenStreetMap using the osm_to_gdf() function, you can also download specific types of data from OpenStreetMap, e.g., roads, buildings, bars, or water features. A complete list of OSM map features can be found at https://wiki.openstreetmap.org/wiki/Map_features. To create a GeoDataFrame of OSM entities from a place name, use the osm_gdf_from_geocode() function (Fig. 4.31):

```
import geemap.osm as osm

Map = geemap.Map(add_google_map=False)
gdf = osm.osm_gdf_from_geocode("New York City")
```

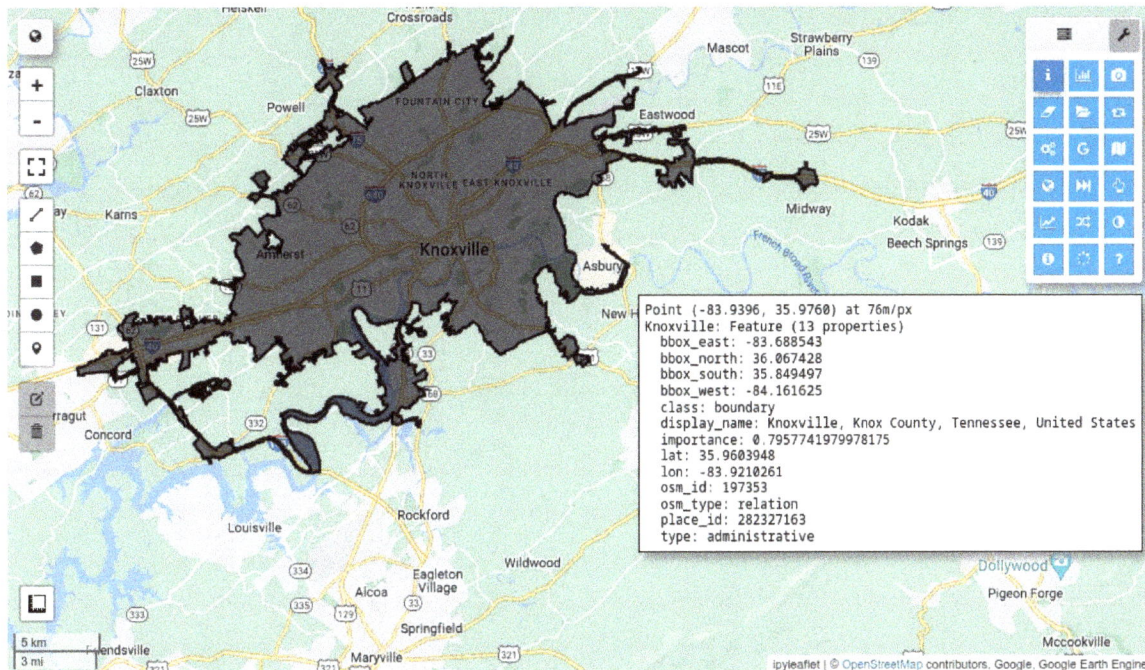

Figure 4.30: Converting OpenStreetMap data to an 'ee.FeatureCollection'.

```
Map.add_gdf(gdf, layer_name="NYC")
Map
```

To create a GeoDataFrame of OSM entities within the boundaries of geocodable places, use the osm_
gdf_from_place() function. Specify the tags parameter to filter the OSM entities by tags. The tags parameter is a dictionary of tags used for finding objects in the selected area. The dictionary keys should be OSM tags, (e.g., building, landuse, highway, etc.) and the dictionary values should be either True to retrieve all items with the given tag, a string to get a single tag-value combination, or a list of strings to get multiple values for the given tag. For example, tags = {'building': True} would return all building footprints in the area. tags = {'amenity':True, 'landuse':['retail','commercial'], 'highway':'bus_stop'} would return all amenities that are landuse=retail, landuse=commercial, and highway=bus_stop. The following example retrieves all buildings within the boundaries of the University of Tennessee, Knoxville, Tennessee (Fig. 4.33):

```
place = "University of Tennessee, Knoxville, TN"
tags = {"building": True}
gdf = osm.osm_gdf_from_place(place, tags)
gdf = gdf[gdf.geometry.type == "Polygon"]
gdf
```

Add the GeoDataFrame to the map (Fig. 4.33):

```
Map = geemap.Map(add_google_map=False)
Map.add_gdf(gdf, layer_name="Buildings")
Map
```

The following example retrieves all bars within a radius of 1,500 meters from the centroid of New York City:

```
gdf = osm.osm_gdf_from_address(
    address="New York City", tags={"amenity": "bar"}, dist=1500
```

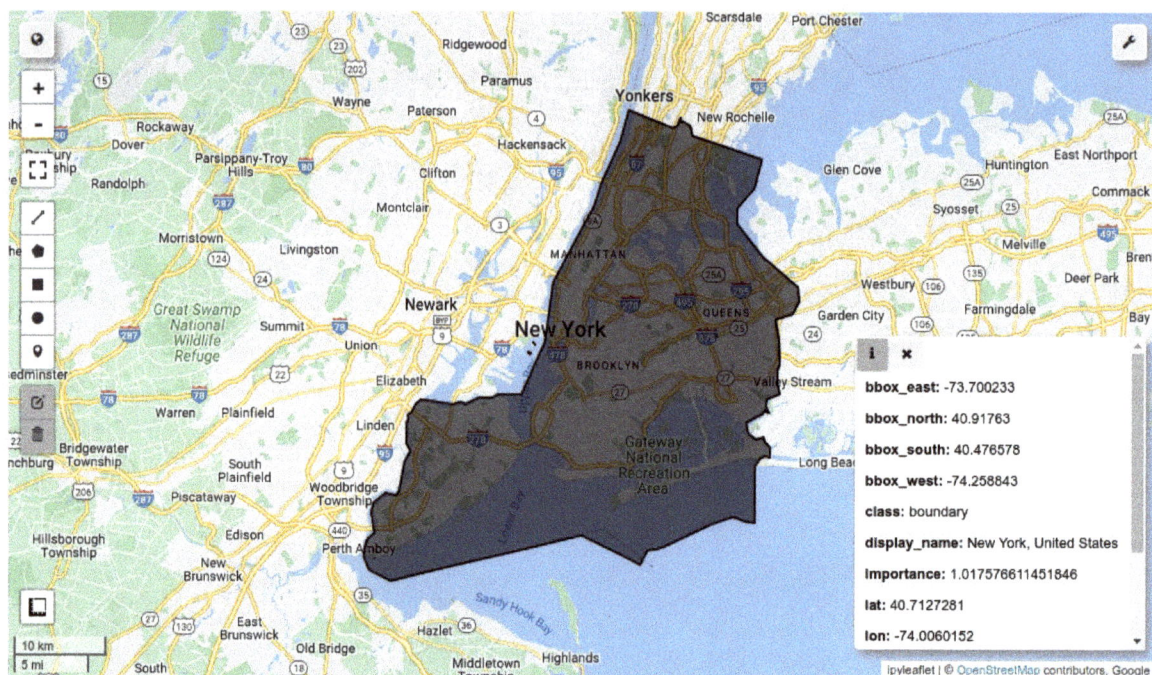

Figure 4.31: Retrieving place(s) by name from OpenStreetMap.

osmid	addr:state	building	ele	gnis:county_name	gnis:feature_id	gnis:import_uuid	gnis:reviewed	name	source	geometry	...	nam
80675037	NaN	university	NaN	NaN	NaN	NaN	NaN	University Printing & Mail	NaN	POLYGON ((-83.92864 35.94639, -83.92866 35.947...	...	
80675040	NaN	university	NaN	NaN	NaN	NaN	NaN	UT Warehouse	NaN	POLYGON ((-83.92919 35.94630, -83.92888 35.946...	...	
80675042	NaN	yes	NaN	NaN	NaN	NaN	NaN	NaN	NaN	POLYGON ((-83.93121 35.94676, -83.93101 35.945...	...	
80675043	NaN	university	NaN	NaN	NaN	NaN	NaN	UT Warehouse	NaN	POLYGON ((-83.93020 35.94693, -83.92997 35.946...	...	

Figure 4.32: Downloading building footprints from OpenStreetMap.

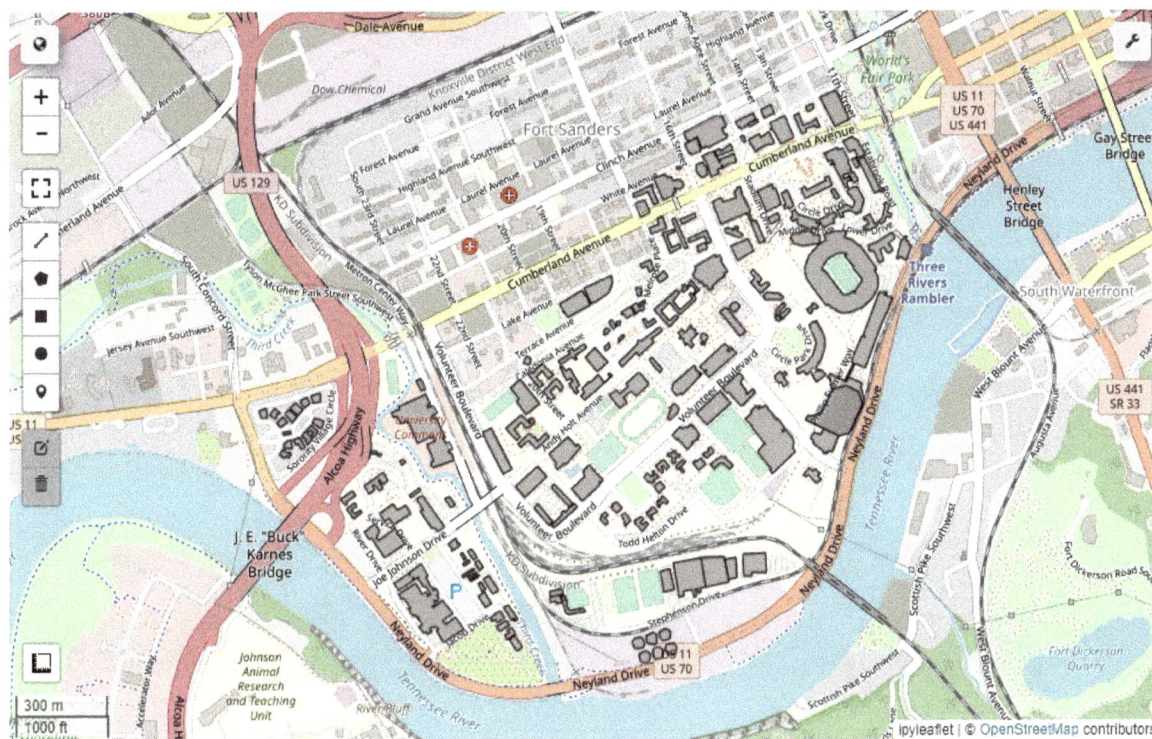

Figure 4.33: Displaying building footprints downloaded from OpenStreetMap.

```
)
gdf
```

Add the GeoDataFrame to the map (Fig. 4.34):

```
Map = geemap.Map(add_google_map=False)
Map.add_gdf(gdf, layer_name="NYC bars")
Map
```

To retrieve OSM entities within a radius of a point, use the `osm_gdf_from_point()` function. The following example retrieves all natural lakes within a radius of 10,500 meters from the point (46.7808, -96.0156):

```
gdf = osm.osm_gdf_from_point(
    center_point=(46.7808, -96.0156),
    tags={"natural": "water"},
    dist=10000,
)
gdf
```

Add the GeoDataFrame to the map (Fig. 4.35):

```
Map = geemap.Map(add_google_map=False)
Map.add_gdf(gdf, layer_name="Lakes")
Map
```

Last but not least, you can use the `add_osm_from_view()` method to retrieve OSM data from the current view of a map. Simply pan and zoom the map to the desired area and call the `add_osm_from_view()` method (Fig. 4.36):

Figure 4.34: Retrieving bars in New York City from OpenStreetMap.

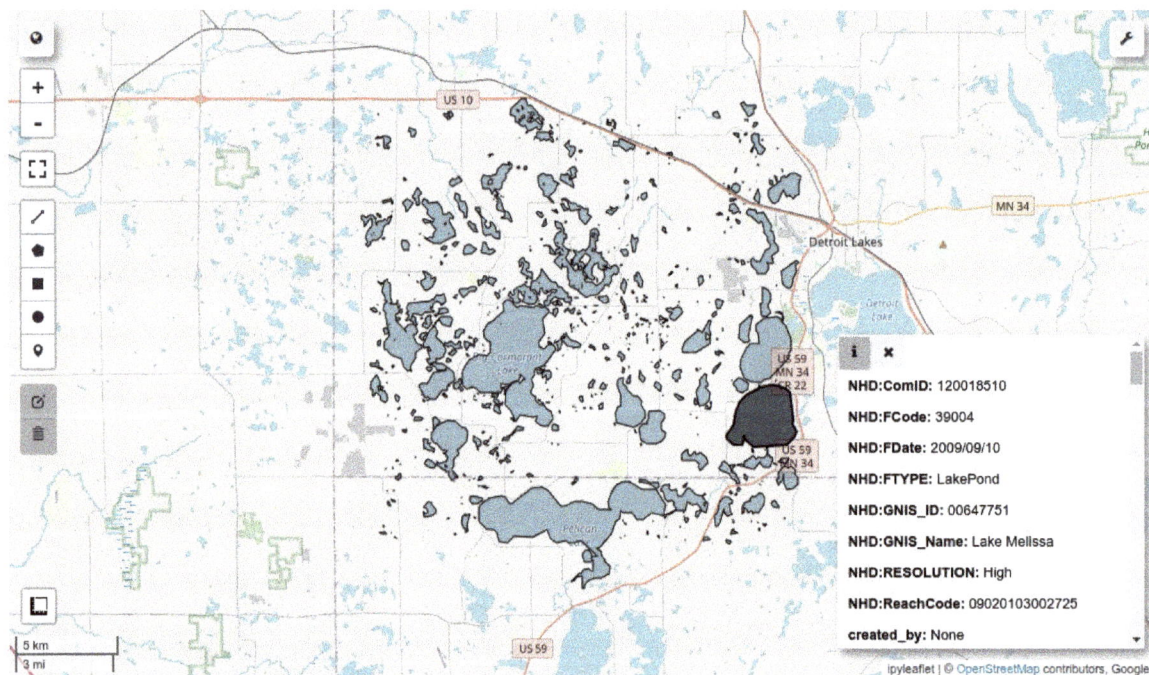

Figure 4.35: Retrieving natural lake features from OpenStreetMap.

```
Map = geemap.Map(center=[40.7500, -73.9854], zoom=16, add_google_map=False)
Map
```

The add_osm_from_view() method is only available for the ipyleaflet plotting backend.

```
Map.add_osm_from_view(tags={"amenity": "bar", "building": True})
```

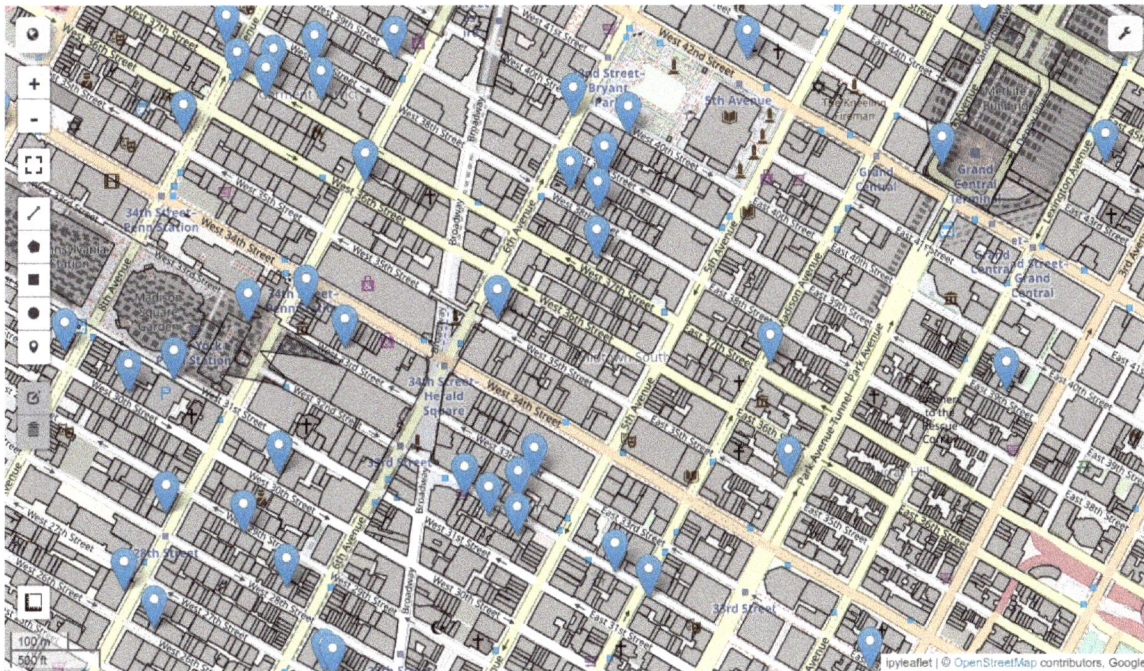

Figure 4.36: Retrieving OSM entities for the current map view.

4.12 Reading PostGIS data

Geemap supports loading data directly from a PostGIS database. To follow the examples in this section, you need to install the sqlalchemy and psycopg2 packages and have PostGIS set up properly on your machine. Watch this video tutorial[107] to set up PostGIS on Windows if needed. **If you are using Google Colab, you can skip this section as Google Colab does not support PostGIS.**

```
mamba install sqlalchemy psycopg2 -c conda-forge
```

Once PostGIS has been properly set up on your machine, you can then load some data into PostGIS. Watch this tutorial[108] to load nyc_data.zip[109] into the PostGIS database. The next step is to connect to the database with Python using the connect_postgis() function. You can directly pass in the username and password to access the database. Alternatively, you can define environment variables. The default environment variables for user and password are SQL_USER and SQL_PASSWORD, respectively. You can also specify user and password parameters to connect to the database by setting use_env_var=False. The following example shows how to connect to the PostGIS database using the default environment variables:

---

[107]video tutorial: youtu.be/LhKj-_-CCfY

[108]tutorial: youtu.be/fROzLrjNDrs

[109]nyc_data.zip: tiny.geemap.org/aqqc

```
con = geemap.connect_postgis(
    database="nyc", host="localhost", user=None, password=None, use_env_var=True
)
```

Create a GeoPandas GeoDataFrame from a SQL query (see Fig. 4.37).

```
sql = 'SELECT * FROM nyc_neighborhoods'
gdf = geemap.read_postgis(sql, con)
gdf
```

	id	geom	boroname	name
0	1	MULTIPOLYGON (((582771.426 4495167.427, 584651...	Brooklyn	Bensonhurst
1	2	MULTIPOLYGON (((585508.753 4509691.267, 586826...	Manhattan	East Village
2	3	MULTIPOLYGON (((583263.278 4509242.626, 583276...	Manhattan	West Village
3	4	MULTIPOLYGON (((597640.009 4520272.720, 597647...	The Bronx	Throggs Neck
4	5	MULTIPOLYGON (((595285.205 4525938.798, 595348...	The Bronx	Wakefield-Williamsbridge
...	...		...	...	...

Figure 4.37: Reading data from a PostGIS database.

Convert the GeoDataFrame to an ee.FeatureCollection and display it on an interactive map:

```
Map = geemap.Map()
Map = geemap.gdf_to_ee(gdf)
Map.addLayer(fc, {}, "NYC EE")
Map.centerObject(fc)
Map
```

Alternatively, you can directly display data from a PostGIS database on an interactive map using the add_gdf_from_postgis() method (Fig. 4.38):

```
Map = geemap.Map()
Map.add_gdf_from_postgis(
    sql, con, layer_name="NYC Neighborhoods", fill_colors=["red", "green", "blue"]
)
Map
```

4.13 Summary

In this chapter, we learned how to deal with local vector and raster data, including GeoTIFF, Cloud Optimized GeoTIFF, Shapefile, GeoJSON, KML, GeoPandas GeoDataFrame, and more. We also learned how to retrieve data from OpenStreetMap and PostGIS with only a few lines of code. Hopefully, at the end of this chapter, you feel comfortable working with geospatial data stored locally or in the cloud. In the next chapter, we will learn how to visualize various types of geospatial data.

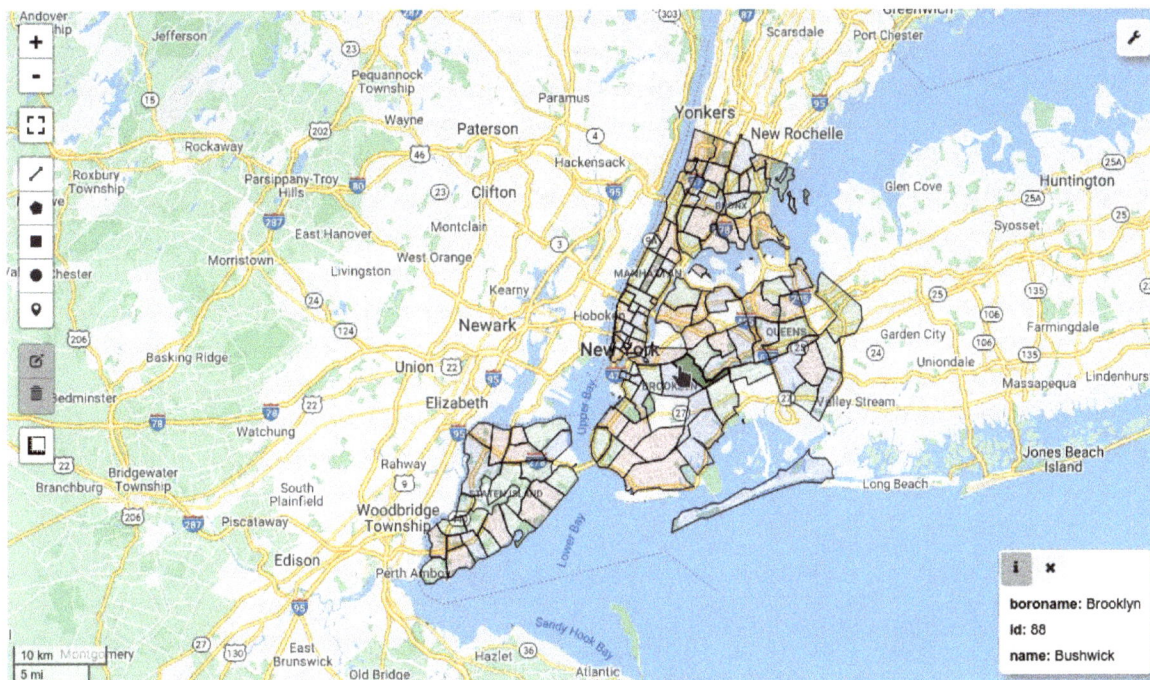

Figure 4.38: Displaying data from a PostGIS database on an interactive map.

5. Visualizing Geospatial Data

5.1 Introduction

There is a saying in the world of geospatial data: "If a picture is worth a thousand words, then a map must be worth a million." Visualizing geospatial data on interactive maps is a great way to get a sense of the data. In this chapter, we will learn how to visualize vector and raster with only a few lines of code. We will also introduce various widgets that can be added to the map to make it more meaningful, such as legends, color bars, split-panel maps, linked maps, timeseries inspector, time slider, and more. Lastly, we will learn how to create choropleth maps with various classification schemes. At the end of this chapter, you should feel comfortable visualizing Earth Engine data and other geospatial data (e.g., NetCDF, LiDAR) on interactive maps.

5.2 Technical requirements

To follow along with this chapter, you will need to have geemap and several optional dependencies installed. If you have already followed Section 1.5 - *Installing geemap*, then you should already have a conda environment with all the necessary packages installed. Otherwise, you can create a new conda environment and install pygis[30] with the following commands, which will automatically install geemap and all the required dependencies:

```
conda create -n gee python
conda activate gee
conda install -c conda-forge mamba
mamba install -c conda-forge pygis
```

Next, launch JupyterLab by typing the following commands in your terminal or Anaconda prompt:

```
jupyter lab
```

Alternatively, you can use geemap with a Google Colab cloud environment without installing anything on your local computer. Click 05_data_viz.ipynb[110] to launch the notebook in Google Colab.

Once in Colab, you can uncomment the following line and run the cell to install pygis, which includes geemap and all the necessary dependencies:

```
# %pip install pygis
```

The installation process may take 2-3 minutes. Once pygis has been installed successfully, click the **RESTART RUNTIME** button that appears at the end of the installation log or go to the **Runtime** menu and select **Restart runtime**. After that, you can start coding.

To begin, import the necessary libraries that will be used in this chapter:

```
import ee
import geemap
```

Initialize the Earth Engine Python API:

---

[110]05_data_viz.ipynb: tiny.geemap.org/ch05

```
geemap.ee_initialize()
```

If this is your first time running the code above, you will need to authenticate Earth Engine first. Follow the instructions in Section 1.7 - *Earth Engine authentication* to authenticate Earth Engine.

5.3 Using the plotting tool

In Chapter 3 - **Using Earth Engine Data**, we introduced the inspector tool for inspecting image pixel values. For multi-spectral and hyperspectral remote sensing images, it would be useful to plot the spectral signatures of selected pixels in addition to the pixel values. Geemap provides a plotting tool for this purpose. You can activate the plotting tool and click on the image to select a pixel or draw a rectangle on the image to select a region. The plotting tool will display the spectral signatures of the selected pixel or the mean spectral signatures of the selected region (Fig. 5.1).

Figure 5.1: The interactive plotting tool.

Let's try out the plotting tool for two datasets, Landsat 7[111] and EO-1 Hyperion[112]. The multi-spectral Landsat 7 image contains seven bands, while the hyperspectral EO-1 Hyperion images contains 198 bands.

```
Map = geemap.Map(center=[40, -100], zoom=4)

landsat7 = ee.Image('LANDSAT/LE7_TOA_5YEAR/1999_2003').select(
    ['B1', 'B2', 'B3', 'B4', 'B5', 'B7']
)

landsat_vis = {'bands': ['B4', 'B3', 'B2'], 'gamma': 1.4}
Map.addLayer(landsat7, landsat_vis, "Landsat")

hyperion = ee.ImageCollection('EO1/HYPERION').filter(
    ee.Filter.date('2016-01-01', '2017-03-01')
)

hyperion_vis = {
    'min': 1000.0,
    'max': 14000.0,
    'gamma': 2.5,
}
Map.addLayer(hyperion, hyperion_vis, 'Hyperion')
Map
```

Click the plotting icon in the toolbar to activate the plotting tool. On the popup dropdown menu, select the desired image layer (e.g., Landsat). Then, click the mouse on the map to select a pixel or click the draw tool to draw a rectangle on the map. The spectral signatures of the selected pixel or the mean spectral signatures of the selected region will be displayed on the right side of the map (Fig. 5.2).

---

[111]Landsat 7: tiny.geemap.org/mwpm
[112]EO-1 Hyperion: tiny.geemap.org/qidr

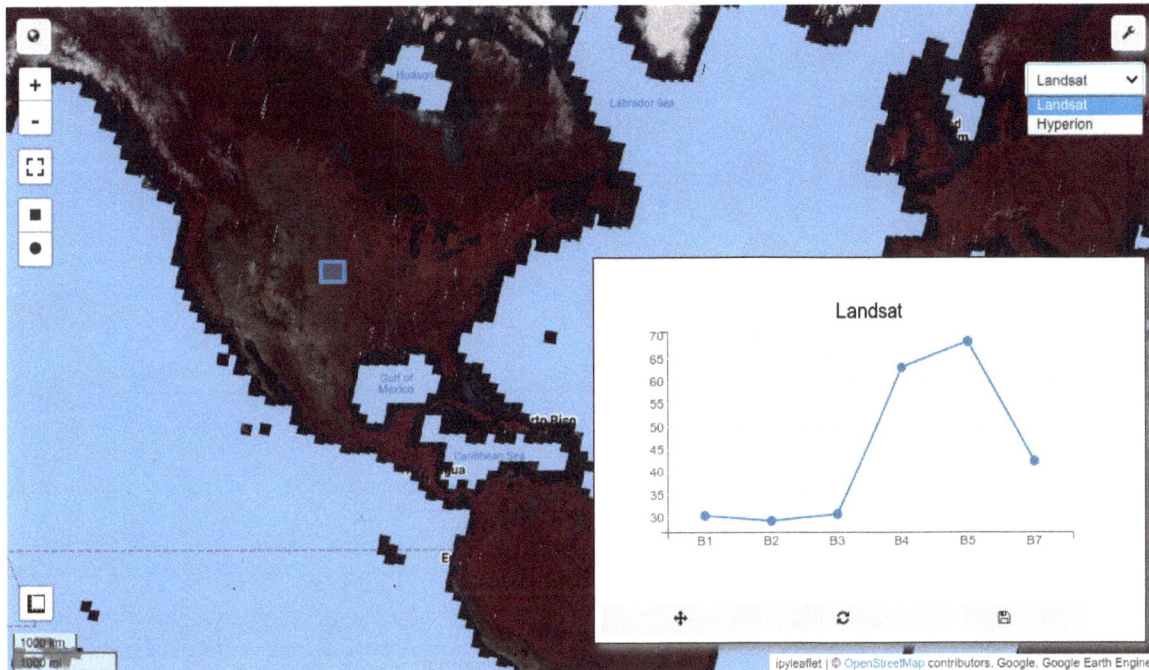

Figure 5.2: Plotting spectral signatures of Landsat 7.

You can customize the plotting options, such as changing the plot type, position, adding a marker cluster, overlaying multiple spectral signatures on the same plot (Fig. 5.3), and so on.

```
Map.set_plot_options(add_marker_cluster=True, overlay=True)
```

Next, let's try out the plotting tool with hyperspectral data. Select Hyperion from the plotting tool popup dropdown menu and click on any Hyperion image to select a pixel, or click the draw tool to draw a rectangle on the image. The line plot consists of 198 data points corresponding to the pixel values of 198 bands at the mouse click location (Fig. 5.4).

5.4   Changing layer opacity

When adding an Earth Engine layer to the map, the opacity parameter of the Map.addLayer() method can be used to control the opacity of the layer. The opacity is a value between 0 and 1, where 0 is fully transparent and 1 is fully opaque, the default opacity being 1. After the layer has been added to the map, the layer opacity can be changed interactively using the layer control widget. First, click the toolbar icon in the upper-right corner of the map. Then, click the sandwich icon next to toolbar icon to activate the layer control widget. Click the checkbox next to the layer name to toggle the visibility of each layer. Alternatively, drag and move the slider bar to change the opacity of the layer (Fig. 5.5). Note that the layer opacity will be updated instantly and reflected on the map. To toggle the visibility of all layers at once, click the checkbox next to the All layers on/off.

```
Map = geemap.Map(center=(40, -100), zoom=4)

dem = ee.Image('USGS/SRTMGL1_003')
states = ee.FeatureCollection("TIGER/2018/States")

vis_params = {
    'min': 0,
    'max': 4000,
```

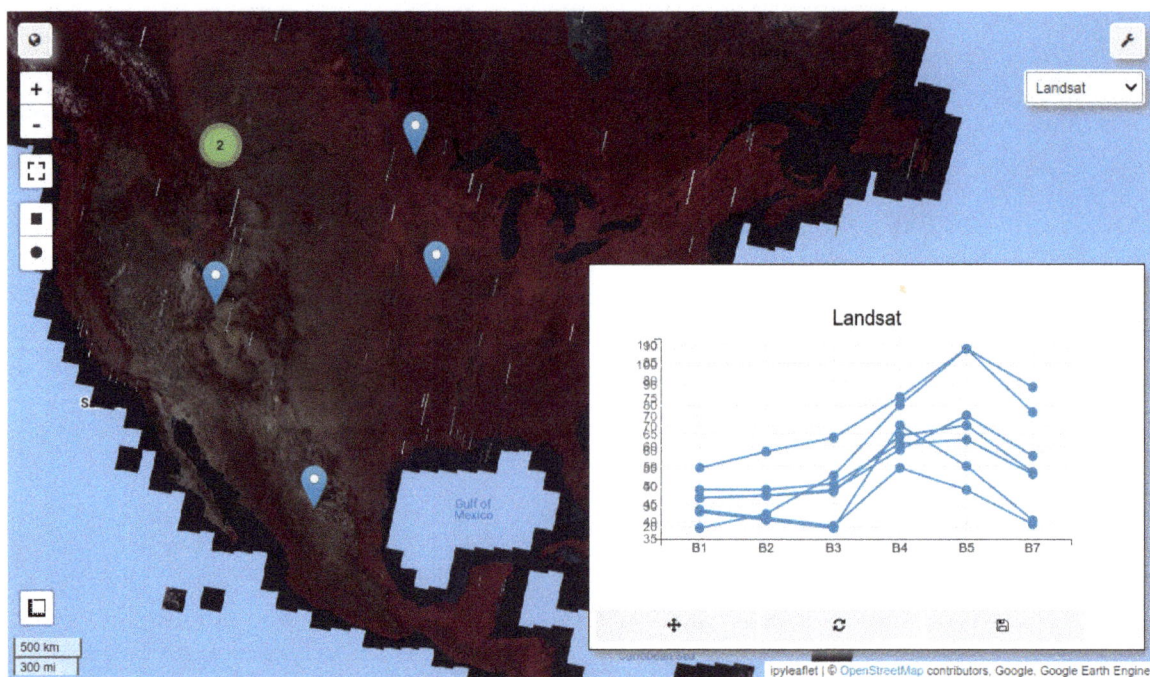

Figure 5.3: Plotting multiple spectral signatures of Landsat 7 on the same plot.

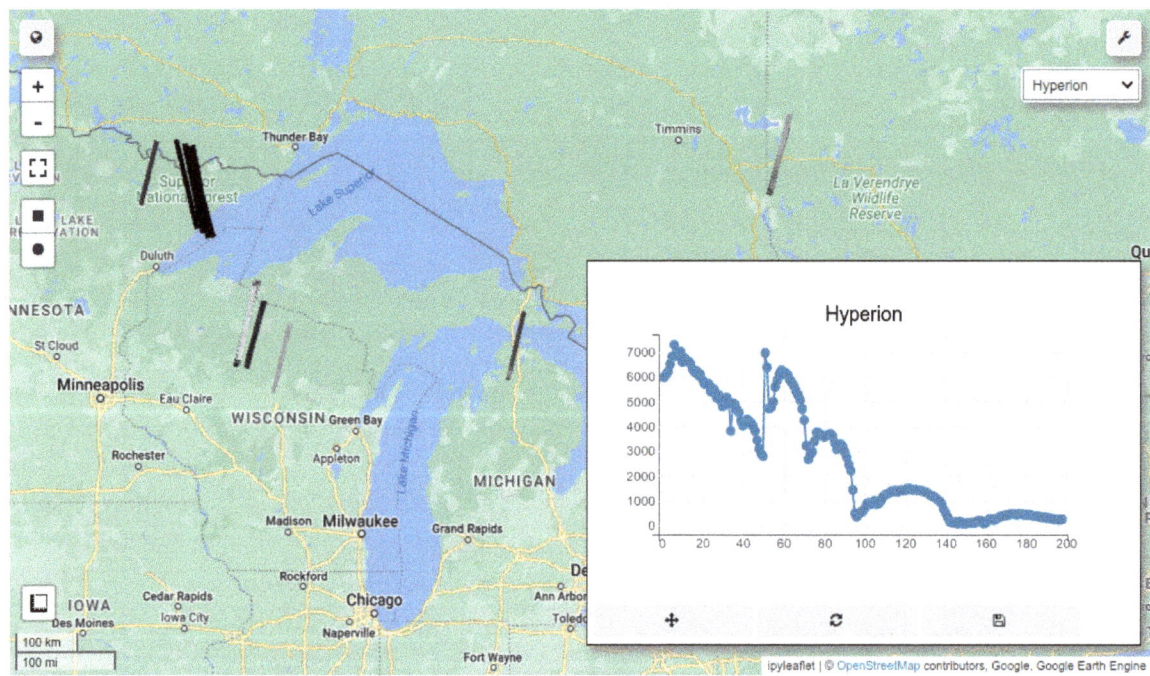

Figure 5.4: Plotting spectral signatures of EO-1 Hyperion hyperspectral images.

```
        'palette': ['006633', 'E5FFCC', '662A00', 'D8D8D8', 'F5F5F5'],
}

Map.addLayer(dem, vis_params, 'SRTM DEM', True, 1)
Map.addLayer(states, {}, "US States", True)

Map
```

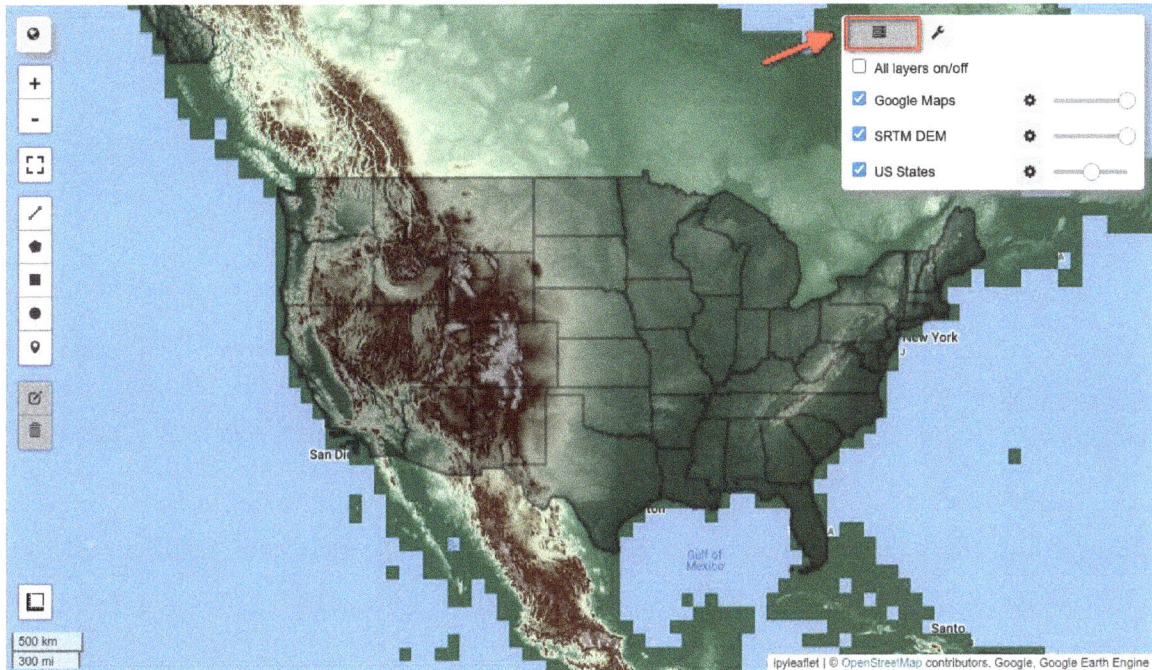

Figure 5.5: Changing layer opacity interactively

5.5 Visualizing raster data

In Section 3.3.1.2 - *Visualizing Earth Engine images*, we introduced the visualization parameters of the Map.addLayer() method. One can fine-tune the visualization parameters programmatically to achieve desirable visualizations. In this section, we will introduce geemap's interactive graphical user interface (GUI) for visualizing Earth Engine raster data interactively without coding.

Single-band images

To visualize a single-band image, click the toolbar icon in the upper-right corner of the map. Then, click the sandwich icon next to toolbar icon to activate the layer control widget. Next, click the gear icon next to the layer name to open the raster visualization GUI. Similar to the visualization parameters of the Map.addLayer() method, the raster visualization GUI provides a number of options for customizing the visualization parameters interactively, such as range, opacity, gamma, palette, etc. Instead of entering hexadecimal color values for the palette manually, you can click the colormap dropdown list to select a predefined colormap, which will automatically update the palette field. There are over 200 colormaps to choose from. A colorbar preview will also be displayed in the lower-right corner of the map automatically. To keep the colorbar after the visualization GUI is closed, click the checkbox next to the Legend option.

Let's try out the visualization GUI for a single-band image. Click the gear icon on the layer control widget to activate the visualization GUI. Note that colorbar will be created based on the palette selected in the visualization parameters (Fig. 5.6).

```
Map = geemap.Map(center=[12, 69], zoom=3)
dem = ee.Image('USGS/SRTMGL1_003')
vis_params = {
    'min': 0,
    'max': 4000,
    'palette': ['006633', 'E5FFCC', '662A00', 'D8D8D8', 'F5F5F5'],
}
Map.addLayer(dem, vis_params, 'SRTM DEM')
Map
```

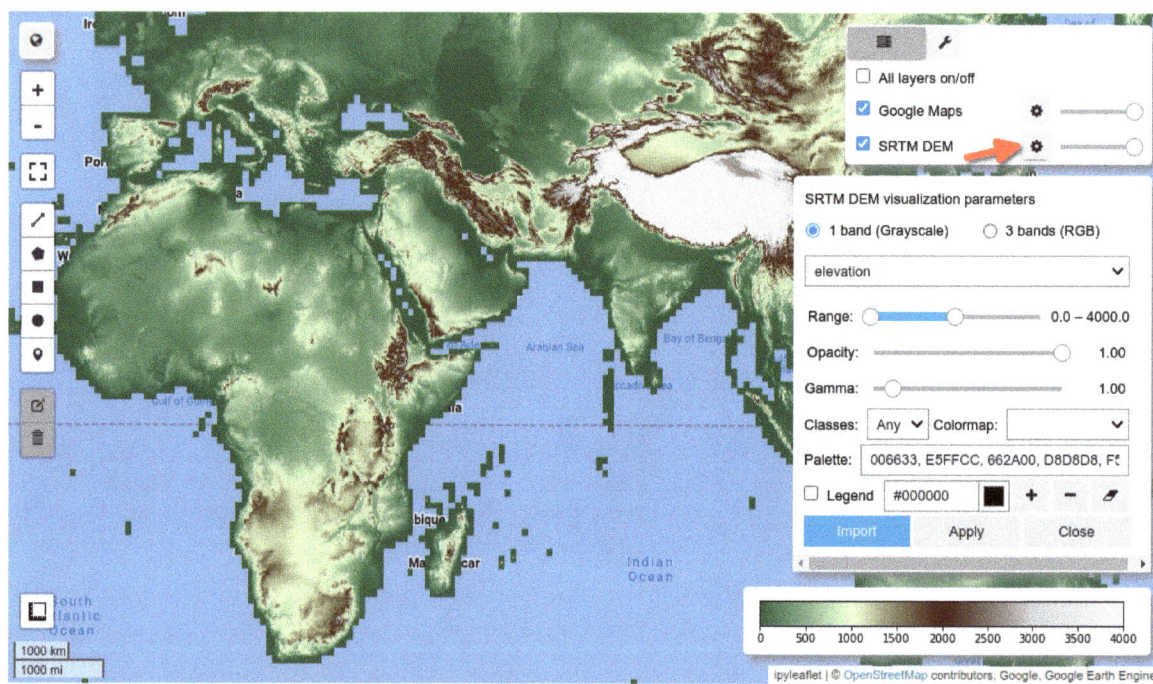

Figure 5.6: Visualizing a single-band SRTM image.

You may drag and move the sliders to change visualization parameters such as range, opacity, and gamma. Click the colormap dropdown list and select a different colormap (e.g., terrain). Click the legend checkbox if you want to keep the legend on the map after closing the GUI. Once you are done with adjusting the visualization parameters, click the Apply button to apply the changes. The image with the new visualization parameters will be displayed on the map (Fig. 5.7).

If you are running Jupyter Notebook (not JupyterLab or Google Colab), you may click the Import button to import the visualization parameters into the Jupyter Notebook. A new cell will be created below the map and populated with the visualization parameters like this:

```
vis_params = {
    'bands': ['elevation'],
    'palette': ['333399', ' 00b2b2', ' 99eb85', ' ccbe7d', ' 997c76', ' ffffff'],
    'min': 0.0,
    'max': 6000.0,
    'opacity': 1.0,
    'gamma': 1.0,
}
```

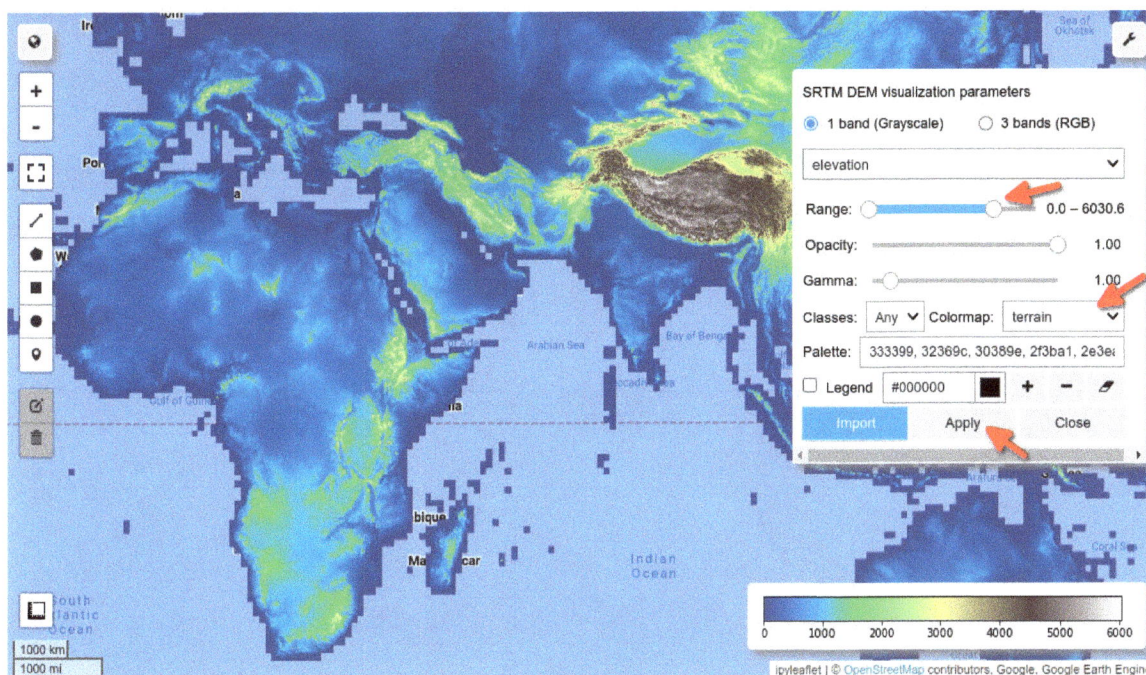

Figure 5.7: Changing image visualization parameters interactively.

This allows you to reuse the parameters for visualizing images programmatically without using the visualization GUI. Once you are satisfied with the visualization, click the `Close` button to close the GUI.

Multi-band images

Next, let's try out the visualization GUI for a multi-band image (e.g., Landsat 7). Activate the visualization GUI and choose an RGB band combination (e.g., `B5/B4/B3`) by selecting a band from the band dropdown list. Adjust other parameters (e.g., range, opacity, gamma) as needed. Note that the `palette` option is not applicable for multi-band images. Once you are done with adjusting the visualization parameters, click the `Apply` button to apply the changes. The image with the new visualization parameters will be displayed on the map (Fig. 5.8).

```
Map = geemap.Map()
landsat7 = ee.Image('LANDSAT/LE7_TOA_5YEAR/1999_2003')
vis_params = {
    'min': 20,
    'max': 200,
    'gamma': 2,
    'bands': ['B4', 'B3', 'B2'],
}
Map.addLayer(landsat7, vis_params, 'Landsat 7')
Map
```

5.6 Visualizing vector data

In Section 3.3 - *Visualizing Feature Collections*, we introduced two ways to visualize Earth Engine vector data programmatically, including the `ee.Image().paint()` and `collection.style()` methods. In this section, we will learn how to visualize vector data interactively without coding.

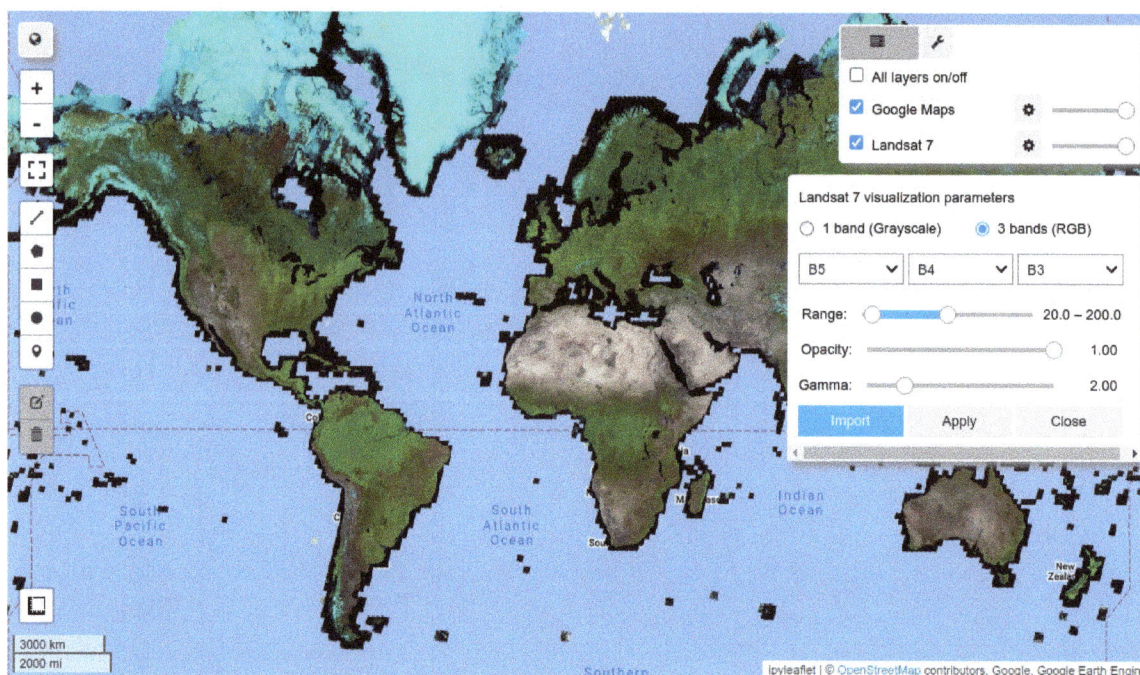

Figure 5.8: Visualizing a multi-band Landsat image.

Let's add some vector data to the map. We will use the TIGER: US Census States 2018[113] dataset.

```
Map = geemap.Map()
states = ee.FeatureCollection("TIGER/2018/States")
Map.addLayer(states, {}, "US States")
Map
```

Open the layer control widget and click the gear icon next to the layer name of the vector dataset (e.g., US States). The visualization GUI will be displayed below the layer control widget. Adjust visualization parameters (e.g., color, opacity, line width, line type, fill color) and layer name as needed. Then, click the Apply button to apply the changes. The vector dataset with the new visualization parameters will be displayed on the map (Fig. 5.9). By default, the vector dataset will be symbolized by black outline and filled with black color at 66% opacity.

To display the vector dataset by outline only, choose an outline color and set the fill opacity to 0. Then, click the Apply button to apply the changes (Fig. 5.10).

If you are running Jupyter Notebook rather than JupyterLab or Google Colab, you may click the Import button to import the visualization parameters into the Jupyter Notebook. A new cell will be created below the map and populated with the visualization parameters like this:

```
vis_params = {
    'color': 'ff0000ff',
    'width': 2,
    'lineType': 'solid',
    'fillColor': '00000000',
}
```

The visualization parameters can be used to re-create the same visualization programmatically without

---

[113]TIGER: US Census States 2018: tiny.geemap.org/qudx

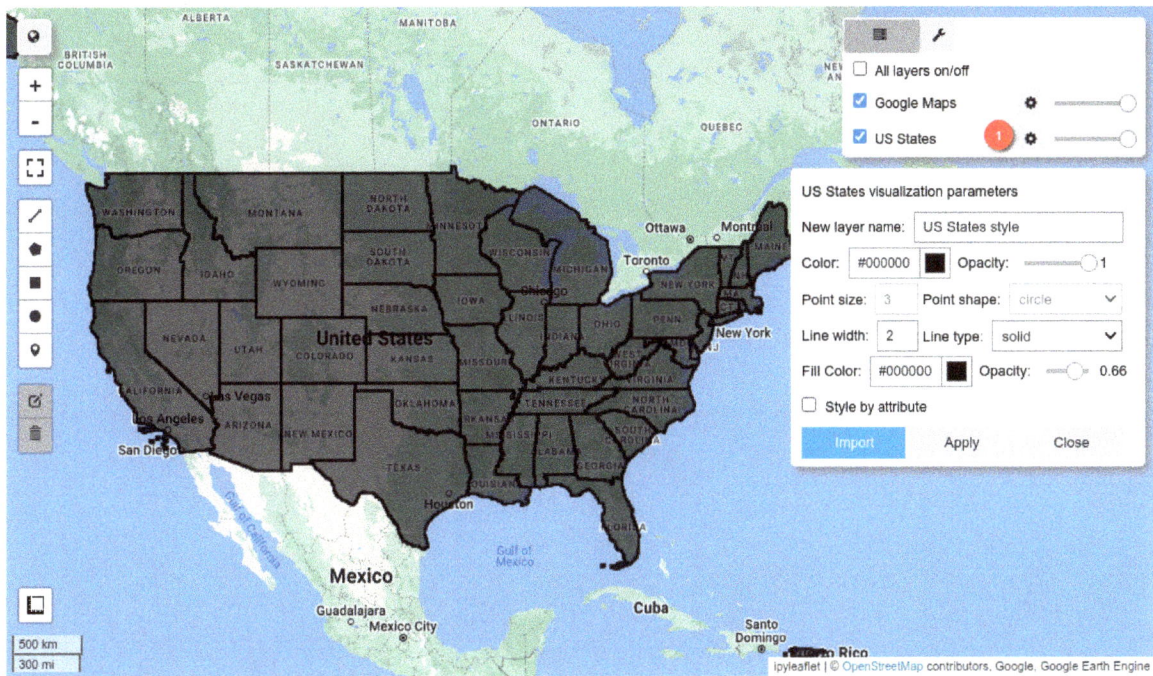

Figure 5.9: The visualization GUI for Earth Engine vector data.

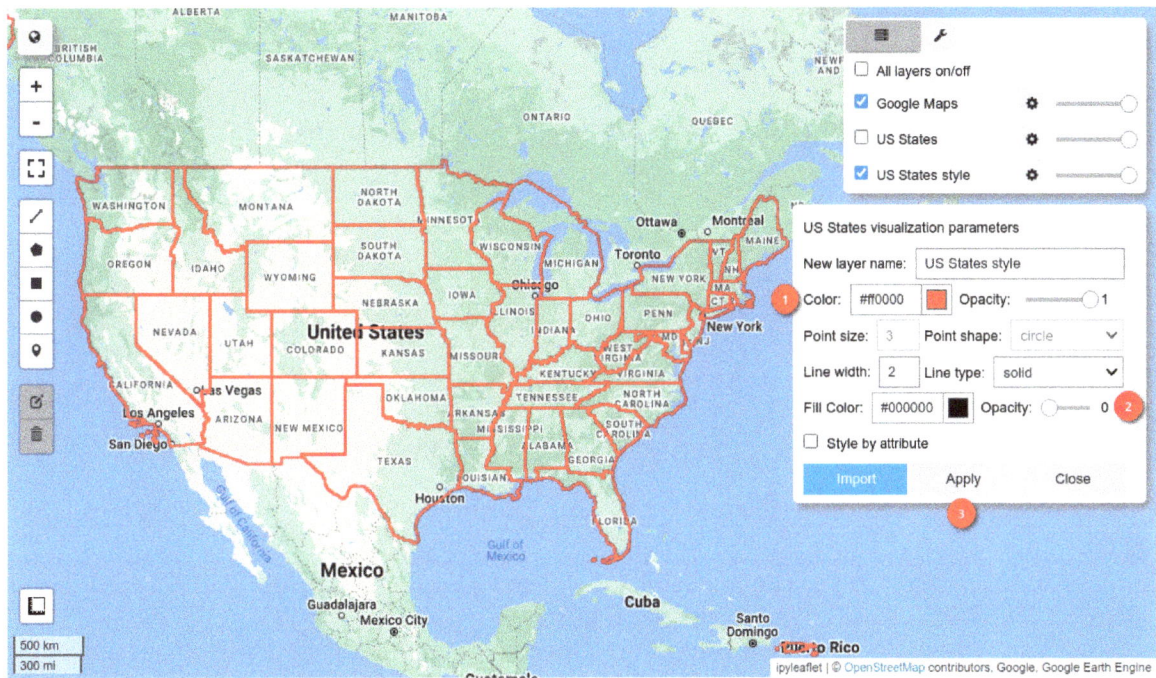

Figure 5.10: Displaying vector data with outline.

using the visualization GUI. For example:

```
Map = geemap.Map(center=[40, -100], zoom=4)
states = ee.FeatureCollection("TIGER/2018/States")
Map.addLayer(states.style(**vis_params), {}, "US States")
Map
```

To create a choropleth map (i.e., symbolizing the vector dataset by attribute values), check the `Style by attribute` option to display more options. Then, select a field name (e.g., `ALAND` represents the land area) from the dropdown list; choose the number of classes (e.g., 5); and select a colormap (e.g., `viridis`). Finally, click the `Apply` button to apply the changes. The choropleth map will be displayed on the map (Fig. 5.11). The yellow color indicates states with a larger land area while the blue color indicates states with a smaller land area.

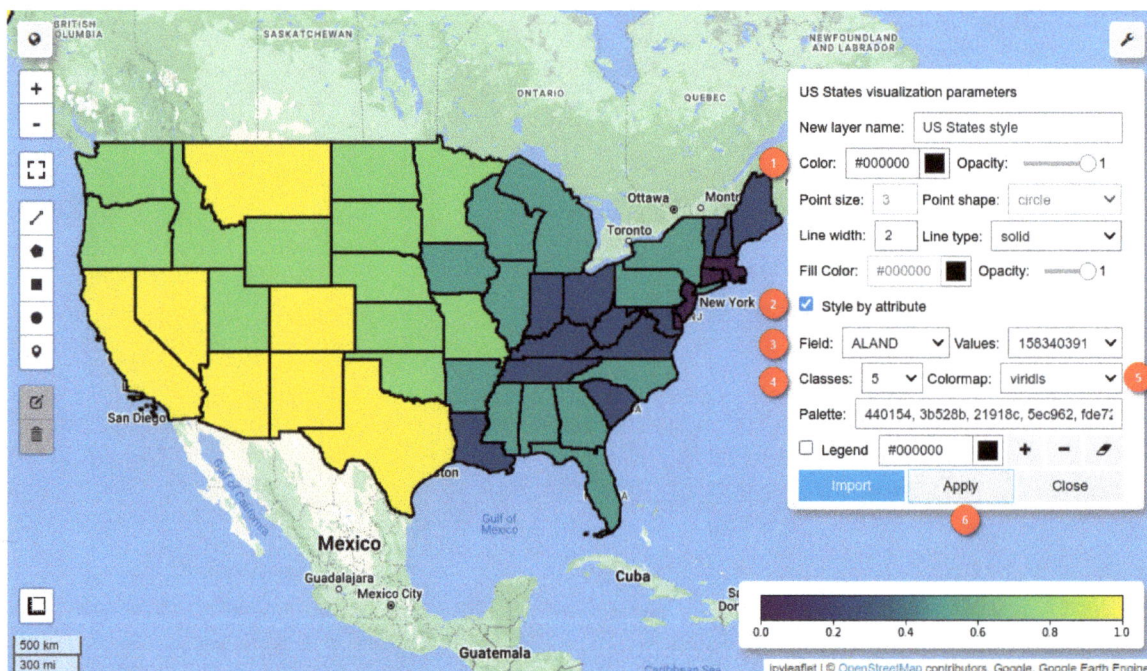

Figure 5.11: Vector data styling by attribute.

Once you are satisfied with the visualization, click the `Close` button to close the visualization GUI.

5.7 Creating legends

In this section, we will learn how to add a legend programmatically to an interactive map.

Built-in legends

Geemap has several built-in legends, making it easy to add a legend to the map with only one line of code. First, let's check what built-in legends are available:

```
from geemap.legends import builtin_legends

for legend in builtin_legends:
    print(legend)
```

```
NLCD
ESA_WorldCover
ESRI_LandCover
ESRI_LandCover_TS
Dynamic_World
NWI
MODIS/051/MCD12Q1
MODIS/006/MCD12Q1
GLOBCOVER
JAXA/PALSAR
Oxford
AAFC/ACI
COPERNICUS/CORINE/V20/100m
COPERNICUS/Landcover/100m/Proba-V/Global
USDA/NASS/CDL
ALOS_landforms
```

Some commonly used legends include `NLCD`, `ESA_WorldCover`, and `ESRI_LandCover`. To add a built-in legend to the map, simply use the `Map.add_legend()` method with the `builtin_legend` parameter. For example, to add the `NLCD` legend to the map, use the following code:

```
Map.add_legend(builtin_legend='NLCD')
```

Let's create an interactive map with the National Land Cover Database (NLCD) 2019[114] dataset and add the NLCD legend to the map (Fig. 5.12). Note that you can specify the legend title and size (e.g., `width`, `height`). The default `height` is `400px`. A scroll bar will be added to the legend automatically when the legend height exceeds the specified height. To show all legend items without the scroll bar, set the `height` to a larger value, e.g., `465px`.

```
Map = geemap.Map(center=[40, -100], zoom=4)
Map.add_basemap('HYBRID')

nlcd = ee.Image('USGS/NLCD_RELEASES/2019_REL/NLCD/2019')
landcover = nlcd.select('landcover')

Map.addLayer(landcover, {}, 'NLCD Land Cover 2019')
Map.add_legend(
    title="NLCD Land Cover Classification", builtin_legend='NLCD', height='465px'
)
Map
```

Custom legends

There are two ways to add a custom legend to the map. The first way is to provide a list of legend keys and colors to the `Map.add_legend()` method through the `legend_keys` and `legend_colors` parameters. Each color can be specified as a hexadecimal string (e.g., '#ff0000') or as a tuple of RGB values (e.g., (255, 0, 0)). For example, to add a legend with the following keys and colors to the map (see Fig. 5.13):

```
Map = geemap.Map(add_google_map=False)

labels = ['One', 'Two', 'Three', 'Four', 'ect']

# colors can be defined using either hex code or RGB (0-255, 0-255, 0-255)
colors = ['#8DD3C7', '#FFFFB3', '#BEBADA', '#FB8072', '#80B1D3']
# legend_colors = [(255, 0, 0), (127, 255, 0), (127, 18, 25), (36, 70, 180), (96, 68 123)]

Map.add_legend(
```

---

[114]National Land Cover Database (NLCD) 2019: tiny.geemap.org/uxwr

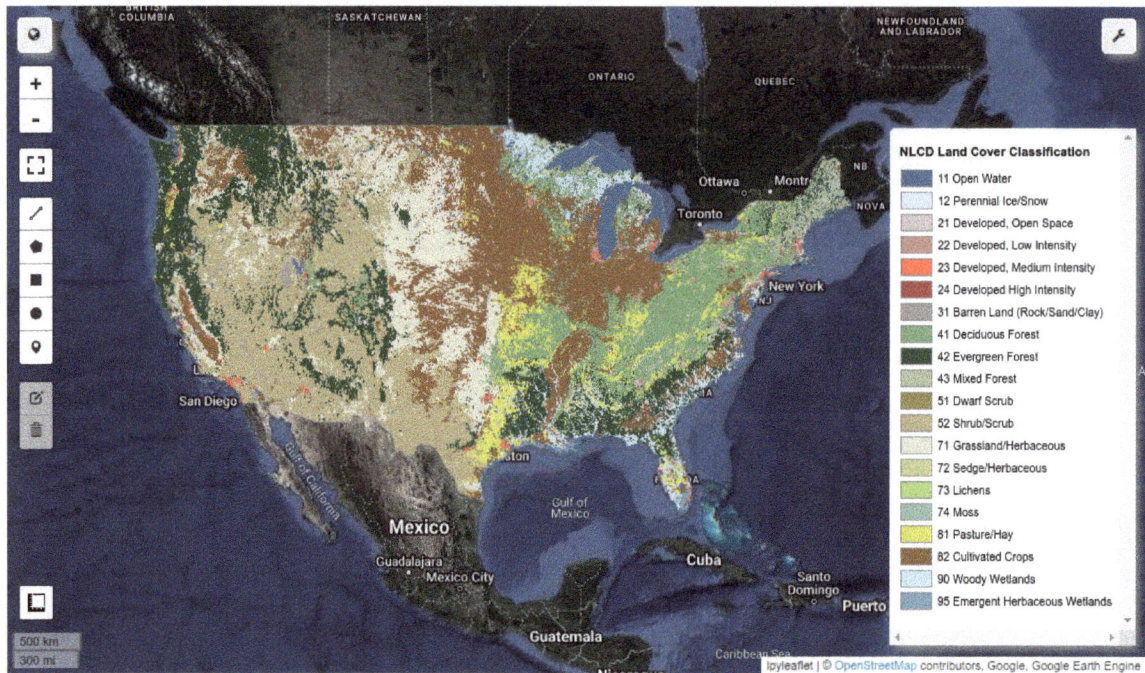

Figure 5.12: Adding the National Land Cover Database (NLCD) legend to the map.

```
    labels=labels, colors=colors, position='bottomright'
)
Map
```

The second way to add a custom legend to a map is to provide a legend dictionary to the Map.add_legend() method through the legend_dict parameter, which is a dictionary of legend keys and colors. For example, to add a custom legend for the NLCD land cover dataset:

```
Map = geemap.Map(center=[40, -100], zoom=4)

legend_dict = {
    '11 Open Water': '466b9f',
    '12 Perennial Ice/Snow': 'd1def8',
    '21 Developed, Open Space': 'dec5c5',
    '22 Developed, Low Intensity': 'd99282',
    '23 Developed, Medium Intensity': 'eb0000',
    '24 Developed High Intensity': 'ab0000',
    '31 Barren Land (Rock/Sand/Clay)': 'b3ac9f',
    '41 Deciduous Forest': '68ab5f',
    '42 Evergreen Forest': '1c5f2c',
    '43 Mixed Forest': 'b5c58f',
    '51 Dwarf Scrub': 'af963c',
    '52 Shrub/Scrub': 'ccb879',
    '71 Grassland/Herbaceous': 'dfdfc2',
    '72 Sedge/Herbaceous': 'd1d182',
    '73 Lichens': 'a3cc51',
    '74 Moss': '82ba9e',
    '81 Pasture/Hay': 'dcd939',
    '82 Cultivated Crops': 'ab6c28',
    '90 Woody Wetlands': 'b8d9eb',
    '95 Emergent Herbaceous Wetlands': '6c9fb8',
}
```

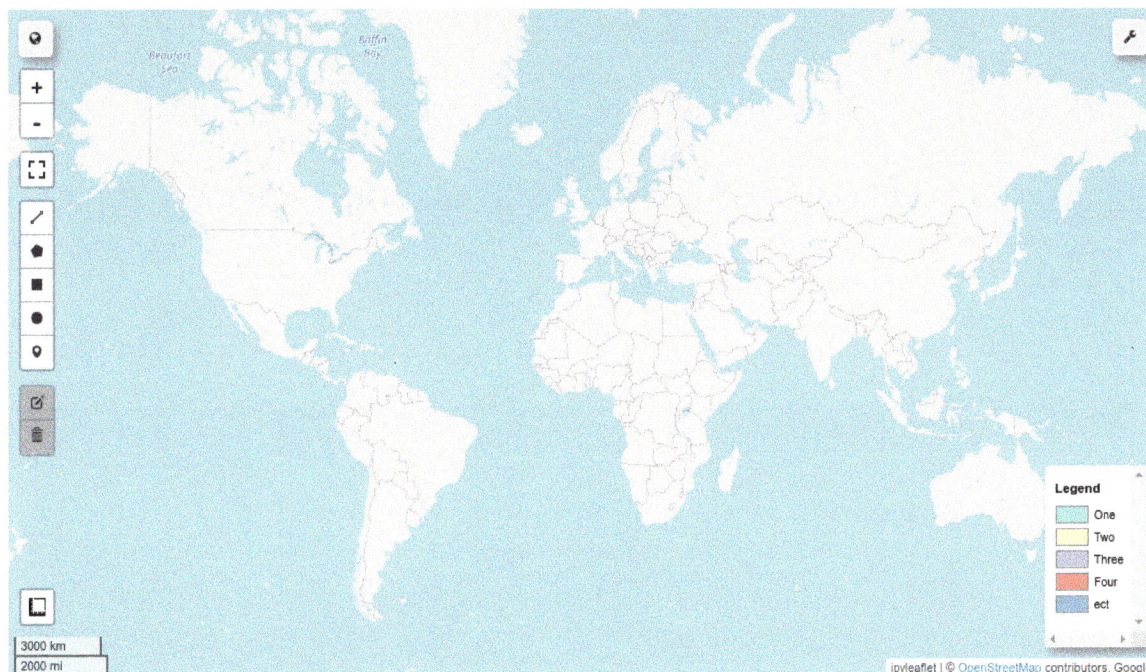

Figure 5.13: Adding a custom legend to the map.

```
nlcd = ee.Image('USGS/NLCD_RELEASES/2019_REL/NLCD/2019')
landcover = nlcd.select('landcover')

Map.addLayer(landcover, {}, 'NLCD Land Cover 2019')
Map.add_legend(title="NLCD Land Cover Classification", legend_dict=legend_dict)
Map
```

The output map should be the same as the map shown in Fig. 5.12.

Earth Engine class table

Many Earth Engine thematic datasets (e.g., land cover datasets) have a class table with three columns, including the Value, Color, and Description columns. For example, let's look at the ESA WorldCover 10m v100[115] class table under the **Bands** tab (Fig. 5.14).

Geemap provides a quick way to create a legend from an Earth Engine class table. Simply copy the class table with the header from the Earth Engine webpage and wrap it with triple quotes and then use the legend_from_ee() function to create a legend. For example, to add a legend to the map with the ESA WorldCover class table (Fig. 5.15):

```
Map = geemap.Map()

dataset = ee.ImageCollection("ESA/WorldCover/v100").first()
Map.addLayer(dataset, {'bands': ['Map']}, "Landcover")

ee_class_table = """
Value   Color   Description
10  006400  Trees
20  ffbb22  Shrubland
```

---

[115]ESA WorldCover 10m v100: tiny.geemap.org/vnue

Description   Bands   Terms of Use   Citations

Resolution
10 meters

Bands

Name	Description
Map	Landcover class

Map Class Table

Value	Color	Description
10	006400	Trees
20	ffbb22	Shrubland
30	ffff4c	Grassland
40	f096ff	Cropland
50	fa0000	Built-up
60	b4b4b4	Barren / sparse vegetation
70	f0f0f0	Snow and ice
80	0064c8	Open water
90	0096a0	Herbaceous wetland
95	00cf75	Mangroves
100	fae6a0	Moss and lichen

Figure 5.14: The Earth Engine class table for the ESA WorldCover.

```
30  ffff4c  Grassland
40  f096ff  Cropland
50  fa0000  Built-up
60  b4b4b4  Barren / sparse vegetation
70  f0f0f0  Snow and ice
80  0064c8  Open water
90  0096a0  Herbaceous wetland
95  00cf75  Mangroves
100 fae6a0  Moss and lichen
"""

legend_dict = geemap.legend_from_ee(ee_class_table)
Map.add_legend(title="ESA Land Cover", legend_dict=legend_dict)
Map
```

5.8   Creating color bars

In the previous section, we learned how to add a legend for discrete data. In this section, we will learn how to add a color bar for continuous data (elevation, temperature, precipitation, etc.) to the map. Let's use the NASA SRTM Digital Elevation dataset[116] as an example:

```
Map = geemap.Map()
dem = ee.Image('USGS/SRTMGL1_003')
vis_params = {
    'min': 0,
    'max': 4000,
    'palette': ['006633', 'E5FFCC', '662A00', 'D8D8D8', 'F5F5F5'],
```

---

[116]NASA SRTM Digital Elevation dataset: tiny.geemap.org/ultl

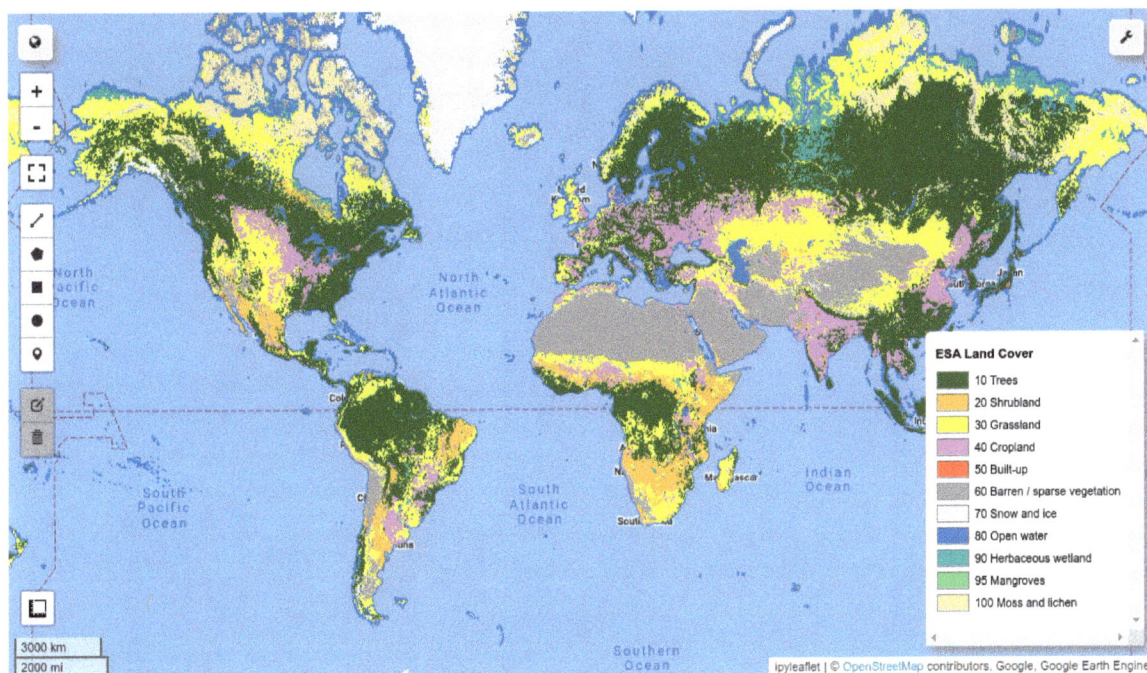

Figure 5.15: Adding a legend for the ESA global land cover dataset.

```
}
Map.addLayer(dem, vis_params, 'SRTM DEM')
Map
```

The DEM dataset is visualized using the provided visualization parameters (i.e., vis_params), which include min, max, and palette. These three parameters are the key parameters for defining the colorbar. The min and max parameters define the minimum and maximum values of the colorbar. The palette parameter defines the color gradient of the colorbar. Let's create our first colorbar using the Map.add_colorbar() method. We can directly pass the vis_params dictionary to the Map.add_colorbar() method. The label parameter adds a title to the colorbar. The layer_name parameter is used to tie the visibility of the colorbar to the corresponding Earth Engine layer. If the layer is toggled off, the colorbar will also become invisible. The following code will add a horizontal colorbar to the map.

```
Map.add_colorbar(vis_params, label="Elevation (m)", layer_name="SRTM DEM")
Map
```

The resulting map should look like Fig. 5.16.

The default orientation of the colorbar is horizontal. To create a vertical colorbar, we can set the orientation parameter. Note that the previously added colorbar with the same layer name will be replaced with the new colorbar.

```
Map.add_colorbar(
    vis_params, label="Elevation (m)", layer_name="SRTM DEM", orientation="vertical"
)
```

The resulting map should look like Fig. 5.17.

To make the colorbar background transparent, set the transparent_bg parameter to True:

```
Map.add_colorbar(
```

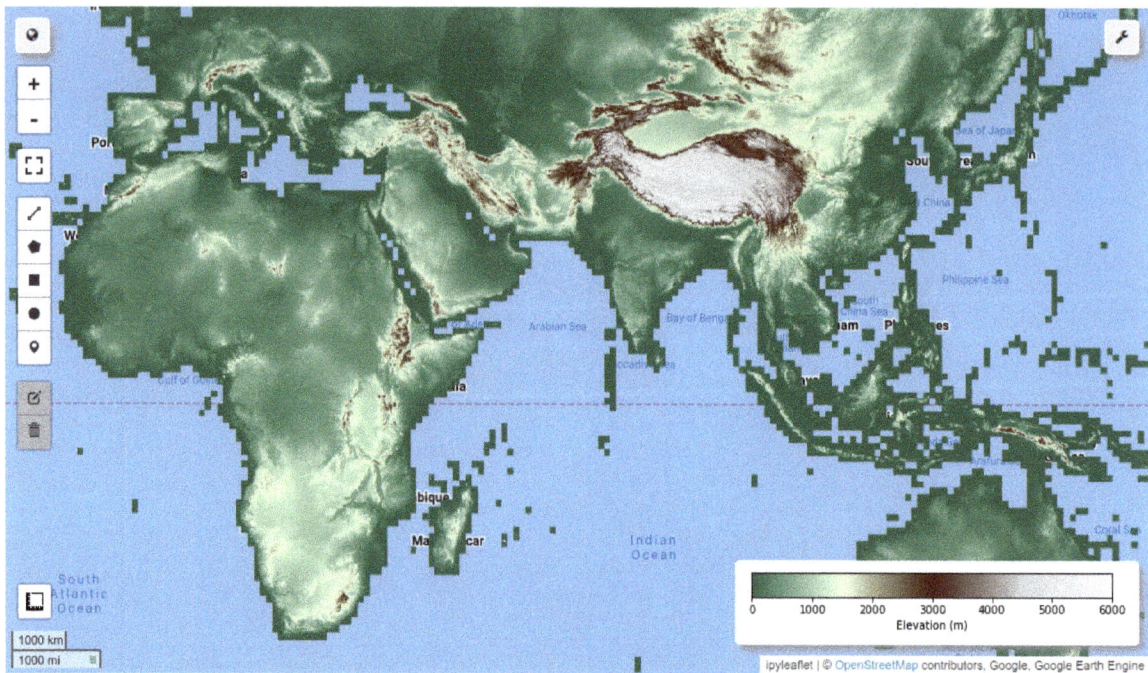

Figure 5.16: Adding a colorbar for the NASA SRTM Digital Elevation dataset.

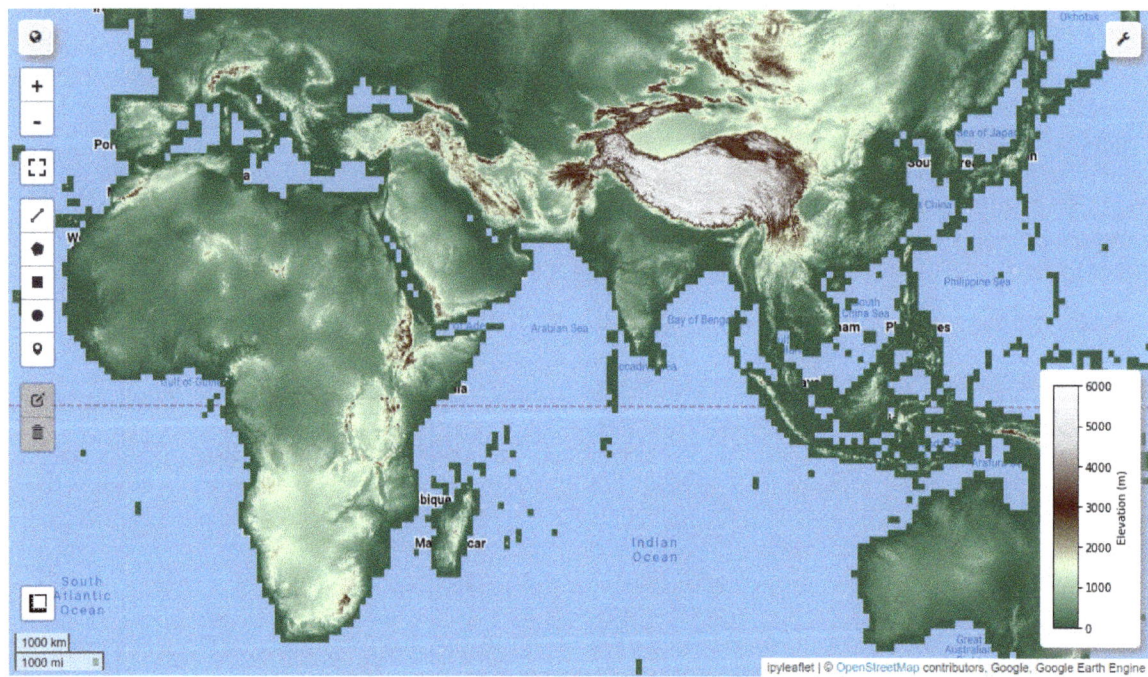

Figure 5.17: Adding a vertical colorbar.

```
    vis_params,
    label="Elevation (m)",
    layer_name="SRTM DEM",
    orientation="vertical",
    transparent_bg=True,
)
```

The resulting map should look like Fig. 5.18.

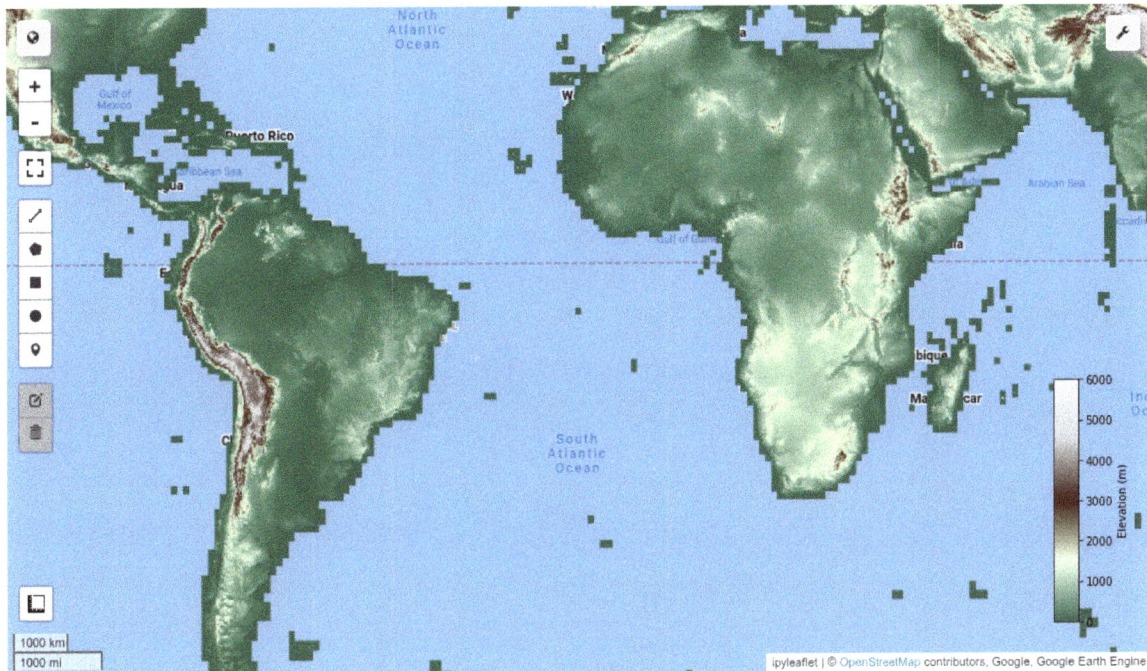

Figure 5.18: Adding a vertical colorbar with a transparent background.

5.9   Displaying labels

Labeling is an easy way to add descriptive text to features on your map. You can turn labels on or off. You can also make labels draggable or lock them so their locations stay fixed as you zoom or pan on your map. Let's add some features to the map, e.g., the US Census States dataset:

```
Map = geemap.Map(center=[40, -100], zoom=4, add_google_map=False)
states = ee.FeatureCollection("TIGER/2018/States")
style = {'color': 'black', 'fillColor': "00000000"}
Map.addLayer(states.style(**style), {}, "US States")
Map
```

To display labels for each feature, we can use the `Map.add_labels()` method. The US Census States dataset has a column called `STUSPS` that contains the US Postal Service state abbreviation of each state. You can specify the `STUSPS` column as the label column. You can also control the font size, color, and font family of the labels. Lastly, you can set the `draggable` parameter to `True` to make the labels draggable (Fig. 5.19).

```
Map.add_labels(
    data=states,
    column="STUSPS",
    font_size="12pt",
    font_color="blue",
```

```
    font_family="arial",
    font_weight="bold",
    draggable=True,
)
```

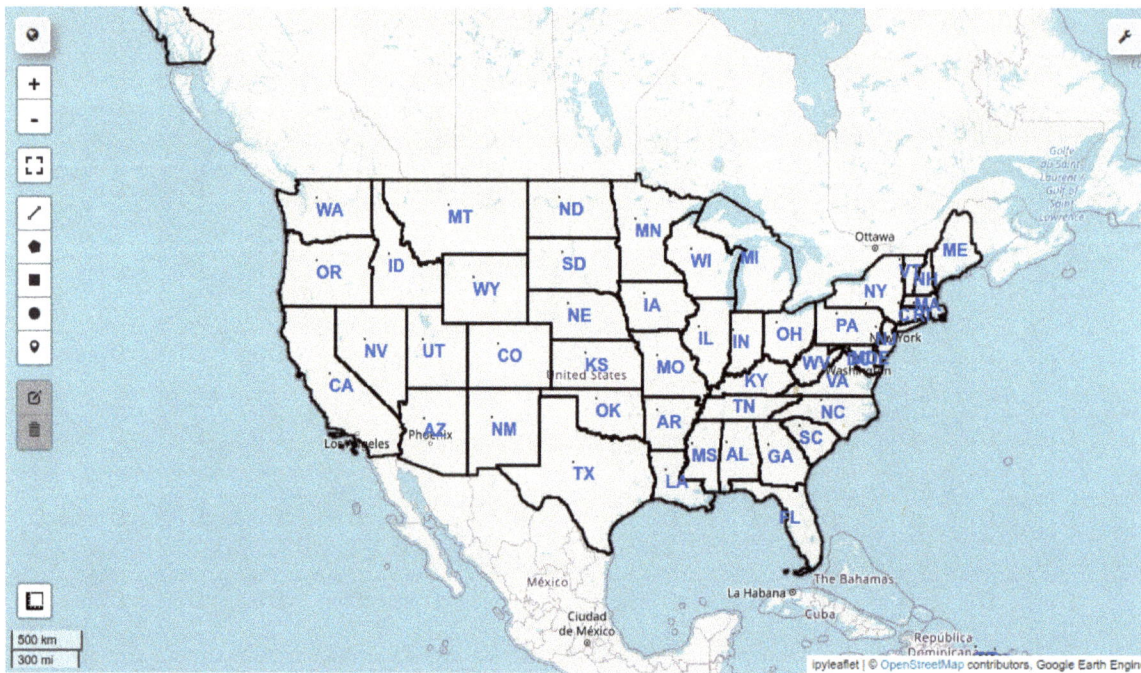

Figure 5.19: Adding labels to the map.

Use the layer control to toggle the labels on and off. To remove the labels from the map, use the `Map.remove_labels()` method.

```
Map.remove_labels()
```

By default, labels are placed around the center of each feature. To precisely control the location of each label, you can provide a Pandas DataFrame with `longitude` and `latitude` columns. For example, let's generate the centroid of each state using the `vector_centroids()` method. The resulting FeatureCollection can then be converted to a Pandas dataframe using the `ee_to_df()` method:

```
centroids = geemap.vector_centroids(states)
df = geemap.ee_to_df(centroids)
df
```

The resulting dataframe should look like (Fig. 5.20).

Run the following cell to add labels to the map. The resulting map should look like Fig. 5.19.

```
Map.add_labels(
    data=df,
    column="STUSPS",
    font_size="12pt",
    font_color="blue",
    font_family="arial",
    font_weight="bold",
    x='longitude',
    y='latitude',
)
```

	latitude	longitude	STATENS	GEOID	AWATER	LSAD	STUSPS	STATEFP	FUNCSTAT	INTPTLAT	DIVISION	REGION	NAME	INTPTLON	MTFCC
0	18.059360	-64.838759	01802710	78	1550236201	00	VI	78	A	+18.3267480	0	9	United States Virgin Islands	-064.9712508	G4000
1	16.796171	145.597019	01779809	69	4644252461	00	MP	69	A	+14.9367835	0	9	Commonwealth of the Northern Mariana Islands	+145.6010210	G4000
2	13.442713	144.769378	01802705	66	934337453	00	GU	66	A	+13.4382886	0	9	Guam	+144.7729493	G4000
3	-13.963954	-170.082423	01802701	60	1307243754	00	AS	60	A	-14.2671590	0	9	American Samoa	-170.6682674	G4000
4	18.216460	-66.414736	01779808	72	4922382562	00	PR	72	A	+18.2176480	0	9	Puerto Rico	-066.4107992	G4000
5	41.594008	-71.524721	01219835	44	1323670487	00	RI	44	A	+41.5974187	1	1	Rhode Island	-071.5272723	G4000
6	43.673856	-71.573904	01779794	33	1026675248	00	NH	33	A	+43.6726907	1	1	New Hampshire	-071.5843145	G4000

Figure 5.20: Converting an Earth Engine FeatureCollection to a Pandas DataFrame.

Map

5.10 Image overlay

Besides adding image layers from Earth Engine, you can also add image layers from other sources. For example, you can add an image (e.g., a logo) from a URL and overlay it on the map. The `ImageOverlay` class is used to convert an image to a tile layer so that it can be added to the map. You need to specify the bounding box when the image will be displayed through the `bounds` parameter. For example, let's retrieve an image from `https://i.imgur.com/06Q1fSz.jpg` and overlay it on the map (Fig. 5.21):

```
Map = geemap.Map(center=(25, -115), zoom=5)
url = 'https://i.imgur.com/06Q1fSz.png'
image = geemap.ImageOverlay(url=url, bounds=((13, -130), (32, -100)))
Map.add_layer(image)
Map
```

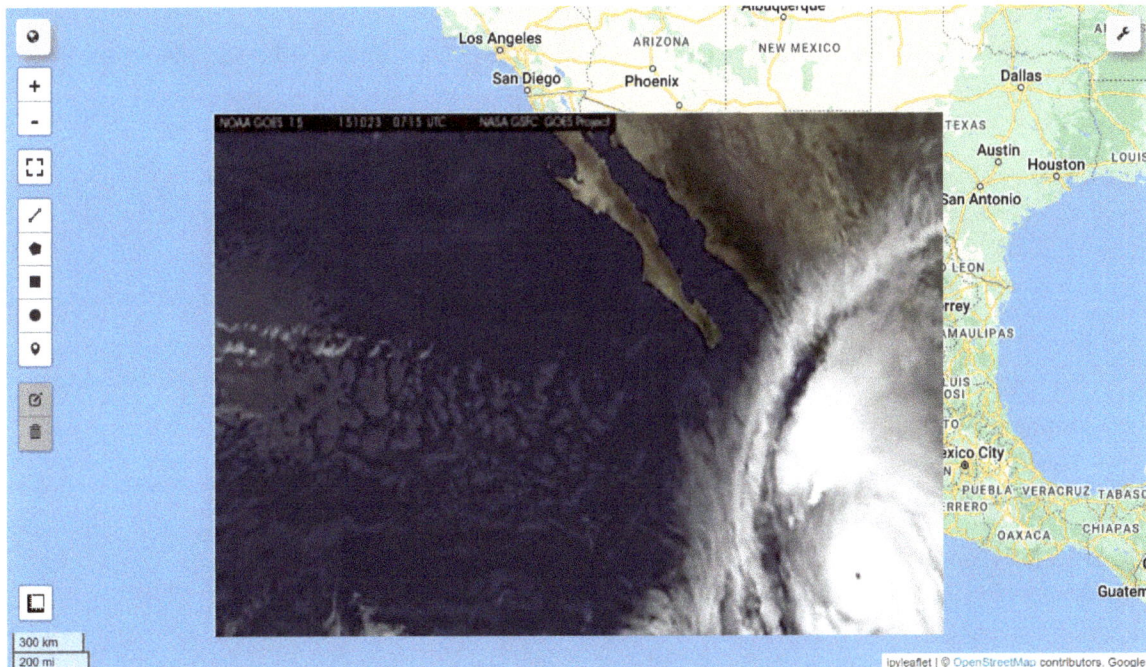

Figure 5.21: Overlaying an image on the map.

You can replace the existing image shown on the map with a new image by passing a new URL to the `image.url` attribute. For example, let's replace the existing image with a new image from `https://i.imgur.com/U0axit9.jpg` (Fig. 5.22):

```
image.url = 'https://i.imgur.com/U0axit9.png'
Map
```

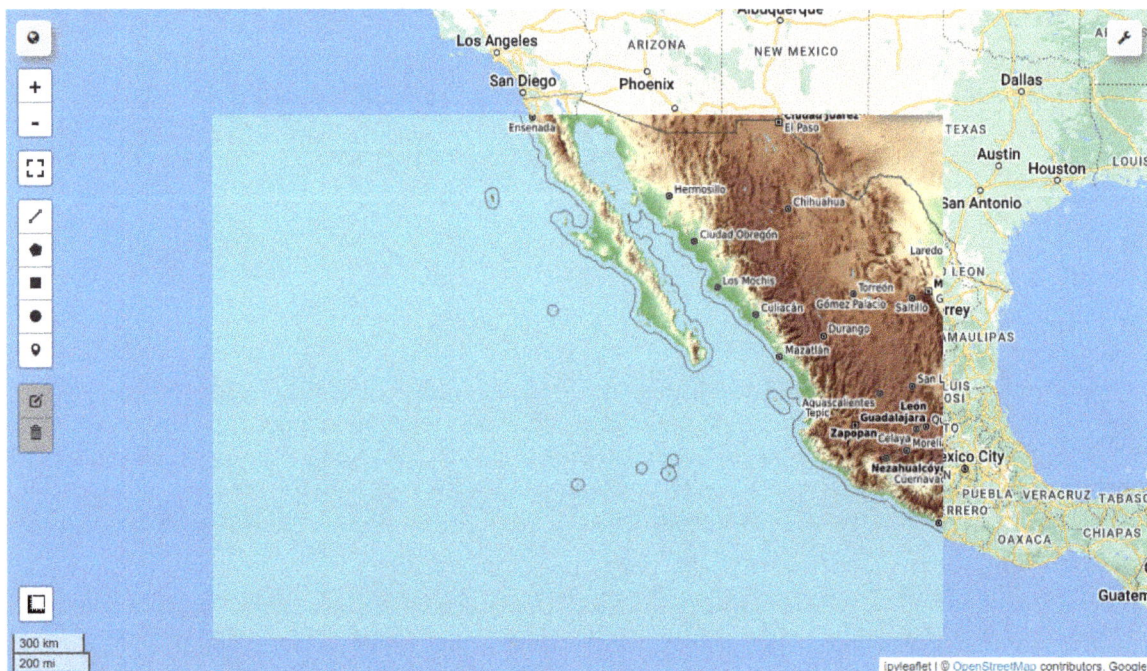

Figure 5.22: Replacing the image overlay with a new image.

The `ImageOverlay` class also works for images stored locally. Let's download a sample image to our computer:

```
url = 'https://i.imgur.com/06Q1fSz.png'
filename = 'hurricane.png'
geemap.download_file(url, filename)
```

Then, simply pass the file path to the `url` parameter of the `ImageOverlay` constructor:

```
Map = geemap.Map(center=(25, -115), zoom=5)
image = geemap.ImageOverlay(url=filename, bounds=((13, -130), (32, -100)))
Map.add_layer(image)
Map
```

The resulting map should look like Fig. 5.21.

5.11   Video overlay

Similar to overlaying an image on the map, you can also overlay a video on the map. Simply pass a URL to the video and a bounding box to the `Map.video_overlay()` method:

```
Map = geemap.Map(center=(25, -115), zoom=5)
url = 'https://labs.mapbox.com/bites/00188/patricia_nasa.webm'
bounds = ((13, -130), (32, -100))
Map.video_overlay(url, bounds)
```

```
Map
```

The resulting map should look like (Fig. 5.23).

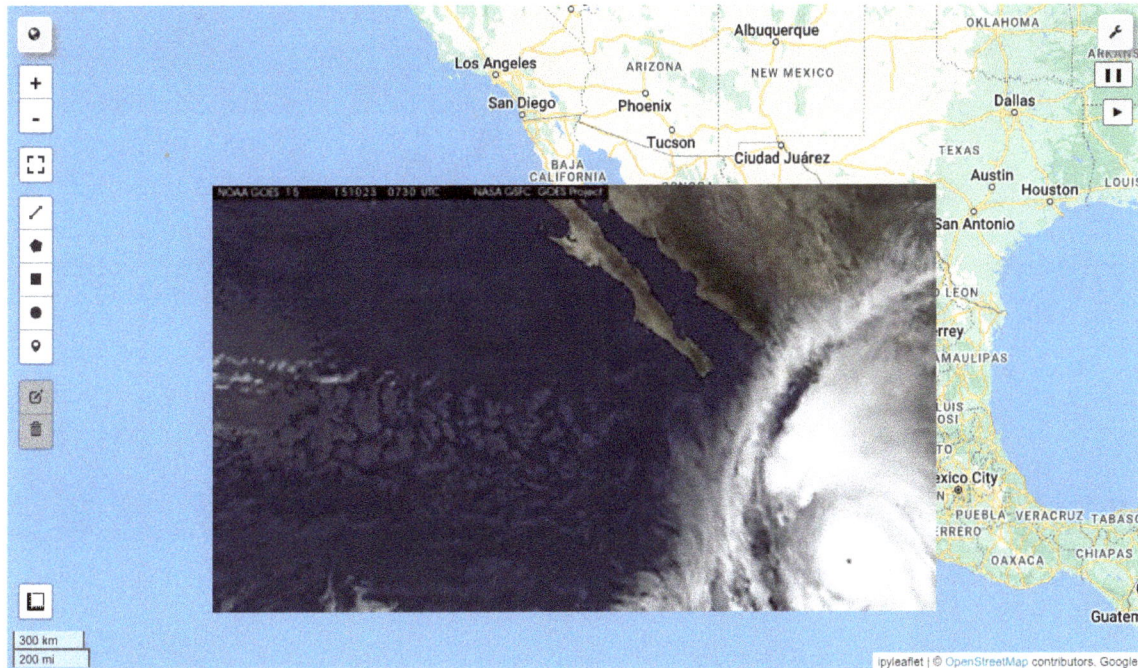

Figure 5.23: Overlaying a video on the map.

5.12   Split-panel maps

In Section 5.4 - *Changing Layer Opacity*, we learned how to change the opacity of a layer. This can be useful for comparing two different layers by changing the opacity layers. In this section, we will introduce the split-panel map that can also be used to compare two layers. A split-panel map is composed of two panels side-by-side with a divider between them. The divider can be dragged and moved to resize the two panels. The split-panel map requires two layers: `left_layer` and `right_layer`. The layer instance can be a string representing a basemap, an HTTP URL to a Cloud Optimized GeoTIFF (COG), or an ipyleaflet/folium TileLayer instance. For example, let's create a split-panel map with a Google satellite imagery layer on the left and a Google terrain layer on the right.

```
Map = geemap.Map()
Map.split_map(left_layer='HYBRID', right_layer='TERRAIN')
Map
```

Any built-in basemaps can be used with the split-panel map by providing the name of the basemap as a string. To see the list of available basemaps, run the follow cell:

```
list(geemap.basemaps.keys())
```

To use Earth Engine layers with the split-panel map, you need to create a TileLayer instance from an Earth Engine dataset using the `ee_tile_layer()` function. The parameters of the `ee_tile_layer()` function are the same as the parameters of the `Map.addLayer()` method. The example below shows how to create a split-panel map with two NLCD land cover layers (Fig. 5.25):

```
Map = geemap.Map(center=(40, -100), zoom=4, height=600)
```

Figure 5.24: A split-panel map with Google satellite imagery layer on the left and Google terrain layer on the right.

```
nlcd_2001 = ee.Image('USGS/NLCD_RELEASES/2019_REL/NLCD/2001').select('landcover')
nlcd_2019 = ee.Image('USGS/NLCD_RELEASES/2019_REL/NLCD/2019').select('landcover')

left_layer = geemap.ee_tile_layer(nlcd_2001, {}, 'NLCD 2001')
right_layer = geemap.ee_tile_layer(nlcd_2019, {}, 'NLCD 2019')

Map.split_map(left_layer, right_layer, add_close_button=True)
Map
```

Zoom and pan the map to a specific region. Then, drag the divider back and forth to see land cover changes between 2001 and 2019. Click the Close button in the lower-right corner of the map to close the split-panel map and return to the default map.

5.13   Linked maps

The split-panel map introduced above can be used to compare two layers side by side. To compare more than two layers simultaneously, you can create linked maps by specifying the number of rows and columns. The layers can come from the same image with different band combinations or a series of images from an image collection. Each layer can have unique visualization parameters. Other parameters of the `linked_maps()` function include: `height`, `center`, `zoom`, `labels`, and `label_position`. The example below creates linked maps of Sentinel-2 imagery with four different band combinations (Fig. 5.26). Dragging and moving one map will automatically update the other linked maps so that all maps are always centered at the same location with the same zoom level. Note that the linked maps might not work properly in Google Colab or JupyterLab. It should work fine in Jupyter Notebook and Visual Studio Code.

```
image = (
    ee.ImageCollection('COPERNICUS/S2')
    .filterDate('2018-09-01', '2018-09-30')
```

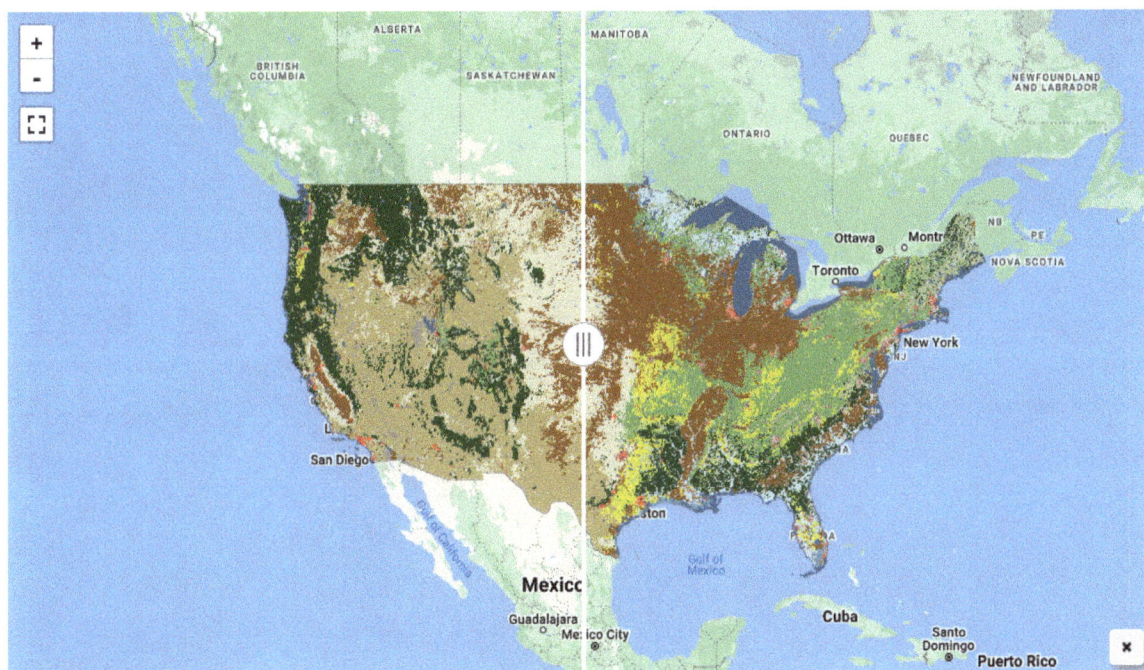

Figure 5.25: A split-panel map with two NLCD land cover layers.

```python
    .map(lambda img: img.divide(10000))
    .median()
)

vis_params = [
    {'bands': ['B4', 'B3', 'B2'], 'min': 0, 'max': 0.3, 'gamma': 1.3},
    {'bands': ['B8', 'B11', 'B4'], 'min': 0, 'max': 0.3, 'gamma': 1.3},
    {'bands': ['B8', 'B4', 'B3'], 'min': 0, 'max': 0.3, 'gamma': 1.3},
    {'bands': ['B12', 'B12', 'B4'], 'min': 0, 'max': 0.3, 'gamma': 1.3},
]

labels = [
    'Natural Color (B4/B3/B2)',
    'Land/Water (B8/B11/B4)',
    'Color Infrared (B8/B4/B3)',
    'Vegetation (B12/B11/B4)',
]

geemap.linked_maps(
    rows=2,
    cols=2,
    height="300px",
    center=[38.4151, 21.2712],
    zoom=12,
    ee_objects=[image],
    vis_params=vis_params,
    labels=labels,
    label_position="topright",
)
```

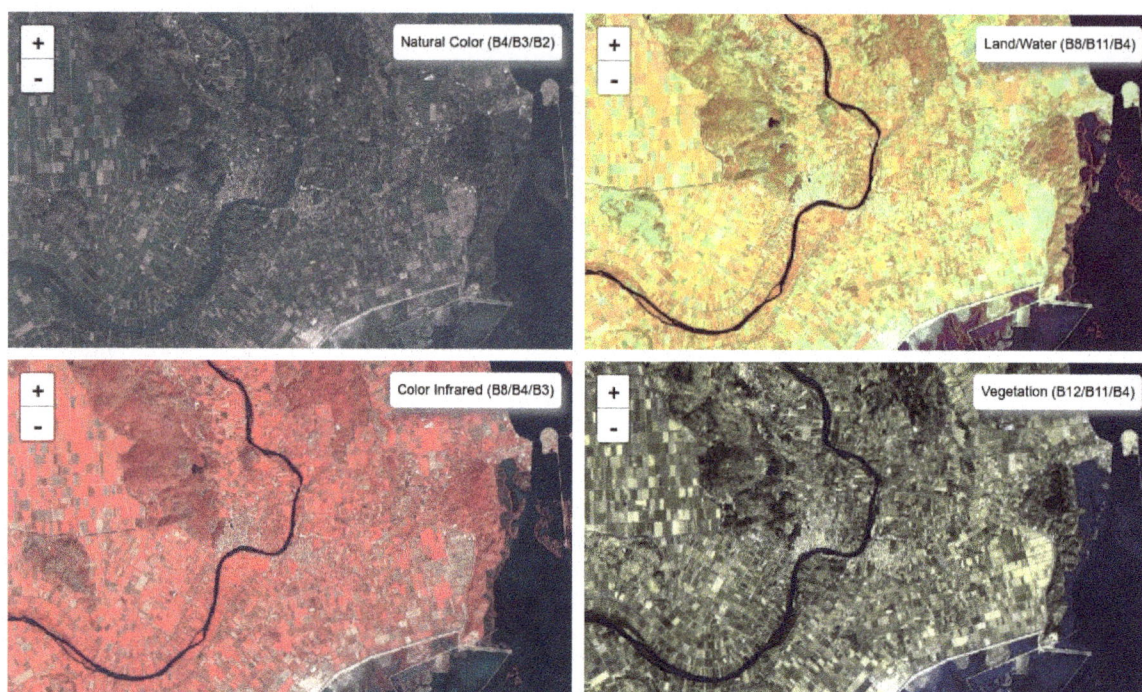

Figure 5.26: Creating linked maps for comparing different band combinations of an image.

5.14   Timeseries inspector

Visualizing image collections

The split-panel map is useful for comparing two layers side by side. To change the layers interactively, you can use the timeseries inspector. Let's take the National Land Cover Database (NLCD) as an example. It is an image collection consisting of NLCD land cover maps from 2001 to 2019. Let's load the datasets and specify visualization parameters for the NLCD maps.

```
Map = geemap.Map(center=[40, -100], zoom=4)
collection = ee.ImageCollection('USGS/NLCD_RELEASES/2019_REL/NLCD').select('landcover')
vis_params = {'bands': ['landcover']}
years = collection.aggregate_array('system:index').getInfo()
years
```

The output should look like this:

```
['2001', '2004', '2006', '2008', '2011', '2013', '2016', '2019']
```

As can be seen above, there are eight time periods of NLCD land cover maps. The follow example creates a split-panel map with the NLCD image collection (Fig. 5.27):

```
Map.ts_inspector(
    left_ts=collection,
    right_ts=collection,
    left_names=years,
    right_names=years,
    left_vis=vis_params,
    right_vis=vis_params,
    width='80px',
)
```

```
Map
```

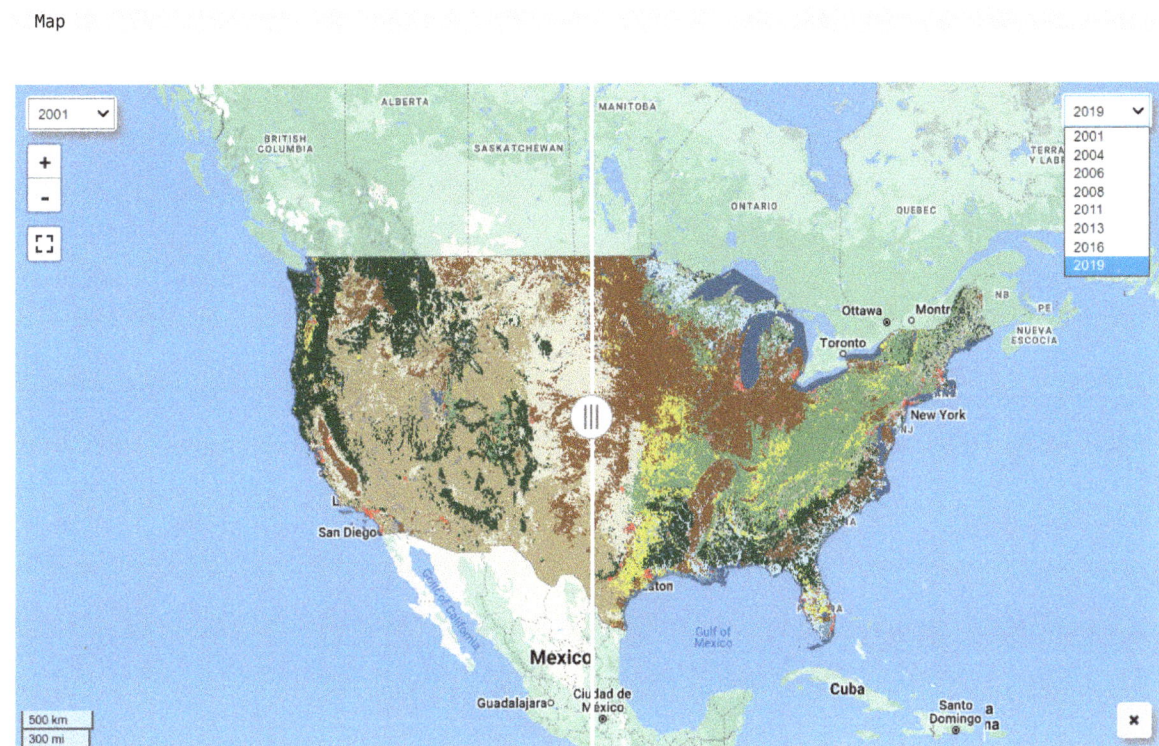

Figure 5.27: A timeseries inspector for visualizing NLCD land cover datasets.

Two dropdown menus appear on the left and right side of the split-panel map. These menus allow you to select the year of the NLCD dataset to be displayed on the left and right side of the split-panel map. When a year is selected on either side, the corresponding NLCD image is displayed on the map. In this example, the same image collection is used on both sides of the split-panel map. They can be different image collections if needed. Similarly, the dropdown labels and visualization parameters can also be different for each side of the split-panel map. The `width` parameter specifies the width of the dropdown menus. Zoom and pan the map to a specific region and then move the divider back and forth to see land cover changes between years. Once you are done, click the Close button in the lower-right corner of the map to close the split-panel map and return to the default map.

Visualizing planet.com imagery

First, you need to sign up a Planet account and get an API key by going to the Planet website - Get Started with Planet APIs[117]. Pass your API key to the environment variable `PLANET_API_KEY`:

```python
import os

os.environ["PLANET_API_KEY"] = "your-api-key"

monthly_tiles = geemap.planet_monthly_tiles()
geemap.ts_inspector(monthly_tiles)

quarterly_tiles = geemap.planet_quarterly_tiles()
geemap.ts_inspector(quarterly_tiles)

tiles = geemap.planet_tiles()
```

---

[117]Get Started with Planet APIs: tiny.geemap.org/ubby

```
geemap.ts_inspector(tiles)

Map = geemap.Map()
Map
```

5.15   Time slider

If you want to animate a series of images in an image collection, the time slider is for you. It can animate any image collection by specifying visualization parameters and a time interval in seconds. The smaller the time interval, the faster the animation, and vice versa.

Visualizing vegetation data

The following example uses the time slider to animate MODIS Normalized Difference Vegetation Index (NDVI) data[118].

```
Map = geemap.Map()

collection = (
    ee.ImageCollection('MODIS/MCD43A4_006_NDVI')
    .filter(ee.Filter.date('2018-06-01', '2018-07-01'))
    .select("NDVI")
)
vis_params = {
    'min': 0.0,
    'max': 1.0,
    'palette': 'ndvi',
}

Map.add_time_slider(collection, vis_params, time_interval=2)
Map
```

The time slider appears in the lower-right corner of the map (Fig. 5.28). It has three buttons: play, pause, and close. The play button starts the animation. The pause button pauses the animation. The close button stops the animation and removes the time slider. Removing the time slider will also remove the corresponding image layers from the map.

Visualizing weather data

The example below uses the time slider to animate Global Forecast System (GFS) temperature data[119] (Fig. 5.29). You can change the opacity of layers by specifying the opacity parameter for the Map.add_ time_slider() method. You can also provide a list of labels to be used for the time slider. In this case, the list of labels contains image acquisition times. Click the play button to start the animation.

```
Map = geemap.Map()

collection = (
    ee.ImageCollection('NOAA/GFS0P25')
    .filterDate('2018-12-22', '2018-12-23')
    .limit(24)
    .select('temperature_2m_above_ground')
)

vis_params = {
```

---

[118]MODIS Normalized Difference Vegetation Index (NDVI) data: tiny.geemap.org/mxyr

[119]Global Forecast System (GFS) temperature data: tiny.geemap.org/ofcv

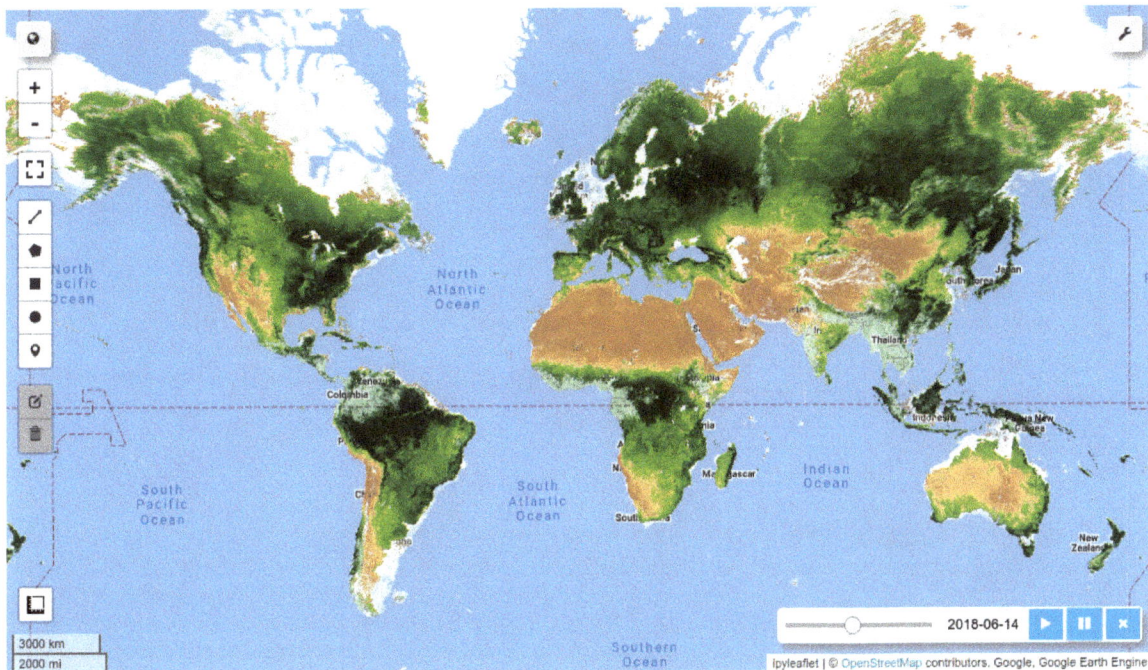

Figure 5.28: Using the time slider to animate MODID NDVI data.

```
    'min': -40.0,
    'max': 35.0,
    'palette': ['blue', 'purple', 'cyan', 'green', 'yellow', 'red'],
}

labels = [str(n).zfill(2) + ":00" for n in range(0, 24)]
Map.add_time_slider(collection, vis_params, labels=labels, time_interval=1, opacity=0.8)
Map
```

Visualizing Sentinel-2 imagery

The following example uses the time slider to animate Sentinel-2 imagery[120] for the San Francisco Bay Area (Fig. 5.30). It filters the image collection to select images with a cloud cover percentage less than 10%. It can be easily adapted to animate any other image collections for a specific region. Again, click the play button to start the animation.

```
Map = geemap.Map(center=[37.75, -122.45], zoom=12)

collection = (
    ee.ImageCollection('COPERNICUS/S2_SR')
    .filterBounds(ee.Geometry.Point([-122.45, 37.75]))
    .filterMetadata('CLOUDY_PIXEL_PERCENTAGE', 'less_than', 10)
)

vis_params = {"min": 0, "max": 4000, "bands": ["B8", "B4", "B3"]}

Map.add_time_slider(collection, vis_params)
Map
```

[120]Sentinel-2 imagery: tiny.geemap.org/xusk

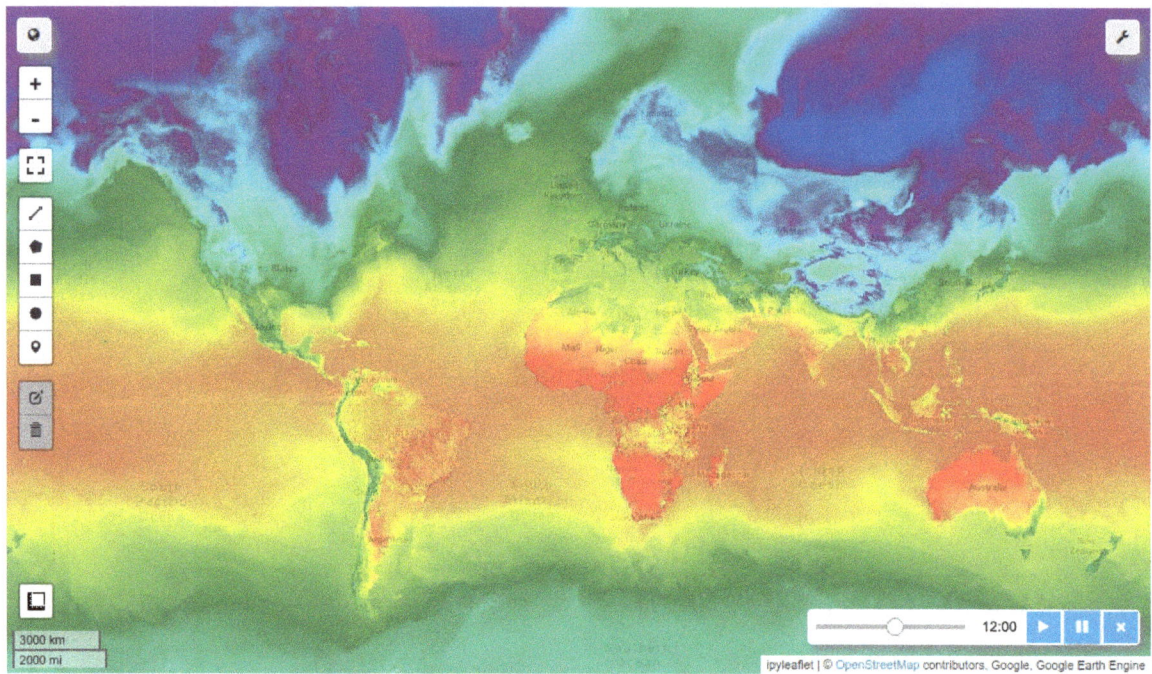

Figure 5.29: Using the time slider to animate GFS temperature data.

Figure 5.30: Using the time slider to animate Sentinel-2 imagery.

5.16 Shaded relief maps

Shaded relief maps are commonly used to visualize digital elevation data. A shaded relief map is a flat map that appears to be three-dimensional images of the physical features on Earth's surface such as mountains, valleys, and plateaus. Compared with hillshades, shaded relief maps can better distinguish landforms and features through the use of colors and shadows. The following example creates a global shaded relief map based on the 30-m NASA SRTM Digital Elevation data[121] using the blend() function, which creates a blended image that is a combination of two images (Fig. 5.31). In this case, the first image is the DEM with visualization parameters, and the second image is the hillshade of the DEM.

```
import geemap.colormaps as cm

Map = geemap.Map()

dem = ee.Image("USGS/SRTMGL1_003")
hillshade = ee.Terrain.hillshade(dem)

vis = {'min': 0, 'max': 6000, 'palette': cm.palettes.terrain}
blend = geemap.blend(top_layer=dem, top_vis=vis)

Map.addLayer(hillshade, {}, 'Hillshade')
Map.addLayer(blend, {}, 'Shaded relief')

Map.add_colorbar(vis, label='Elevation (m)')
Map.setCenter(91.4206, 27.3225, zoom=9)
Map
```

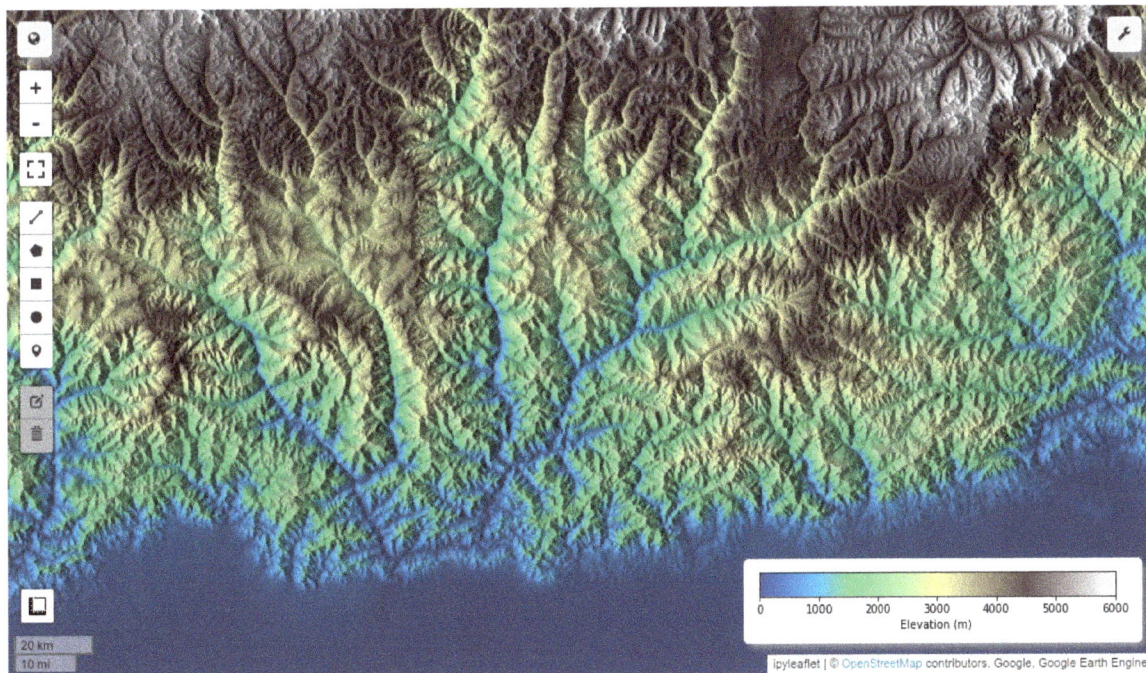

Figure 5.31: Creating a shaded relief map based on a DEM.

To compare the shaded relief map and the hillshade, we can create a split-panel map to inspect the difference between the two (Fig. 5.32). The left_layer is the shaded relief map, and the right_layer is the hillshade.

[121]NASA SRTM Digital Elevation data: tiny.geemap.org/hclo

```
left_layer = geemap.ee_tile_layer(blend, {}, "Shaded relief")
right_layer = geemap.ee_tile_layer(hillshade, {}, "Hillshade")
Map.split_map(left_layer, right_layer)
```

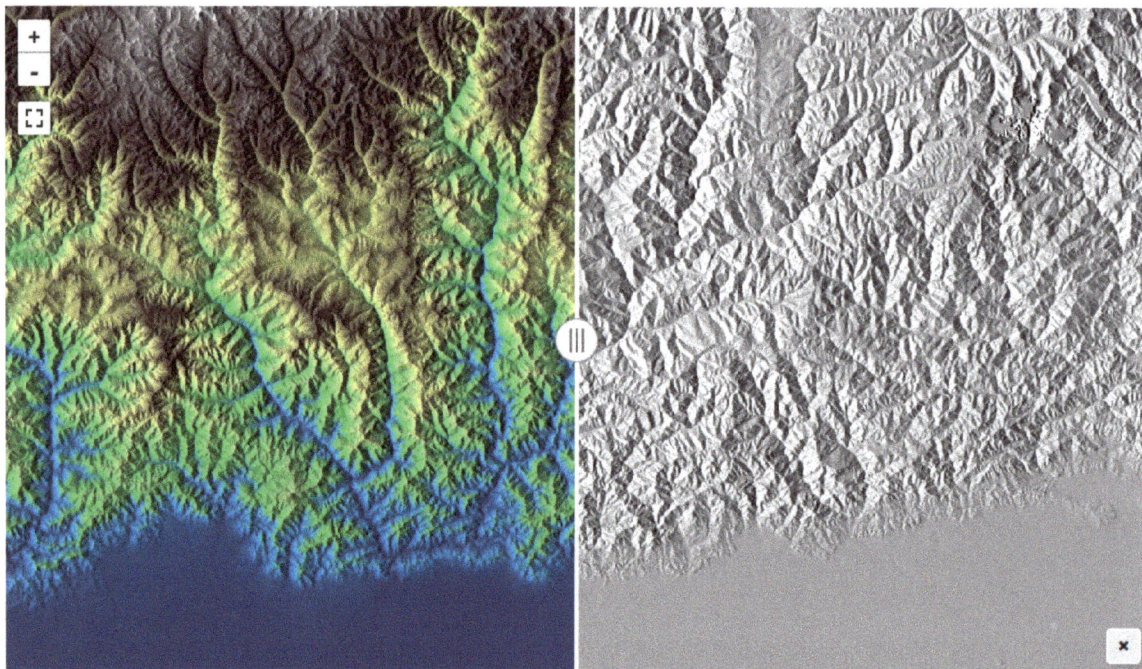

Figure 5.32: Creating a split-panel map for comparing the shaded relief map and hillshade.

Zoom and pan the map to a specific region. Then drag and move the divider to compare the two image layers. Click the close button in the lower-right corner to remove the split-panel map.

Aside from creating a shaded relief map based on a DEM, the `blend()` function can also be used to blend any two images. The following example blends the NLCD land cover with the SRTM, resulting in a nice-looking land cover map (Fig. 5.33).

```
Map = geemap.Map()
nlcd = ee.Image("USGS/NLCD_RELEASES/2019_REL/NLCD/2019").select('landcover')
nlcd_vis = {'bands': ['landcover']}
blend = geemap.blend(nlcd, dem, top_vis=nlcd_vis, expression='a*b')
Map.addLayer(blend, {}, 'Blend NLCD')
Map.add_legend(builtin_legend='NLCD', title='NLCD Land Cover')
Map.setCenter(-118.1310, 35.6816, 10)
Map
```

5.17 Elevation contours

To generate elevation contours, use the `create_contours()` function. The following example creates elevation contours for the 30-m NASA SRTM Digital Elevation data[121] :

```
import geemap.colormaps as cm
```

First, create the terrain hillshade based on the DEM and add it to the map:

```
Map = geemap.Map()
image = ee.Image("USGS/SRTMGL1_003")
hillshade = ee.Terrain.hillshade(image)
```

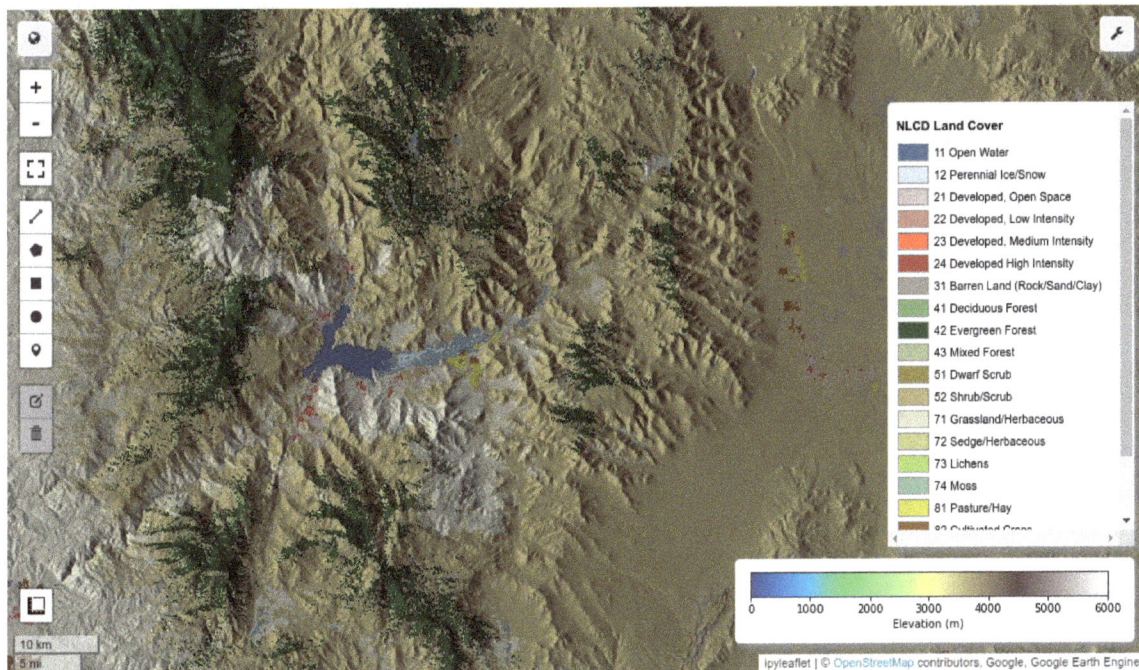

Figure 5.33: Blending NLCD land cover with hillshade.

```
Map.addLayer(hillshade, {}, "Hillshade")
Map
```

Overlay the semi-transparent DEM on the hillshade to create the effect of a shaded relief map with a color bar:

```
vis_params = {'min': 0, "max": 5000, "palette": cm.palettes.dem}
Map.addLayer(image, vis_params, "dem", True, 0.5)
Map.add_colorbar(vis_params, label='Elevation (m)')
```

Lastly, use `create_contours()` function to generate elevation contours with specified minimum elevation, maximum elevation, and contour interval. Specify the `region` parameter to limit the contours to a specific region. Otherwise, the contours will be generated for the entire image. The following code generates elevation contours for the entire image with a contour interval of 100 meters (Fig. 5.34):

```
contours = geemap.create_contours(image, 0, 5000, 100, region=None)
Map.addLayer(contours, {'palette': 'black'}, 'contours')
Map.setCenter(-119.3678, 37.1671, 12)
```

5.18   Visualizing NetCDF data

In this section, we will learn how to visualize NetCDF (network Common Data Form) data. NetCDF (network Common Data Form) is a file format for storing multidimensional scientific data (variables) such as temperature, humidity, pressure, wind speed, and direction. Let's download a sample NetCDF dataset of global wind speed:

```
url = 'https://github.com/gee-community/geemap/raw/master/examples/data/wind_global.nc'
filename = 'wind_global.nc'
geemap.download_file(url, output=filename)
```

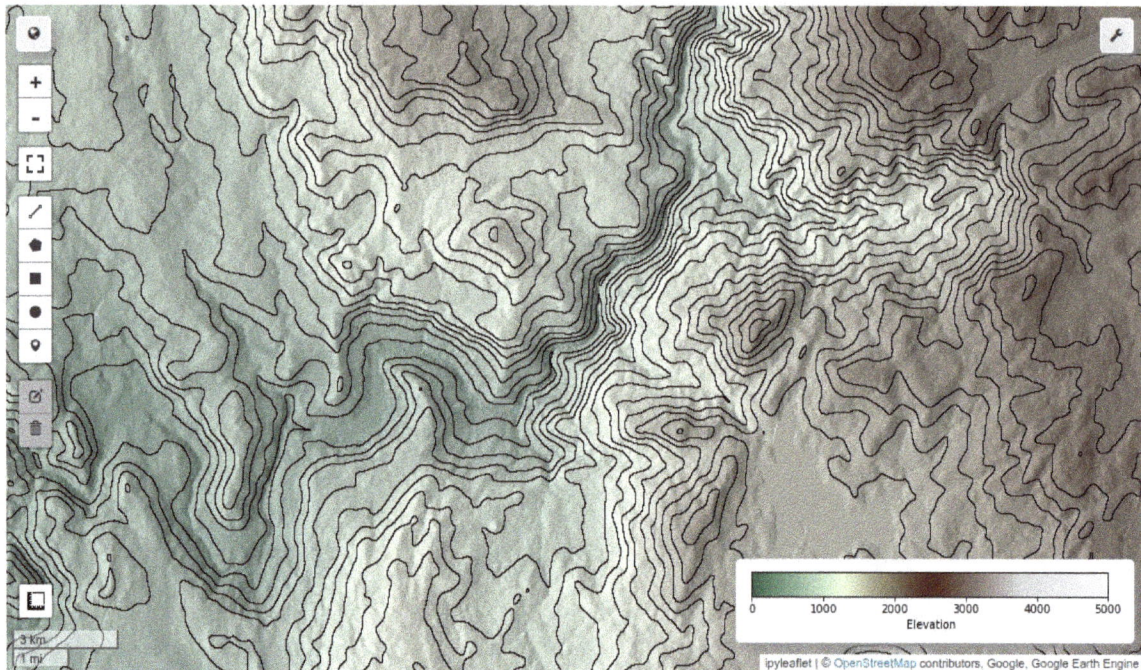

Figure 5.34: Elevation contours of the NASA SRTM Digital Elevation Model.

First, let's read the NetCDF dataset and see what it contains (Fig. 5.35).

```
data = geemap.read_netcdf(filename)
data
```

Figure 5.35: Displaying NetCDF metadata.

There are two data variables contained in this dataset: `u_wind` and `v_wind`. The `u_wind` variable contains the wind speed in the x-direction, and the `v_wind` variable contains the wind speed in the y-direction. The meteorological convention for winds is that U component is positive for a west to east flow (eastward wind) and the V component is positive for south to north flow (northward wind).

Next, let's create an interactive map and use the `Map.add_netcdf()` method to visualize the data. Note that we can also overlay country borders on the map by adding a GeoJSON file to the map (Fig. 5.36). Since geemap's layer control widget does support GeoJSON layers, we can pass the `layers_control=True` parameter to the `Map` constructor to activate ipyleaflet's built-in layer control so that we can toggle the country layer on and off. We also need to set the `shift_lon=True` parameter because the longitude range of the dataset is [0, 360], which needs to be converted to [-180, 180] in order to be displayed on the interactive map. Note that the `Map.add_netcdf()` method does not support Google Colab at the time of writing.

```
Map = geemap.Map(layers_control=True)
Map.add_netcdf(
    filename,
    variables=['v_wind'],
    palette='coolwarm',
    shift_lon=True,
    layer_name='v_wind',
)

geojson = 'https://github.com/gee-community/geemap/raw/master/examples/data/countries.geojson'
Map.add_geojson(geojson, layer_name='Countries')
Map
```

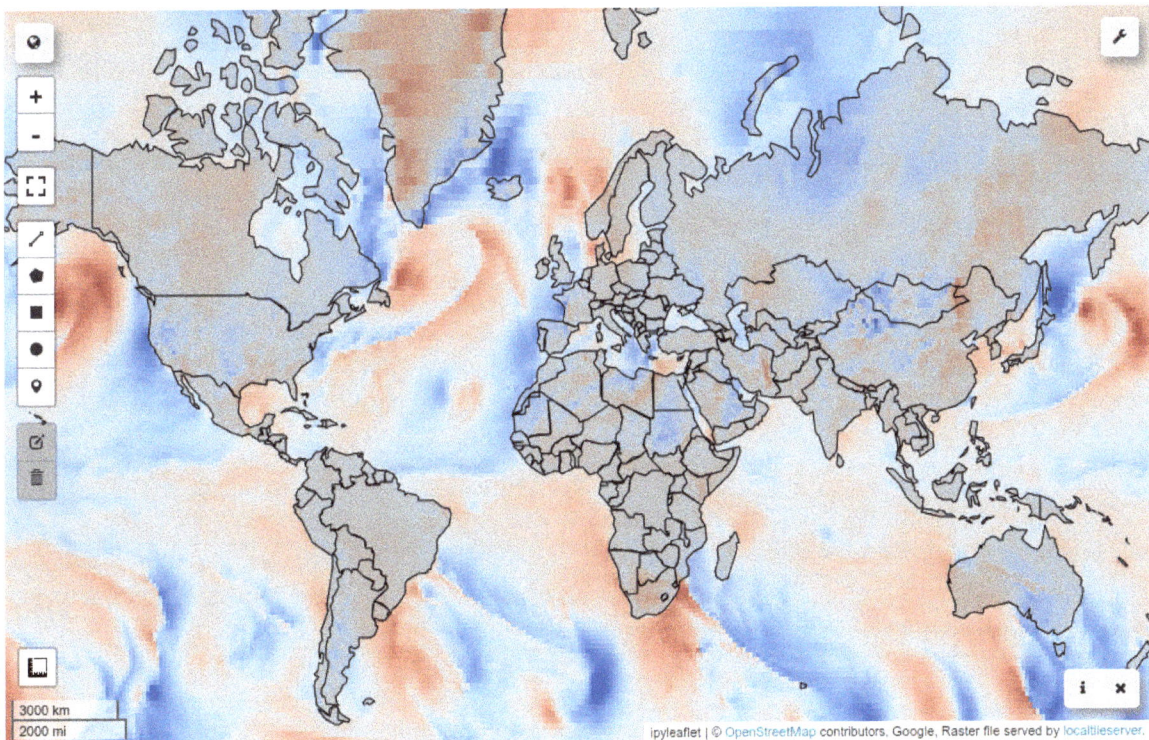

Figure 5.36: Adding NetCDF data of global wind speed to the map.

The `Map.add_netcdf()` method can only visualize one variable at a time. To visualize multiple variables, we can create a velocity map by using the `Map.add_velocity()` method. The zonal speed is the wind speed in the x-direction, and the meridional speed is the wind speed in the y-direction.

```
Map = geemap.Map(layers_control=True)
Map.add_basemap('CartoDB.DarkMatter')
Map.add_velocity(filename, zonal_speed='u_wind', meridional_speed='v_wind')
Map
```

The resulting map should look like (Fig. 5.37).

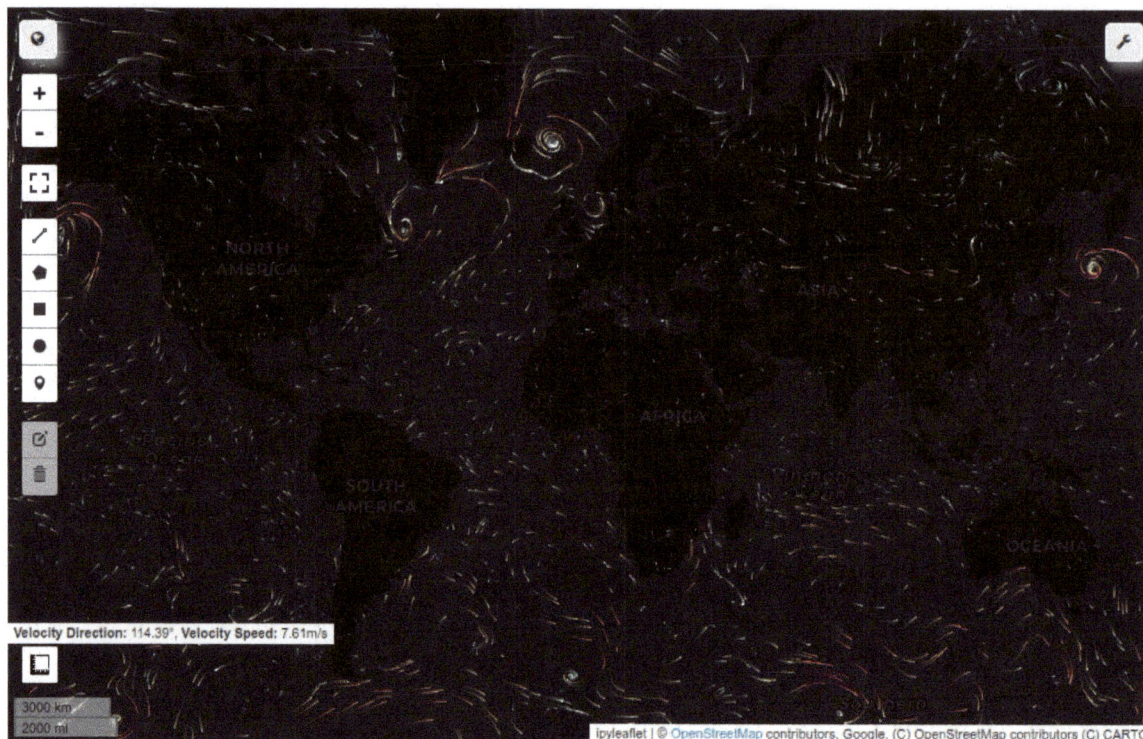

Figure 5.37: Visualizing global wind velocity

5.19   Visualizing LiDAR data

LiDAR stands for Light Detection and Ranging. It is a remote sensing technology that uses a laser to detect and measure the distance between objects and the light that emits them. One of the products of LiDAR data is called a **point cloud**, which is a 3D representation of the points in the LiDAR data. In this section, we will learn how to read and visualize LiDAR data. First, let's download a sample LiDAR dataset[122] (52.1 MB) for Madison, Mississippi.

This section requires additional packages that are not included in the geemap package, including laspy[123] and pyvista-xarray[124]. To install these packages, run the following commands in the terminal and restart the kernel after the installation is completed.

```
%pip install "geemap[lidar]"
```

Once the installation is completed, we can download the sample LiDAR dataset using the download_file() function. The unzip parameter is set to True to unzip the downloaded file.

```
import os

url = (
    'https://drive.google.com/file/d/1H_X1190vL63BoFYa_cVBDxtIa8rG-Usb/view?usp=sharing'
)
```

---

[122]sample LiDAR dataset: tiny.geemap.org/xykc

[123]laspy: laspy.readthedocs.io

[124]pyvista-xarray: tiny.geemap.org/raxz

```
filename = 'madison.las'

if not os.path.exists(filename):
    geemap.download_file(url, 'madison.zip', unzip=True)
```

Once the dataset is downloaded and unzipped, we can use the read_lidar() function to read the Li-DAR data.

```
las = geemap.read_lidar(filename)
```

After the dataset is loaded into memory, we can inspect the metadata, such as the header, point count, and names of the dimensions.

```
las.header
```

```
<LasHeader(1.3, <PointFormat(1, 0 bytes of extra dims)>)>
```

```
las.header.point_count
```

```
4068294
```

```
list(las.point_format.dimension_names)
```

```
['X',
 'Y',
 'Z',
 'intensity',
 'return_number',
 'number_of_returns',
 'scan_direction_flag',
 'edge_of_flight_line',
 'classification',
 'synthetic',
 'key_point',
 'withheld',
 'scan_angle_rank',
 'user_data',
 'point_source_id',
 'gps_time']
```

Based on the list of dimensions names, we can check the values of each dimension, such as the X, Y, Z, and intensity dimensions.

```
las.X
```

```
array([5324343, 5324296, 5323993, ..., 5784049, 5784359, 5784667],
      dtype=int32)
```

```
las.intensity
```

```
array([5324343, 5324296, 5323993, ..., 5784049, 5784359, 5784667],
      dtype=int32)
```

To visualize the LiDAR data, we can use the view_lidar() function. It has four plotting backends, including pyvista[125], ipygany[126], panel[127], and open3d[128]. The default backend is pyvista. Note that the view_lidar() function does not support Google Colab at the time of writing. To visualize the

LiDAR data using the pyvista backend (Fig. 5.38):

```
geemap.view_lidar(filename, cmap='terrain', backend='pyvista', background='gray')
```

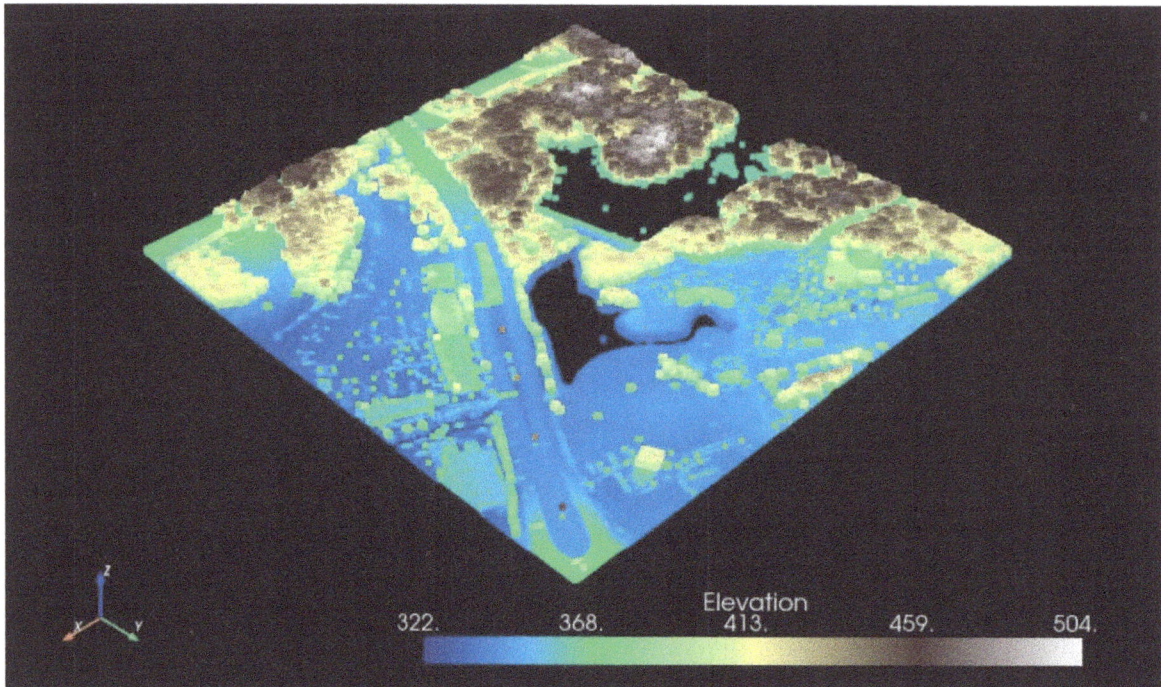

Figure 5.38: Visualizing LiDAR data using the pyvista backend.

To visualize the LiDAR data using the ipygany backend (Fig. 5.39):

```
geemap.view_lidar(filename, backend='ipygany', background='white')
```

5.20   Visualizing raster data in 3D

In this section, we will learn how to visualize raster data in 3D. This is particularly useful for visualizing digital elevation models (DEMs) and other types of raster data. First, let's download a sample DEM from the Shuttle Radar Topography Mission (SRTM):

```
url = 'https://github.com/giswqs/data/raw/main/raster/srtm90.tif'
image = 'srtm90.tif'
if not os.path.exists(image):
    geemap.download_file(url, image)
```

Before we dive into 3D visualization, it's helpful to look at the raster data in two dimensions first. We can plot our raster image using the plot_raster() function from the geemap library. The colormap is set to terrain to give the image a realistic and intuitive color representation.

```
geemap.plot_raster(image, cmap='terrain', figsize=(15, 10))
```

The output is shown in Fig. 5.40.

---

[125]pyvista: github.com/pyvisa/pyvisa

[126]ipygany: github.com/QuantStack/ipygany

[127]panel: panel.holoviz.org

[128]open3d: tiny.geemap.org/fryd

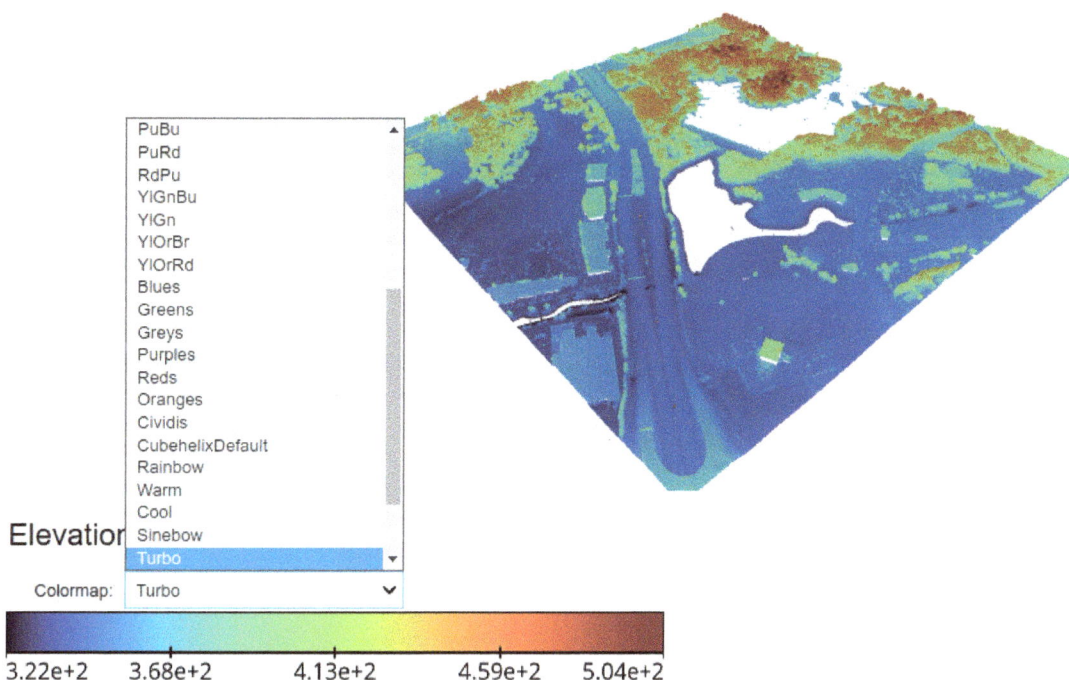

Figure 5.39: Visualizing LiDAR data using the ipygany backend.

This 2D visualization gives us a general view of our geographical area and the elevations across the landscape.

Now that we have an overview of our geographical area in 2D, let's dive into 3D visualization. This can give us a more realistic, visually intuitive representation of the landscape's terrain. We can generate a 3D plot using the `plot_raster_3d()` function, like this:

```
geemap.plot_raster_3d('srtm90.tif', factor=2, cmap='terrain', background='gray')
```

The output should look like (Fig. 5.41). You can click and drag the mouse to rotate the image. You can also use the mouse wheel to zoom in and out.

5.21 Creating choropleth maps

Choropleth maps are thematic maps which use varying colors or patterns within predefined areas to represent the magnitude of a particular quantity, like population density or per-capita income. In this section, we'll walk you through creating various choropleth maps using Python and the geemap library, focusing on various classification schemes.

Firstly, we need a sample dataset for our demonstration. The dataset we'll be using is a GeoJSON file which includes geographical data about different countries. We can directly import it from the geemap library:

```
data = geemap.examples.datasets.countries_geojson
```

Before creating the map, it's crucial to understand what classification schemes are and why they are important. Classification schemes determine how the data range is divided into different classes for representation on the map.

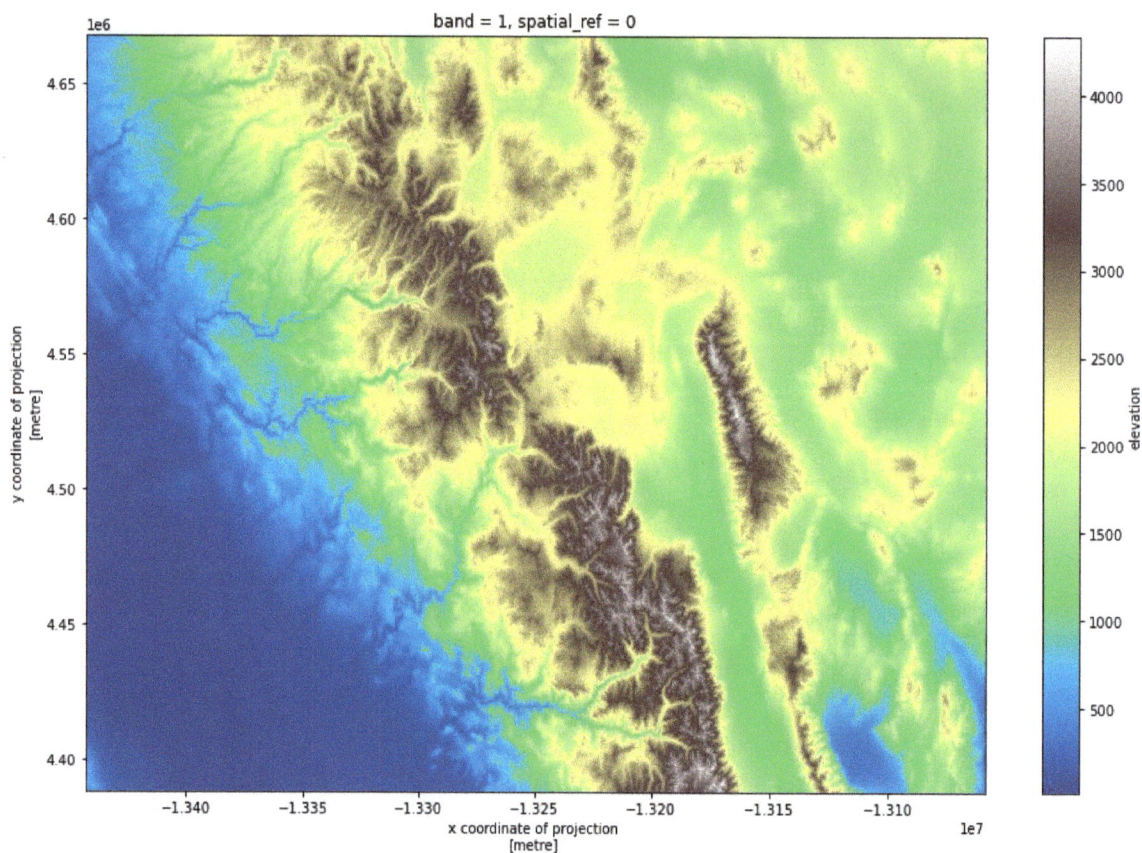

Figure 5.40: Visualizing a raster image in 2D.

The available schemes are as follows:

- BoxPlot
- EqualInterval
- FisherJenks
- FisherJenksSampled
- HeadTailBreaks
- JenksCaspall
- JenksCaspallForced
- JenksCaspallSampled
- MaxP
- MaximumBreaks
- NaturalBreaks
- Quantiles
- Percentiles
- StdMean
- UserDefined

With our data and classification schemes ready, let's create our choropleth map. The geemap library provides an add_data() method that we can use to add our GeoJSON data to the map. This function allows us to specify the column we want to visualize (POP_EST or population estimate in this case), the classification scheme we want to use, and the color map (Blues in this case).

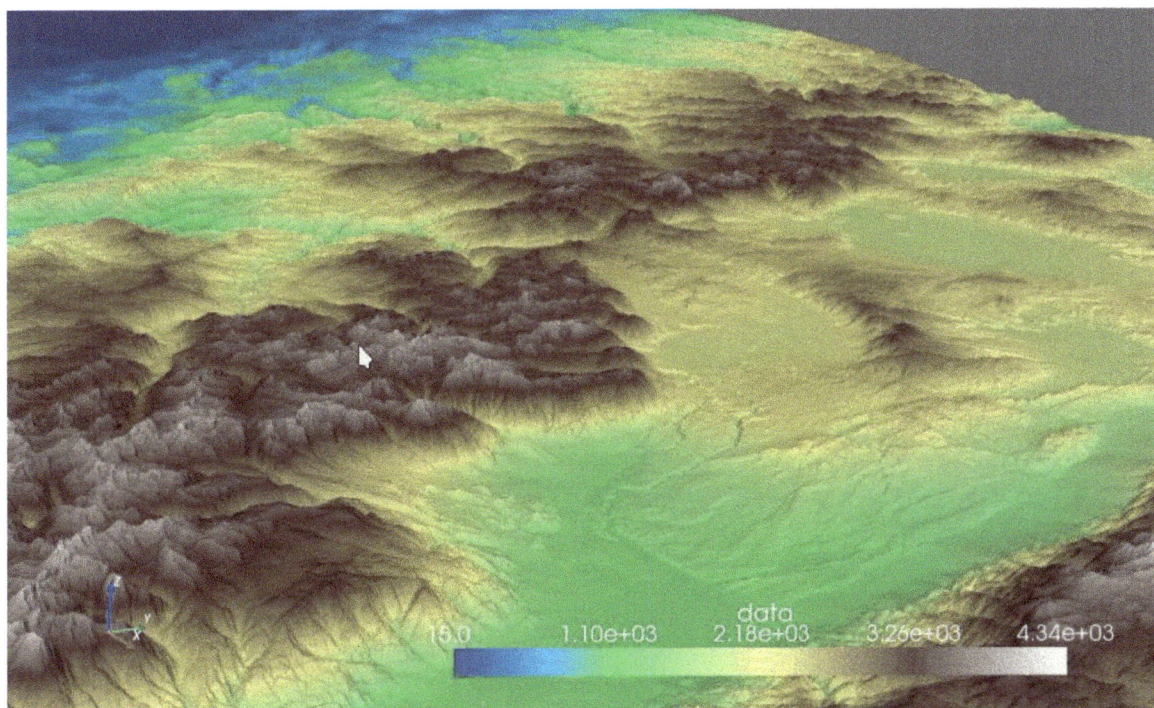

Figure 5.41: Visualizing a raster image in 3D.

Here is how we can create choropleth maps using four different classification schemes, including `Quantiles`, `EqualInterval`, `FisherJenks`, and `JenksCaspall`.

Using the `Quantiles` classification scheme:

```
Map = geemap.Map()
Map.add_data(
    data, column='POP_EST', scheme='Quantiles', cmap='Blues', legend_title='Population'
)
Map
```

The output is shown in Fig. 5.42.

Using the `EqualInterval` classification scheme:

```
Map = geemap.Map()
Map.add_data(
    data,
    column='POP_EST',
    scheme='EqualInterval',
    cmap='Blues',
    legend_title='Population',
)
Map
```

The output is shown in Fig. 5.43.

Using the `FisherJenks` classification scheme:

```
Map = geemap.Map()
Map.add_data(
    data,
```

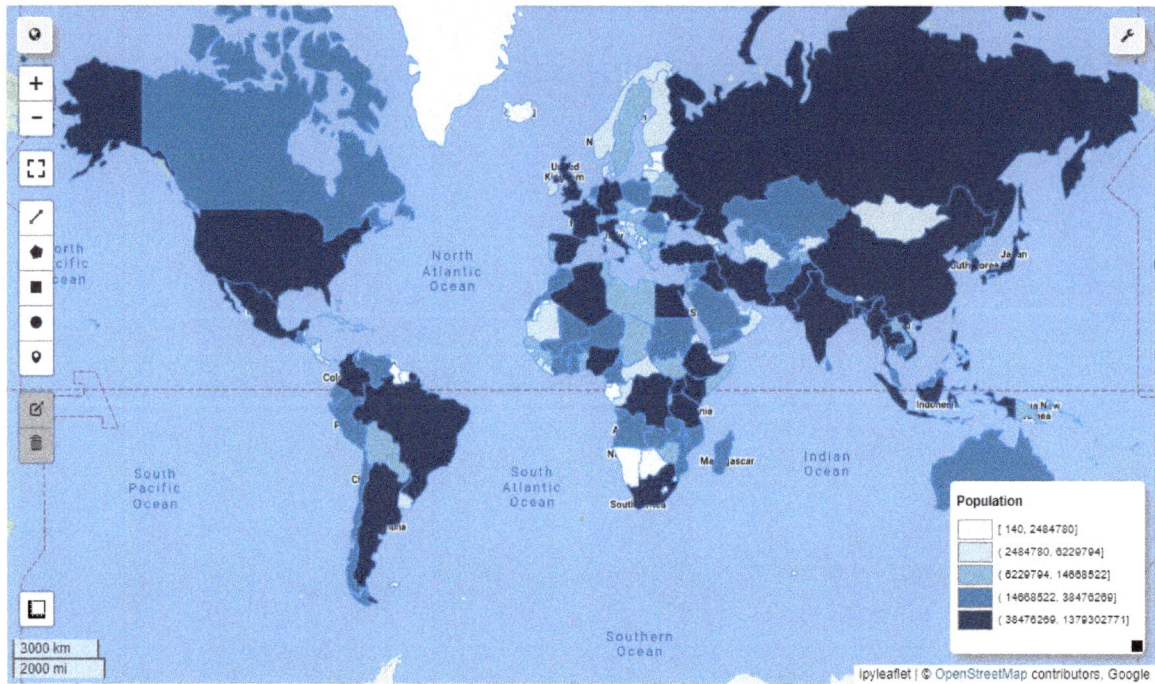

Figure 5.42: A choropleth map with the Quantiles classification scheme.

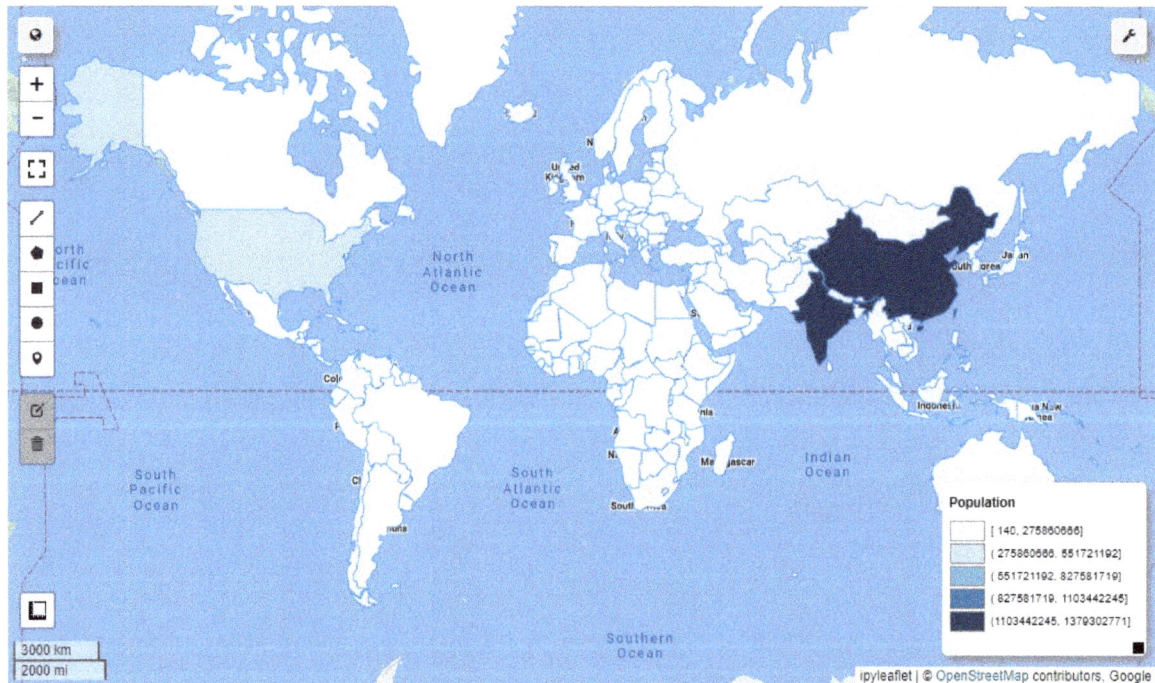

Figure 5.43: A choropleth map with the Equal Interval classification scheme.

```
    column='POP_EST',
    scheme='FisherJenks',
    cmap='Blues',
    legend_title='Population',
)
Map
```

The output is shown in Fig. 5.44.

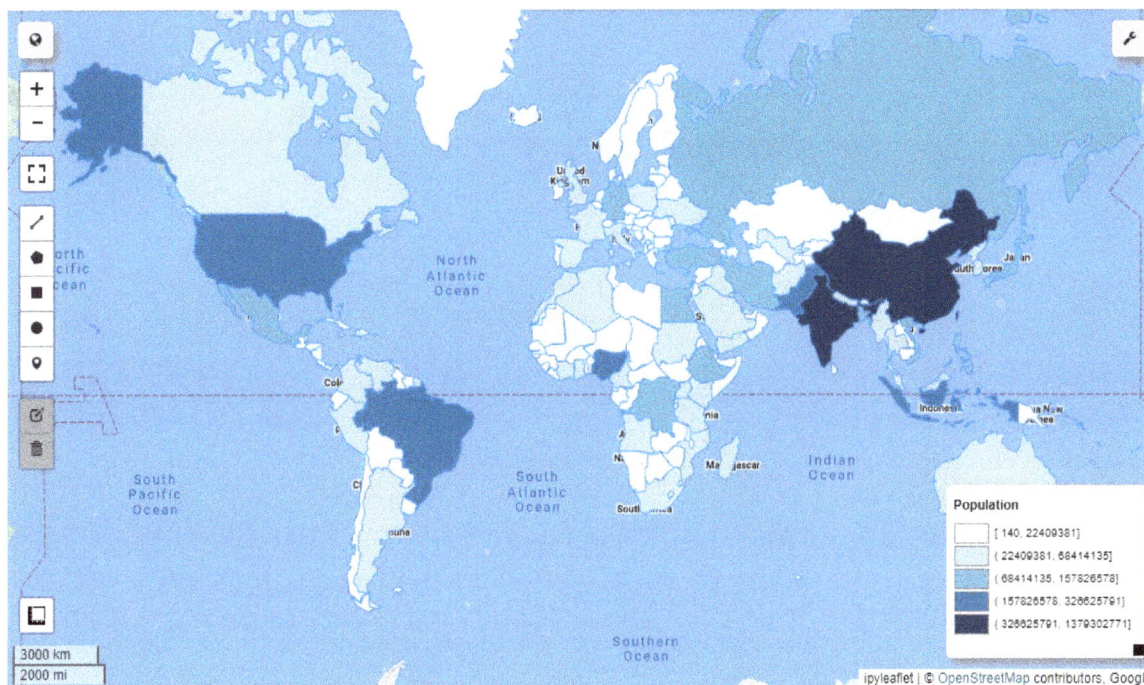

Figure 5.44: A choropleth map with the Fisher Jenks classification scheme.

Using the `JenksCaspall` classification scheme:

```
Map = geemap.Map()
Map.add_data(
    data,
    column='POP_EST',
    scheme='JenksCaspall',
    cmap='Blues',
    legend_title='Population',
)
Map
```

The output is shown in Fig. 5.45.

Creating choropleth maps using Python and the geemap library is a straightforward process. However, the real power of these maps lies in the use of different classification schemes. By trying out various classification schemes, you can find the one that best represents your data and delivers the insights you're looking to glean from your map. Happy mapping!

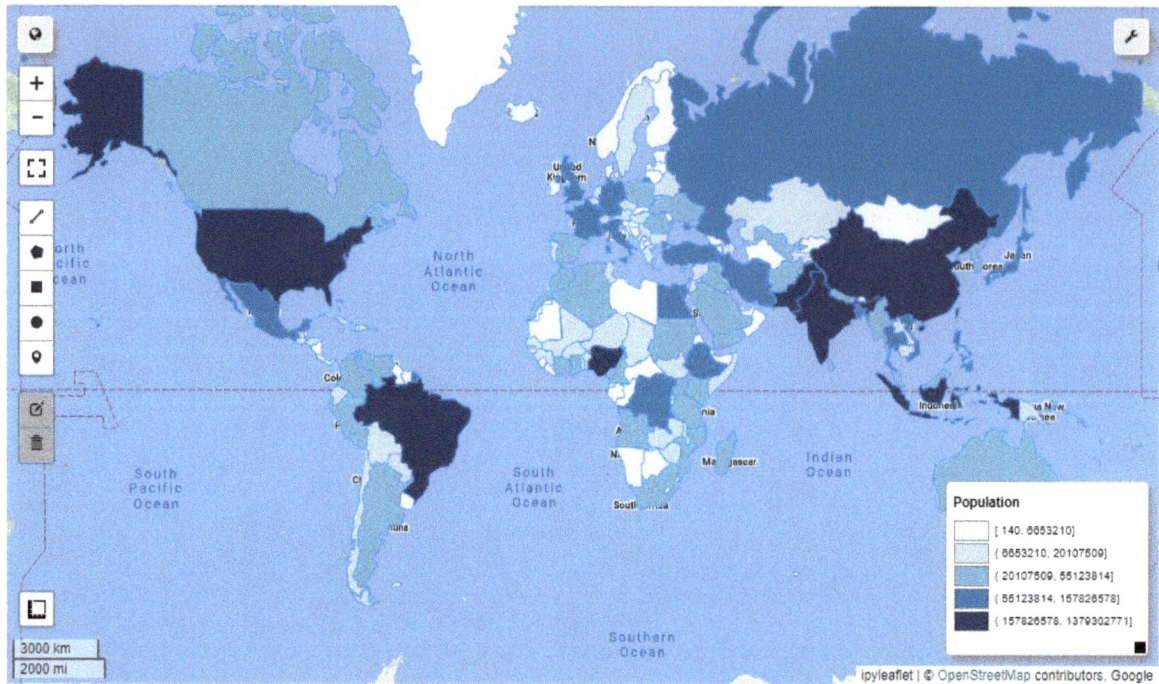

Figure 5.45: A choropleth map with the Jenks Caspall classification scheme.

5.22 Summary

In this chapter, we learned how to visualize different types of Earth Engine data, such as ee.Image and ee.FeatureCollection objects. The visualization techniques introduced in this chapter are essential in exploring and inspecting Earth Engine data. We also learned how to visualize LiDAR data and local raster data in 3D. Lastly, we explored various classification schemes for creating choropleth maps. In the next chapter, we will learn how to perform analysis on Earth Engine data.

6. Analyzing Geospatial Data

6.1 Introduction

In addition to having a multi-petabyte catalog of geospatial datasets, Google Earth Engine can perform planetary-scale geospatial analysis. In this chapter, we will explore how to use Earth Engine to analyze geospatial data at scale, such as descriptive statistics, zonal statistics, and various image processing techniques (e.g., supervised classification, unsupervised classification, and accuracy assessment). We will also learn how to visualize analysis results with interactive charts. By utilizing the continent functions provided by geemap, we can perform planetary-scale geospatial analysis and visualization with a few lines of code.

6.2 Technical requirements

To follow along with this chapter, you will need to have geemap and several optional dependencies installed. If you have already followed Section 1.5 - *Installing geemap*, then you should already have a conda environment with all the necessary packages installed. Otherwise, you can create a new conda environment and install pygis[30] with the following commands, which will automatically install geemap and all the required dependencies:

```
conda create -n gee python
conda activate gee
conda install -c conda-forge mamba
mamba install -c conda-forge pygis
```

Next, launch JupyterLab by typing the following commands in your terminal or Anaconda prompt:

```
jupyter lab
```

Alternatively, you can use geemap with Google Colab without installing anything on your local computer. Click 06_data_analysis.ipynb[129] to launch the notebook in Google Colab.

Once in Colab, you can uncomment the following line and run the cell to install pygis, which includes geemap and all the necessary dependencies:

```
# %pip install pygis
```

The installation process may take 2-3 minutes. Once pygis has been installed successfully, click the **RESTART RUNTIME** button that appears at the end of the installation log or go to the **Runtime** menu and select **Restart runtime**. After that, you can start coding.

To begin, import the necessary libraries that will be used in this chapter:

```
import ee
import geemap
```

Initialize the Earth Engine Python API:

---

[129]06_data_analysis.ipynb: tiny.geemap.org/ch06

```
geemap.ee_initialize()
```

If this is your first time running the code above, you will need to authenticate Earth Engine first. Follow the instructions in Section 1.7 - *Earth Engine authentication* to authenticate Earth Engine.

6.3 Earth Engine data reductions

Reducers can be used to aggregate data over space, time, bands, lists, and other data structures in Earth Engine. For example, given a list of numbers, you might want to calculate some descriptive statistics on the list, such as the minimum, maximum, sum, mean, median, and standard deviation. These operations are what reducers do, taking an input dataset and produce a single output. In this section, you will learn how to use reducers to aggregate various types of Earth Engine data, such as lists, images, and image collections.

List reductions

First, let's create a list of numbers using the ee.List.sequence() function:

```
values = ee.List.sequence(1, 10)
print(values.getInfo())
```

The output should be:

```
[1, 2, 3, 4, 5, 6, 7, 8, 9, 10]
```

Next, use reducers to calculate the descriptive statistics of the list, such as the minimum, maximum, sum, mean, median, and standard deviation. For example, to calculate the number of values in the list, use the ee.Reducer.count() function:

```
count = values.reduce(ee.Reducer.count())
print(count.getInfo())  # 10
```

To find out the minimum value in the list, use the ee.Reducer.min() function:

```
min_value = values.reduce(ee.Reducer.min())
print(min_value.getInfo())  # 1
```

To find out the maximum value in the list, use the ee.Reducer.max() function:

```
max_value = values.reduce(ee.Reducer.max())
print(max_value.getInfo())  # 10
```

To find out the minimum and maximum values in the list in one round, use ee.Reducer.minMax():

```
min_max_value = values.reduce(ee.Reducer.minMax())
print(min_max_value.getInfo())
```

Note that the result of calling the ee.Reducer.minMax() function is a dictionary, with the keys min and max:

```
{'max': 10, 'min': 1}
```

To calculate the mean of the list, use the ee.Reducer.mean() function:

```
mean_value = values.reduce(ee.Reducer.mean())
print(mean_value.getInfo())  # 5.5
```

To calculate the median of the list, use the `ee.Reducer.median()` function:

```
median_value = values.reduce(ee.Reducer.median())
print(median_value.getInfo())  # 5.5
```

To calculate the sum of the list, use the `ee.Reducer.sum()` function:

```
sum_value = values.reduce(ee.Reducer.sum())
print(sum_value.getInfo())  # 55
```

To calculate the standard deviation of the list, use the `ee.Reducer.stdDev()` function:

```
std_value = values.reduce(ee.Reducer.stdDev())
print(std_value.getInfo())  # 2.8723
```

ImageCollection reductions

Reducing an `ImageCollection` is one of the most common operations in Earth Engine. Assume that you have a series of images as an `ImageCollection` covering an area of interest, and you want to produce a good-quality cloud-free image of the area based on the existing images. In this case, you might want to use the `imageCollection.reduce()` or `imageCollection.median()` methods to reduce the `ImageCollection` to a single image. The output is computed pixel-wise, such that each pixel in the output is composed of the median value of all the images in the collection at that location. The following example illustrates how to reduce an `ImageCollection` of Landsat 8 TOA Reflectance[130] images filtered by path and row (see Fig. 6.1).

```
Map = geemap.Map()

# Load an image collection, filtered so it's not too much data.
collection = (
    ee.ImageCollection('LANDSAT/LC08/C01/T1_TOA')
    .filterDate('2021-01-01', '2021-12-31')
    .filter(ee.Filter.eq('WRS_PATH', 44))
    .filter(ee.Filter.eq('WRS_ROW', 34))
)

# Compute the median in each band, each pixel.
# Band names are B1_median, B2_median, etc.
median = collection.reduce(ee.Reducer.median())

# The output is an Image.  Add it to the map.
vis_param = {'bands': ['B5_median', 'B4_median', 'B3_median'], 'gamma': 2}
Map.setCenter(-122.3355, 37.7924, 8)
Map.addLayer(median, vis_param)
Map
```

Note that calling the `reducer()` function will result in the bands names being suffixed with `_median`, such as `B1_median`, `B2_median`, etc. For basic statistics like min, max, mean, median, etc., `ImageCollection` has shortcut methods like `min()`, `max()`, `mean()`, `median()`, etc. Therefore, we can simplify collection.reduce(ee.Reducer.median()) as `collection.median()`:

```
median = collection.median()
print(median.bandNames().getInfo())
```

The output should look like the following:

```
['B1', 'B2', 'B3', 'B4', 'B5', 'B6', 'B7', 'B8', 'B9', 'B10', 'B11', 'BQA']
```

---

[130]Landsat 8 TOA Reflectance: tiny.geemap.org/weci

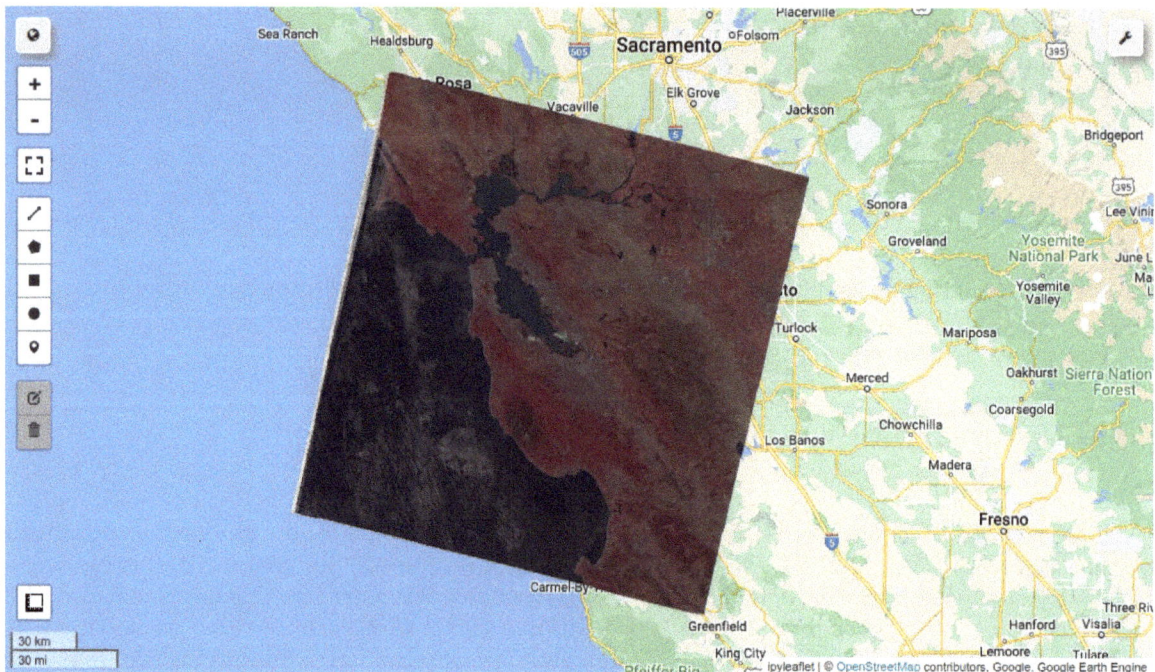

Figure 6.1: A false color composite of the median of Landsat 8 TOA scenes in 2021.

Note that the resultant band names are no longer suffixed with the name of the reducer, i.e., _median.

Image reductions

Reducing an `Image` is similar to reducing an `ImageCollection`, except the bands of the image are input to the reducer rather than the images in the collection. The output is a single image aggregated over the input bands of the image. The following example illustrates how to reduce the RGB bands of a Landsat 8 TOA Reflectance[130] image to a single image with each pixel representing the maximum value of the three bands:

```
Map = geemap.Map()
image = ee.Image('LANDSAT/LC08/C01/T1/LC08_044034_20140318').select(['B4', 'B3', 'B2'])
maxValue = image.reduce(ee.Reducer.max())
Map.centerObject(image, 8)
Map.addLayer(image, {}, 'Original image')
Map.addLayer(maxValue, {'max': 13000}, 'Maximum value image')
Map
```

Activate the Inspector tool and click on the image to see the values of the three RGB bands and the maximum value (see Fig. 6.2).

FeatureCollection reductions

To reduce properties of features in a `FeatureCollection`, use the `collection.reduceColumns()` method. The following code uses the US census blocks data as an example. First, filter the feature collection to only include blocks in Benton County, Oregon (see Fig. 6.3).

```
Map = geemap.Map()
census = ee.FeatureCollection('TIGER/2010/Blocks')
benton = census.filter(
```

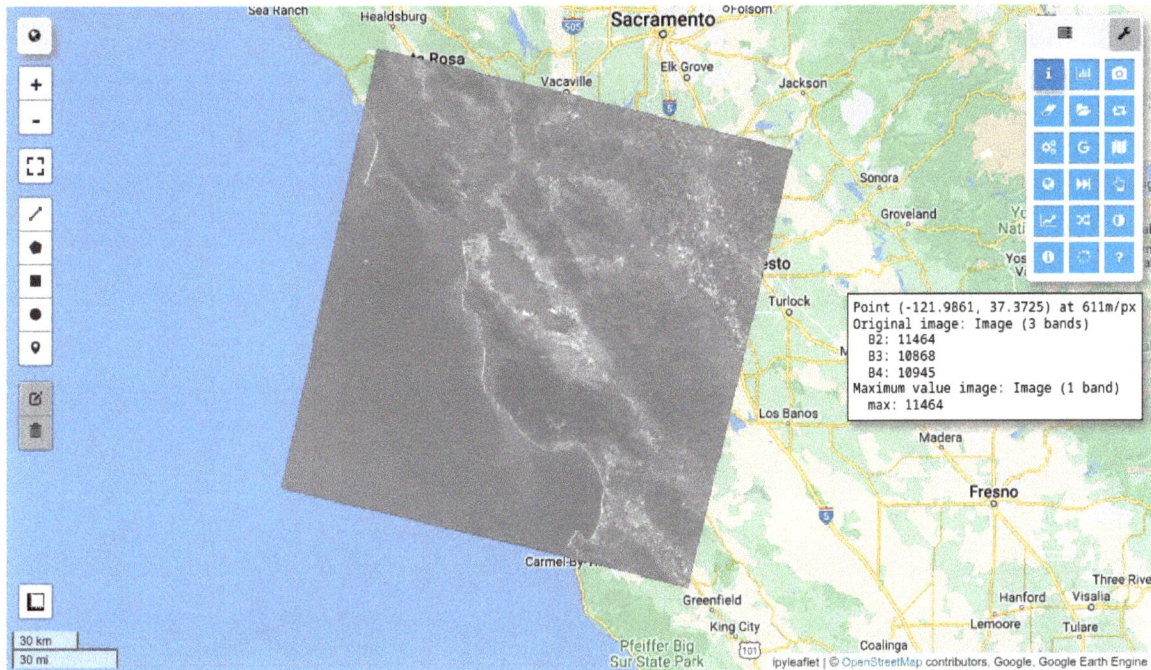

Figure 6.2: A maximum value image of the red, green, and blue bands of a Landsat 8 TOA image in 2014.

```
    ee.Filter.And(ee.Filter.eq('statefp10', '41'), ee.Filter.eq('countyfp10', '003'))
)
Map.setCenter(-123.27, 44.57, 13)
Map.addLayer(benton)
Map
```

The variables of interest are the total population (`pop10`) and total housing units (`housing10`). You can get their sum(s) by supplying a summing `reducer` argument to the `reduceColumns()` method and printing the result:

```
# Compute sums of the specified properties.
properties = ['pop10', 'housing10']
sums = benton.filter(ee.Filter.notNull(properties)).reduceColumns(
    **{'reducer': ee.Reducer.sum().repeat(2), 'selectors': properties}
)
sums
```

The output is a `Dictionary` representing the aggregated property according to the specified reducer:

```
{'sum': [85579, 36245]}
```

Note that the example above uses the `ee.Filter.notNull()` method to only select features with non-null values. This is a convenient way to filter out features with null values that might affect the result. Also note that the use of the `repeat(2)` method call in the reducer argument is necessary to account for the fact that two columns are being reduced. Alternatively, you can use the `collection.aggregate_sum()` method to sum the values of a specific property. For example, to sum the total population and housing units of all city blocks in Benton County, Oregon:

```
print(benton.aggregate_sum('pop10'))   # 85579
print(benton.aggregate_sum('housing10'))  # 36245
```

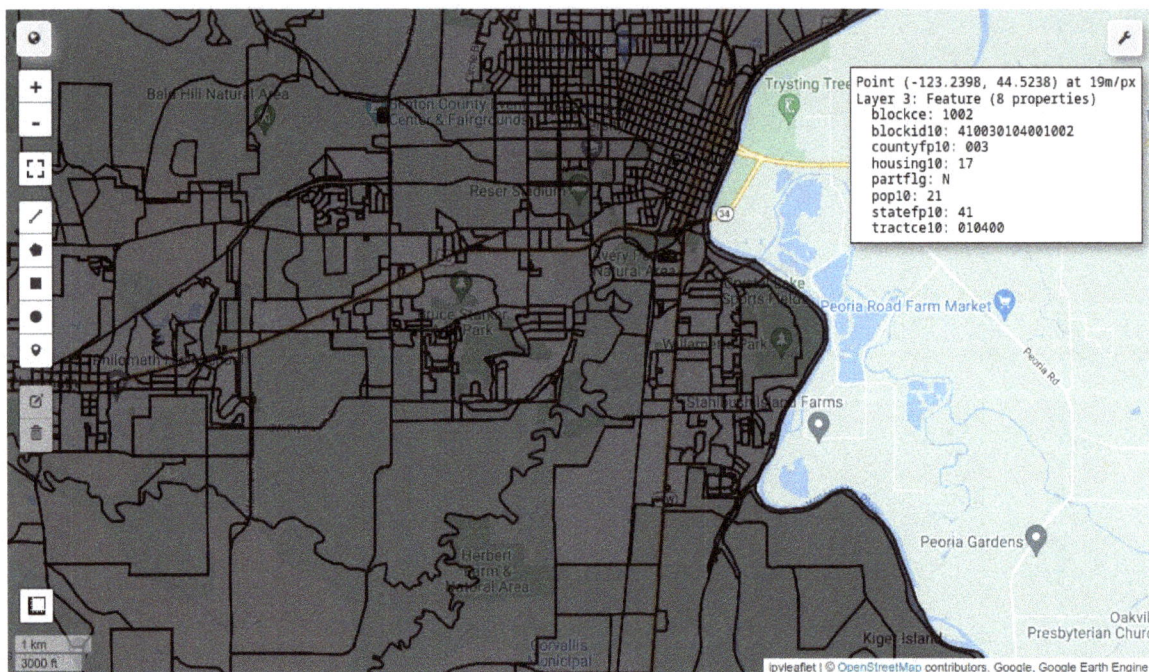

Figure 6.3: US census blocks in Benton County, Oregon.

Other aggregating methods available for ee.FeatureCollection include aggregate_array(), aggregate_min(), aggregate_max(), aggregate_mean(), aggregate_count(), and aggregate_sum(). Besides, the aggregate_stats() method can be handy for computing descriptive statistics of a selected property of a FeatureCollection:

```
benton.aggregate_stats('pop10')
```

```
{'max': 909,
 'mean': 28.72742531050688,
 'min': 0,
 'sample_sd': 65.02307972064482,
 'sample_var': 4228.000896357331,
 'sum': 85579,
 'sum_sq': 15049451,
 'total_count': 2979,
 'total_sd': 65.01216522963706,
 'total_var': 4226.58162784563,
 'valid_count': 2979,
 'weight_sum': 2979,
 'weighted_sum': 85579}
```

6.4 Image descriptive statistics

When visualizing an ee.Image, it is often necessary to set visualization parameters such as the min and max values of the image to visualize. If you are not sure about the value range of the image, you can simply add the image to the map without setting any visualization parameters and then use the Inspector tool to get a good estimate of the range of the image, which can then be used to set the visualization parameters. Alternatively, you can use the image_stats() function to calculate the descriptive statistics of the image, which can give you precise numbers to be used for visualization parameters. First, let's add a Landsat image to the map (see Fig. 6.4).

```
Map = geemap.Map()

centroid = ee.Geometry.Point([-122.4439, 37.7538])
image = ee.ImageCollection('LANDSAT/LC08/C01/T1_SR').filterBounds(centroid).first()
vis = {'min': 0, 'max': 3000, 'bands': ['B5', 'B4', 'B3']}

Map.centerObject(centroid, 8)
Map.addLayer(image, vis, "Landsat-8")
Map
```

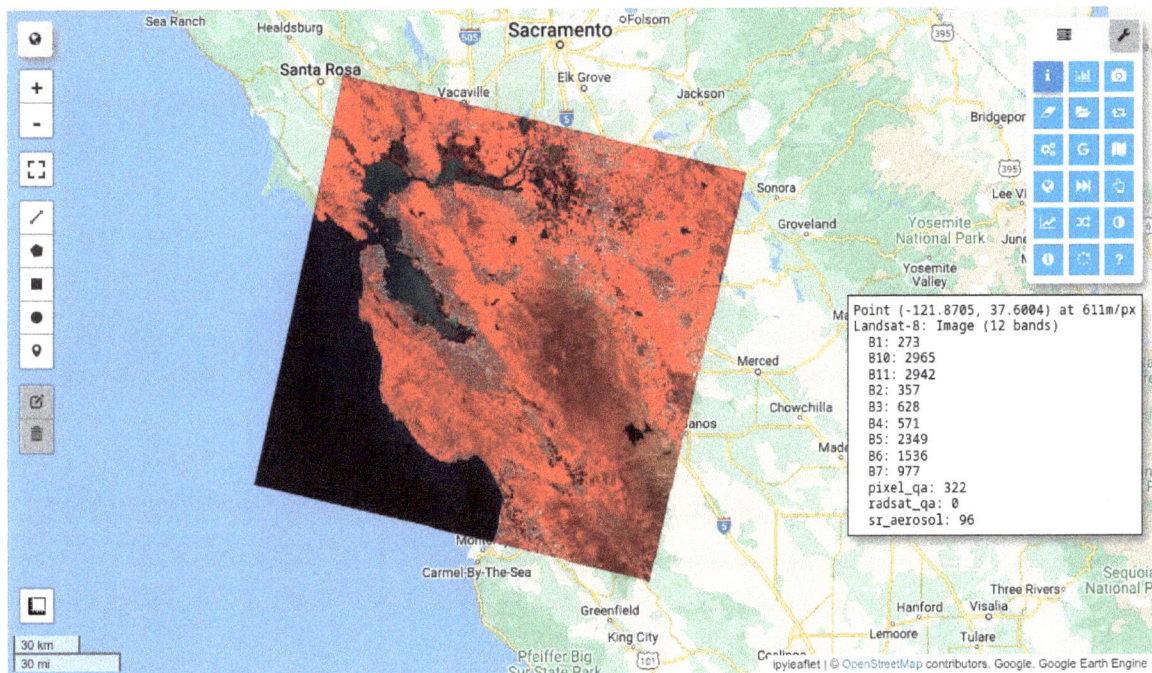

Figure 6.4: A false color composite of a Landsat 8 image covering central California.

To see the image properties, use the `image.propertyNames()` method:

```
image.propertyNames()

['IMAGE_QUALITY_TIRS',
 'CLOUD_COVER',
 'system:id',
 'EARTH_SUN_DISTANCE',
 'LANDSAT_ID',
 'system:footprint',
 'system:version',
 'CLOUD_COVER_LAND',
 ...
]
```

Note that the `image.propertyNames()` method only print out the names of the properties, not the values. To get the value of a specific property, use the `image.get()` method:

```
image.get('CLOUD_COVER')   # 0.05
```

The `image.get()` method can only get the value of a single property. To get image property names and values all at once, use the `image_props()` function:

```
props = geemap.image_props(image)
```

```
props
```

```
{'CLOUD_COVER': 0.05,
 'CLOUD_COVER_LAND': 0.06,
 'EARTH_SUN_DISTANCE': 1.001791,
 'ESPA_VERSION': '2_23_0_1b',
 'GEOMETRIC_RMSE_MODEL': 6.678,
 'GEOMETRIC_RMSE_MODEL_X': 4.663,
 'GEOMETRIC_RMSE_MODEL_Y': 4.78,
 'IMAGE_DATE': '2013-04-09',
 'IMAGE_QUALITY_OLI': 9,
 'IMAGE_QUALITY_TIRS': 9,
 'LANDSAT_ID': 'LC08_L1TP_044034_20130409_20170310_01_T1',
 ...
}
```

The `image_props()` function returns a `Dictionary` of the image properties. The keys of the dictionary are the property names and the values are the property values. Note that the property names are ordered alphabetically, which is a bit different from the order in which they are returned by the `image.propertyNames()` method.

To compute the descriptive statistics of the image, use the `image_stats()` function:

```
stats = geemap.image_stats(image, scale=30)
stats
```

```
{'max': {'B1': 13619,
  'B10': 3157,
  'B11': 3129,
  'B2': 13927,
  'B3': 15027,
  'B4': 15006,
  'B5': 14734,
  'B6': 15394,
  'B7': 15875,
  'pixel_qa': 834,
  'radsat_qa': 254,
  'sr_aerosol': 228},
 'mean': {...},
 'min': {...},
 'std': {...},
 'sum': {...},
}
```

The `image_stats()` function returns a `Dictionary` describing the descriptive statistics of the image. The keys of the dictionary include statistics such as `max`, `min`, `mean`, `std`, and `sum`. The values of the dictionary are the corresponding statistics for each band of the image.

6.5   Zonal statistics with Earth Engine

Zonal statistics

To get the statistics in each zone of an `Image`, also known as zonal statistics, you can use the `zonal_stats()` function. The following adds three sample datasets to the map, including the NASA SRTM, a 5-year Landsat TOA composite, and the US state boundaries (see Fig. 6.5).

```
Map = geemap.Map(center=[40, -100], zoom=4)

# Add NASA SRTM
dem = ee.Image('USGS/SRTMGL1_003')
```

```python
dem_vis = {
    'min': 0,
    'max': 4000,
    'palette': ['006633', 'E5FFCC', '662A00', 'D8D8D8', 'F5F5F5'],
}
Map.addLayer(dem, dem_vis, 'SRTM DEM')

# Add 5-year Landsat TOA composite
landsat = ee.Image('LANDSAT/LE7_TOA_5YEAR/1999_2003')
landsat_vis = {'bands': ['B4', 'B3', 'B2'], 'gamma': 1.4}
Map.addLayer(landsat, landsat_vis, "Landsat", False)

# Add US Census States
states = ee.FeatureCollection("TIGER/2018/States")
style = {'fillColor': '00000000'}
Map.addLayer(states.style(**style), {}, 'US States')
Map
```

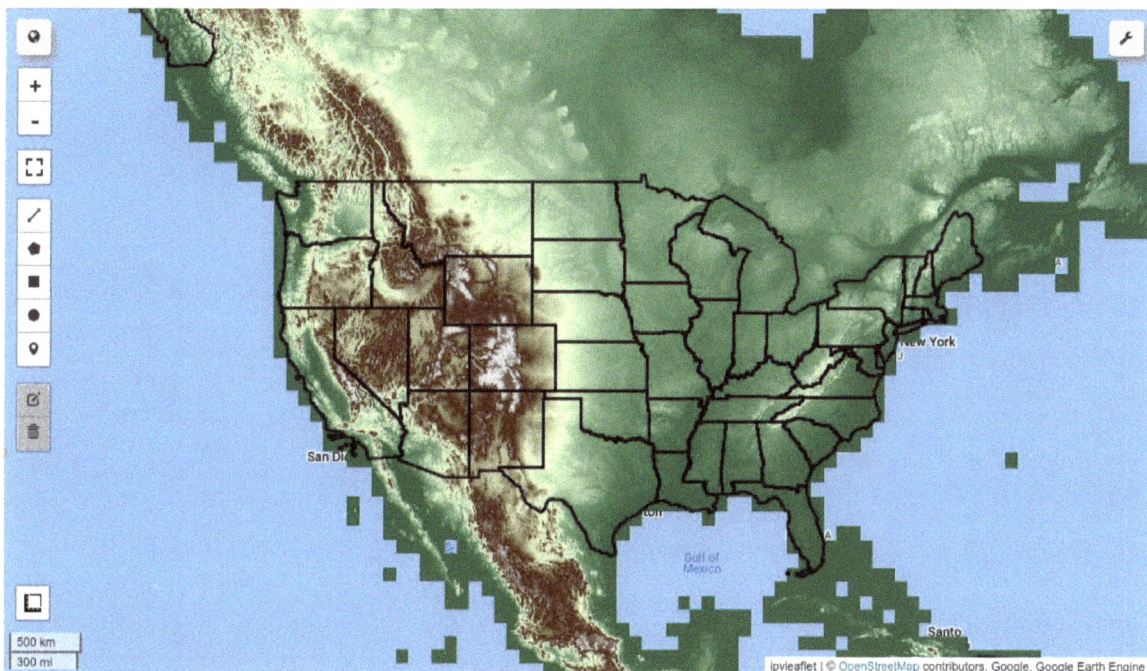

Figure 6.5: Computing the mean elevation of each state in the US.

To calculate the mean elevation of each US state, use the `zonal_stats()` function. The input image in this example is the NASA SRTM, and the input zone layer is a `FeatureCollection` of US states. The output file format can be in the `csv`, `shp`, `geojson`, `kml`, or `kmz` format. To return the result as a `FeatureCollection`, set the `return_fc` parameter to `True`. The supported statistics include `MEAN`, `MAXIMUM`, `MINIMUM`, `MEDIAN`, `STD`, `MIN_MAX`, `VARIANCE`, and `SUM`. Under the hood, the `zonal_stats()` function calls the `image.reduceRegions()` method to compute the statistics. The following code computes the mean elevation of each state using a scale of 1,000 meters and saves the result as a CSV file:

```python
out_dem_stats = 'dem_stats.csv'
geemap.zonal_stats(
    dem, states, out_dem_stats, statistics_type='MEAN', scale=1000, return_fc=False
)
```

The content of the CSV file should look like Fig. 6.6.

mean	STATENS	GEOID	AWATER	LSAD	STUSPS	STATEFP	FUNCSTAT	INTPTLAT	DIVISION	REGION	NAME
133.11131	1779780	9	1815617571	0	CT	9	A	41.57986	1	1	Connecticut
217.09862	1779787	23	11746549764	0	ME	23	A	45.40928	1	1	Maine
131.0798	606926	25	7129925486	0	MA	25	A	42.15652	1	1	Massachusetts
65.26633	1779795	34	3544860246	0	NJ	34	A	40.10727	2	1	New Jersey
383.06976	1779798	42	3394589990	0	PA	42	A	40.9025	2	1	Pennsylvania
313.5933	1779796	36	19246994695	0	NY	36	A	42.9134	2	1	New York
191.29839	1779784	17	6214824948	0	IL	17	A	40.10288	3	2	Illinois
310.04827	1779806	55	29344951758	0	WI	55	A	44.63091	3	2	Wisconsin

Figure 6.6: Zonal statistics of the mean elevation of each state in the US.

Note the mean column in the input CSV file which contains the mean elevation of each state. Similarly, you can use the `zonal_stats()` function to compute the zonal statistics of a multi-band image:

```
out_landsat_stats = 'landsat_stats.csv'
geemap.zonal_stats(
    landsat,
    states,
    out_landsat_stats,
    statistics_type='MEAN',
    scale=1000,
    return_fc=False,
)
```

The content of the CSV file should look like Fig. 6.7.

B1	B2	B3	B4	B5	B6_VCID_2	B7	STATENS	GEOID	NAME
26.4	21.1	16.4	56.4	32.9	189.8	14.4	1779780	9	Connecticut
24.4	19.5	13.3	61.6	28.6	188.7	11.8	1779787	23	Maine
26.4	20.5	14.9	50.1	27.0	188.9	12.1	606926	25	Massachusetts
28.3	23.2	18.6	52.2	33.7	192.7	16.5	1779795	34	New Jersey
26.6	22.1	17.3	68.5	40.7	192.1	18.0	1779798	42	Pennsylvania
26.0	21.0	15.4	63.6	34.4	191.0	14.8	1779796	36	New York
27.8	23.9	21.3	59.7	47.7	194.8	25.2	1779784	17	Illinois
26.6	21.5	17.1	55.5	35.7	189.5	16.6	1779806	55	Wisconsin

Figure 6.7: Zonal statistics of the mean band values of each state in the US.

Note that the first seven columns in the CSV file contain the mean band value of each state.

Zonal statistics by group

To compute zonal statistics by group (e.g., the land cover composition of each zone), use the `zonal_stats_by_group()` function. The following example shows how to calculate the area of each land cover type in each US state. First, let's add relevant datasets to the map (see Fig. 6.8).

```
Map = geemap.Map(center=[40, -100], zoom=4)

# Add NLCD data
dataset = ee.Image('USGS/NLCD_RELEASES/2019_REL/NLCD/2019')
landcover = dataset.select('landcover')
Map.addLayer(landcover, {}, 'NLCD 2019')

# Add US census states
states = ee.FeatureCollection("TIGER/2018/States")
style = {'fillColor': '00000000'}
Map.addLayer(states.style(**style), {}, 'US States')

# Add NLCD legend
Map.add_legend(title='NLCD Land Cover', builtin_legend='NLCD')
Map
```

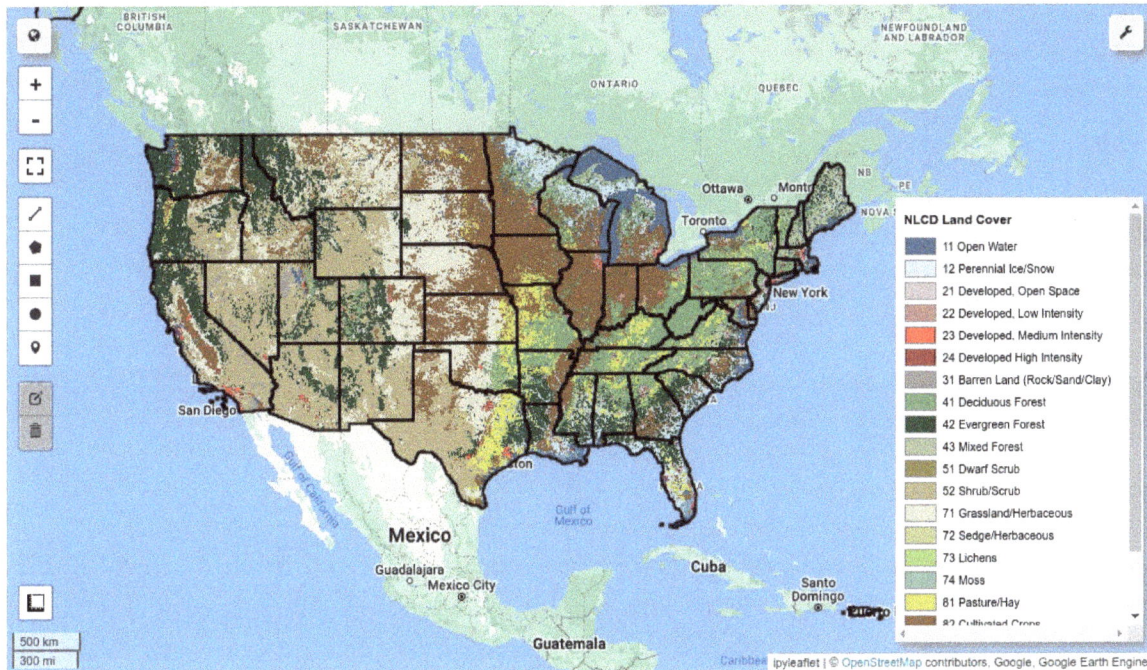

Figure 6.8: Calculating the land cover composition of each state in the US.

Next, use the `zonal_stats_by_group()` function to compute the land cover composition of US states. The input image is the NLCD 2019 land cover image, and the input zone layer is the US states. The output file format can be in the `csv`, `shp`, `geojson`, `kml`, or `kmz` formats. To return the result as a `FeatureCollection`, set the `return_fc` parameter to `True`. The `statistics_type` parameter specifies the statistic to compute. It can be either `SUM` (the total area of each land cover in each zone) or `PERCENTAGE` (the percentage of each land cover type in each zone). The default areal unit is square meters. Specify the `denominator` parameter to convert the area to other areal units. For example, setting `denominator=1e6` to convert the area from square meters to square kilometers. The following code computes the total area of each land cover type in each US state and saves the result as a CSV file:

```
nlcd_stats = 'nlcd_stats.csv'

geemap.zonal_stats_by_group(
    landcover,
    states,
    nlcd_stats,
    statistics_type='SUM',
    denominator=1e6,
    decimal_places=2,
)
```

The content of the CSV file should look like Fig. 6.9.

Class_81	Class_71	Class_82	Class_95	Class_41	Class_52	Class_31	Class_42	NAME
464.02	54.8	219.86	84.62	6390.41	27.27	25.89	139.98	Connecticut
2243.57	1230.16	1065.44	770.27	12842.87	1610.55	108.94	18030.04	Maine
741.49	176.77	273.68	340.79	4599.49	123.78	151.36	1756.69	Massachusetts
864.59	102.44	1983.88	929.36	4117.03	60.7	94.38	769.87	New Jersey
16149.9	551.85	11839.33	155.35	59331.78	506.33	218.69	1323.84	Pennsylvania
17556.42	442.47	11240.54	640.18	49387.58	517.93	207.5	9763.94	New York

Figure 6.9: The total area of each land cover type in each US state.

To compute the percentage area of each land cover type in each US state, use the `statistics_type` parameter and set it to PERCENTAGE:

```
nlcd_stats = 'nlcd_stats_pct.csv'

geemap.zonal_stats_by_group(
    landcover,
    states,
    nlcd_stats,
    statistics_type='PERCENTAGE',
    denominator=1e6,
    decimal_places=2,
)
```

The content of the CSV file should look like Fig. 6.10.

Class_81	Class_71	Class_82	Class_95	Class_41	Class_52	Class_31	Class_42	NAME
0.032	0.004	0.015	0.006	0.445	0.002	0.002	0.010	Connecticut
0.024	0.013	0.012	0.008	0.140	0.018	0.001	0.197	Maine
0.027	0.006	0.010	0.012	0.168	0.005	0.006	0.064	Massachusetts
0.038	0.005	0.088	0.041	0.182	0.003	0.004	0.034	New Jersey
0.135	0.005	0.099	0.001	0.497	0.004	0.002	0.011	Pennsylvania
0.124	0.003	0.080	0.005	0.350	0.004	0.001	0.069	New York

Figure 6.10: The percentage area of each land cover type in each US state.

Zonal statistics with two images

The `zonal_stats()` function introduced above can be used to compute zonal statistics between an `ee.Image` and an `ee.FeatureCollection`. If the zone layer is an image, you might want to convert the image to an `ee.FeatureCollection` before using the `zonal_stats()` function. However, if the image covers a large area and contains complicated zonal polygons, the conversion process might be time-consuming or even impossible. In this case, you can use the `image_stats_by_zone()` function to compute the zonal statistics between two images. Under the hood, the function loops through each zone in the zone image and computes the zonal statistics of the input image within the zone. Once all the zones are processed, the function returns the zonal statistics as a Pandas DataFrame. The following example shows how to compute the mean elevation of each land cover type in the US. First, let's add the USGS 3DEP 10-m DEM[131] to the map:

```
Map = geemap.Map(center=[40, -100], zoom=4)
dem = ee.Image('USGS/3DEP/10m')
vis = {'min': 0, 'max': 4000, 'palette': 'terrain'}
Map.addLayer(dem, vis, 'DEM')
Map
```

Then, add NLCD 2019 land cover data and legend to the map:

```
landcover = ee.Image("USGS/NLCD_RELEASES/2019_REL/NLCD/2019").select('landcover')
Map.addLayer(landcover, {}, 'NLCD 2019')
Map.add_legend(title='NLCD Land Cover Classification', builtin_legend='NLCD')
```

Use the `image_stats_by_zone()` function to compute the mean elevation of each land cover type in the US without converting the NLCD image to an `ee.FeatureCollection` first. The result is returned as a Pandas DataFrame if the `out_csv` parameter is not specified:

```
stats = geemap.image_stats_by_zone(dem, landcover, reducer='MEAN')
stats
```

---

[131] USGS 3DEP 10-m DEM: tiny.geemap.org/ydzj

The resulting DataFrame should look like the following:

index	zone	stac
0	11	164.747176
1	12	2057.655823
2	21	200.148143
3	22	234.370185
...	...	...
15	95	315.732807

To save the resulting Pandas DataFrame as a CSV file:

```
stats.to_csv('mean.csv', index=False)
```

Besides the MEAN reducer, the image_stats_by_zone() function also supports several other reducers, including MAXIMUM, MINIMUM, MEDIAN, MODE, STD, SUM, and VARIANCE. The following code computes the standard deviation of elevation of each land cover type in the US. Specify the out_csv parameter to save the result as a CSV file:

```
geemap.image_stats_by_zone(dem, landcover, out_csv="std.csv", reducer='STD')
```

6.6   Coordinate grids and fishnets

Creating coordinate grids

When computing zonal statistics or performing other spatial analyses, you might want to create a coordinate grid to represent the zones so that you can analyse the data by latitude and longitude. The following examples show you how to create three types of coordinate grids: a latitude grid, a longitude grid, and a coordinate grid.

To create a latitude grid, use the create_latitude_grid() function:

```
lat_grid = geemap.latitude_grid(step=5.0, west=-180, east=180, south=-85, north=85)
```

Specify the region of interest (ROI) by setting the west, east, south, and north parameters. The step parameter specifies the latitude interval. The default value is 1.0. The following code adds the latitude grid to the map (see Fig. 6.11).

```
Map = geemap.Map()
style = {'fillColor': '00000000'}
Map.addLayer(lat_grid.style(**style), {}, 'Latitude Grid')
Map
```

Note that the latitude grid is a ee.FeatureCollection object. To inspect the attributes of the grid, use the ee_to_df() function to convert the ee.FeatureCollection to a Pandas DataFrame:

```
df = geemap.ee_to_df(lat_grid)
df
```

The resulting DataFrame should look like the following:

index	east	north	west	south
0	180	-80	-180	-85

...continued on next page

index	east	north	west	south
1	180	-75	-180	-80
2	180	-70	-180	75
...	...	...	...	...
33	180	85	-180	80

To create a longitude grid, use the `create_longitude_grid()` function:

```
lon_grid = geemap.longitude_grid(step=5.0, west=-180, east=180, south=-85, north=85)
```

To add the longitude grid to the map (see Fig. 6.12):

```
Map = geemap.Map()
style = {'fillColor': '00000000'}
Map.addLayer(lon_grid.style(**style), {}, 'Longitude Grid')
Map
```

To create a coordinate grid, use the `latlon_grid()` function:

```
grid = geemap.latlon_grid(
    lat_step=10, lon_step=10, west=-180, east=180, south=-85, north=85
)
```

Note that you can specify different latitude and longitude intervals.

To add the coordinate grid to the map (see Fig. 6.13):

```
Map = geemap.Map()
style = {'fillColor': '00000000'}
Map.addLayer(grid.style(**style), {}, 'Coordinate Grid')
Map
```

Creating fishnets

The `latlon_grid()` function introduced in the previous section creates a rectangular coordinate grid by specifying the coordinates of the upper left and lower right corners of the region of interest. To create a coordinate grid from a polygon feature without knowing the coordinates of the upper left and lower right corners, use the `fishnet()` function.

First, create an interactive map and draw a rectangle on the map:

```
Map = geemap.Map()
Map
```

The following code adds a rectangle to the map if no rectangle is drawn on the map:

```
roi = Map.user_roi

if roi is None:
    roi = ee.Geometry.BBox(-112.8089, 33.7306, -88.5951, 46.6244)
    Map.addLayer(roi, {}, 'ROI')
    Map.user_roi = None

Map.centerObject(roi)
```

Next, use the `fishnet()` function to create a coordinate grid from the drawn rectangle. Note that the input feature can be a drawn polygon on the map or a file path to a vector dataset. Specify horizontal and vertical intervals in degrees or the number of rows and columns for the fishnet. Specify the `delta`

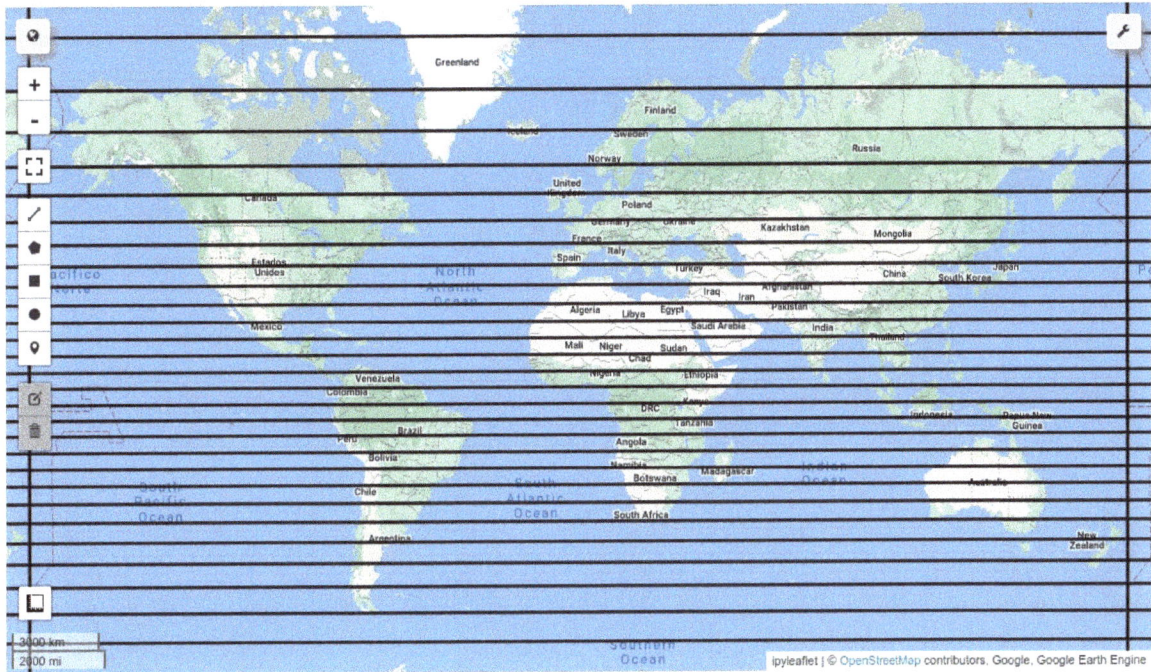

Figure 6.11: An evenly spaced latitude grid.

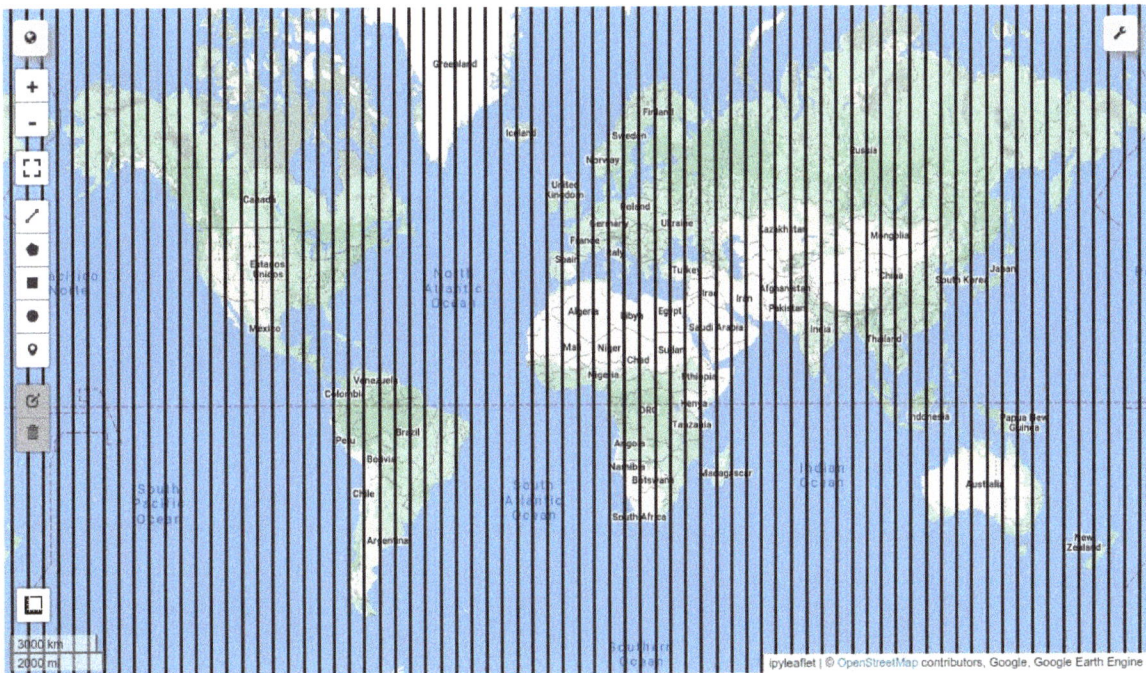

Figure 6.12: An evenly spaced longitude grid.

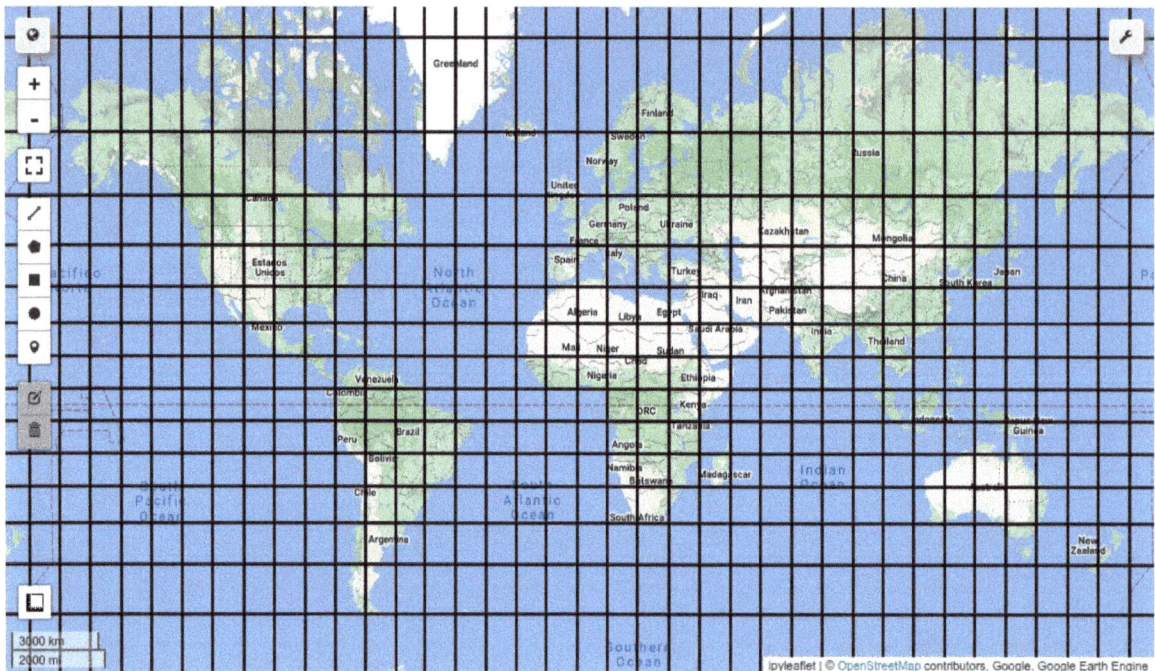

Figure 6.13: A rectangular coordinate grid.

parameter in degrees to make sure the fishnet fully covers the input feature. The following code creates
a fishnet of rectangular cells with a 2-degree horizontal and vertical interval (see Fig. 6.14).

```
fishnet = geemap.fishnet(roi, h_interval=2.0, v_interval=2.0, delta=1)
style = {'color': 'blue', 'fillColor': '00000000'}
Map.addLayer(fishnet.style(**style), {}, 'Fishnet')
```

Besides using a rectangle to create a fishnet, you can also use a polygon feature to create a fishnet. The
following code creates a fishnet based on a polygon feature.

First, create an interactive map and draw a polygon on the map:

```
Map = geemap.Map()
Map
```

The following code adds a polygon to the map if no polygon is drawn on the map:

```
roi = Map.user_roi

if roi is None:
    roi = ee.Geometry.Polygon(
        [
            [
                [-64.602356, -1.127399],
                [-68.821106, -12.625598],
                [-60.647278, -22.498601],
                [-47.815247, -21.111406],
                [-43.860168, -8.913564],
                [-54.582825, -0.775886],
                [-60.823059, 0.454555],
                [-64.602356, -1.127399],
            ]
        ]
    )
```

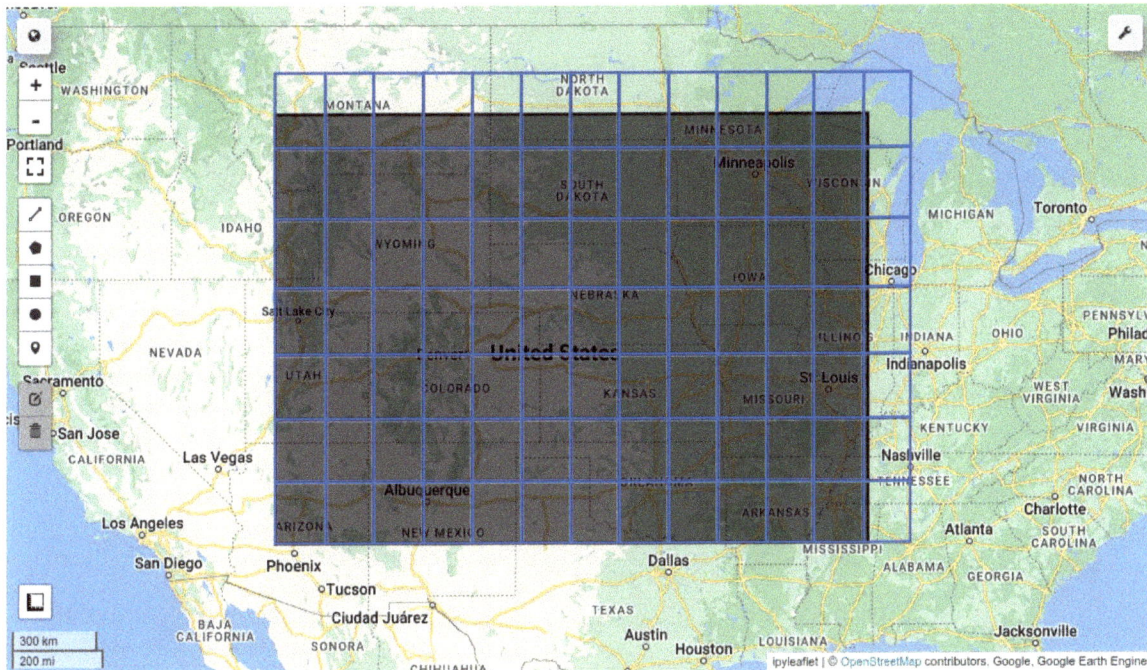

Figure 6.14: Creating a fishnet based on a rectangle.

```
    )
    Map.addLayer(roi, {}, 'ROI')

Map.centerObject(roi)
Map
```

Next, use the `fishnet()` function to create a coordinate grid from the drawn polygon. Instead of specifying the horizontal and vertical intervals, you can specify the number of rows and columns for the fishnet. The following code creates a fishnet of rectangular cells with 6 rows and 8 columns (see Fig. 6.15).

```
fishnet = geemap.fishnet(roi, rows=6, cols=8, delta=1)
style = {'color': 'blue', 'fillColor': '00000000'}
Map.addLayer(fishnet.style(**style), {}, 'Fishnet')
```

6.7 Extracting pixel values

In this section, you will learn how to extract pixel values from an image or image collection. For an `ee.ImageCollection`, you need to convert it to an `ee.Image` first using the `collection.toBands()` method. The input features can be points or lines. For polygon features, use the `zonal_stats()` function introduced in the previous section.

Extracting values to points

In this example, we will extract elevation values and Landsat 7 pixel values for US cities. First, create an interactive map and add the elevation data and Landsat 7 imagery to the map:

```
Map = geemap.Map(center=[40, -100], zoom=4)
```

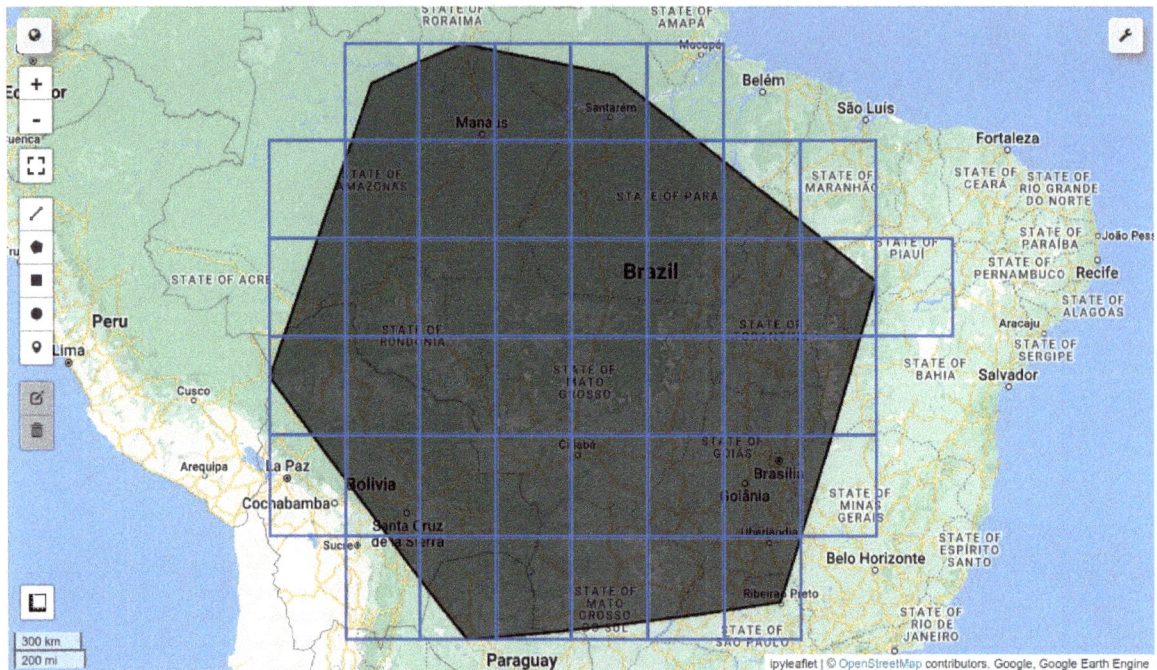

Figure 6.15: Creating a fishnet based on a polygon.

```
dem = ee.Image('USGS/SRTMGL1_003')
landsat7 = ee.Image('LANDSAT/LE7_TOA_5YEAR/1999_2003')

vis_params = {
    'min': 0,
    'max': 4000,
    'palette': ['006633', 'E5FFCC', '662A00', 'D8D8D8', 'F5F5F5'],
}

Map.addLayer(
    landsat7,
    {'bands': ['B4', 'B3', 'B2'], 'min': 20, 'max': 200, 'gamma': 2},
    'Landsat 7',
)
Map.addLayer(dem, vis_params, 'SRTM DEM', True, 1)
Map
```

Download the shapefile of US cities from GitHub:

```
in_shp = 'us_cities.shp'
url = 'https://github.com/giswqs/data/raw/main/us/us_cities.zip'
geemap.download_file(url)
```

Convert the shapefile to an ee.FeatureCollection object and add it to the map:

```
in_fc = geemap.shp_to_ee(in_shp)
Map.addLayer(in_fc, {}, 'Cities')
```

Extract pixel values using the extract_values_to_points() function. The output format can be shapefile, GeoJSON, KML, or CSV. If the out_fc parameter is not specified, the result will be returned as an ee.FeatureCollection object. The following code extracts elevation values of the US cities and saves the result as a shapefile:

```
geemap.extract_values_to_points(in_fc, dem, out_fc="dem.shp")
```

Convert the shapefile to a GeoPandas GeoDataFrame and display the first five rows (see Fig. 6.16):

```
geemap.shp_to_gdf("dem.shp")
```

	STATE	ELEV_IN_FT	id	POP_2010	first	geometry
0	ND	1611	53	40888	491.0	POINT (-101.29627 48.23251)
1	ND	830	101	52838	257.0	POINT (-97.03285 47.92526)
2	ND	1407	153	15427	426.0	POINT (-98.70844 46.91054)
3	ND	902	177	105549	276.0	POINT (-96.78980 46.87719)
4	ND	2411	192	17787	733.0	POINT (-102.78962 46.87918)
...	...	...	...	...	...	...
3661	NY	49	38128	2565635	18.0	POINT (-73.94956 40.65009)

Figure 6.16: Converting a shapefile to a GeoPandas GeoDataFrame.

You can also extract pixel values from a multi-band image. For an ee.ImageCollection, you can convert it to a multi-band image using the collection.toBands() method. The following code extracts Landsat 7 pixel values of the US cities and saves the result as a CSV file:

```
geemap.extract_values_to_points(in_fc, landsat7, 'landsat.csv')
```

Convert the CSV file to a Pandas DataFrame and display the first five rows (see Fig. 6.17):

```
geemap.csv_to_df('landsat.csv')
```

	B1	B2	B3	B4	B5	B6_VCID_2	B7	system:index	STATE	ELEV_IN_FT	POP_2010	id
0	42	42	44	49	52	203	46	0	ND	1611	40888	53
1	38	35	36	53	53	203	40	1	ND	830	52838	101
2	38	38	40	49	54	200	45	2	ND	1407	15427	153
3	40	38	39	49	49	206	38	3	ND	902	105549	177
4	39	41	44	50	57	205	48	4	ND	2411	17787	192
...	...	...	...	...	...	...	...	...	...	...	...	...
3661	36	29	27	28	27	197	22	3661	NY	49	2565635	38128

Figure 6.17: Converting a CSV file to a Pandas DataFrame.

Extracting pixel values along a transect

Besides extracting pixel values based on point features, you can also extract pixel values along a line transect. For example, you might want to extract an elevation profile along a river or a mountain range.

First, let's add the SRTM DEM to the map and overlay it on the Google terrain basemap with a 50% transparency:

```
Map = geemap.Map(center=[40, -100], zoom=4)
Map.add_basemap("TERRAIN")

image = ee.Image('USGS/SRTMGL1_003')
vis_params = {
    'min': 0,
    'max': 4000,
    'palette': ['006633', 'E5FFCC', '662A00', 'D8D8D8', 'F5F5F5'],
}
Map.addLayer(image, vis_params, 'SRTM DEM', True, 0.5)
Map
```

Next, pan and zoom to the area of interest and draw a line transect on the map. If no line is drawn, the default line in the following code will be used (see Fig. 6.18).

```
line = Map.user_roi
if line is None:
    line = ee.Geometry.LineString(
        [[-120.2232, 36.3148], [-118.9269, 36.7121], [-117.2022, 36.7562]]
    )
    Map.addLayer(line, {}, "ROI")
Map.centerObject(line)
```

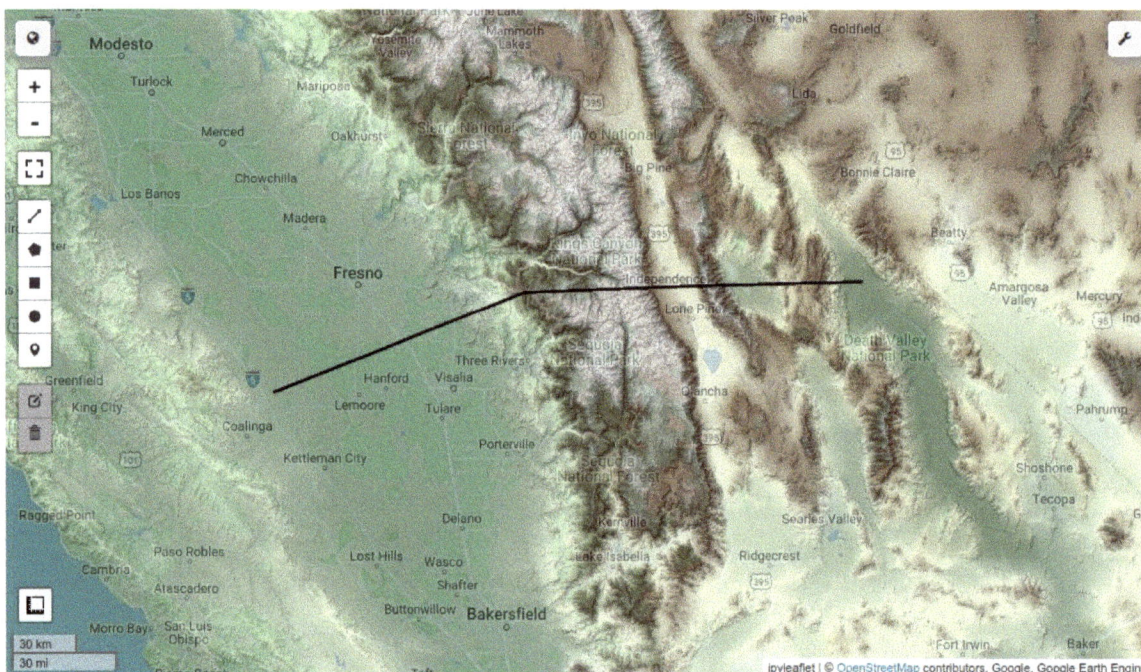

Figure 6.18: Drawing a transect on the map.

With the DEM and the transect, you can extract elevation values along the transect using the `extract_transect()` function. By default, the transect will be split into 100 segments. You can change the number of segments using the `n_segments` parameter. The pixels along each segment will be aggregated to generate the elevation profile with a selected reducer, such as mean, median, min, max, and stdDev. The result will be returned as a Pandas DataFrame with two columns, `distance` and `elevation`. The `distance` column is the distance from the start point of the transect to the end point. The `elevation` column is the elevation value along the transect.

Note that the `elevation` column name is determined by the reducer. For example, if you use the `mean` reducer, the column name will be `mean`. By default, the result is returned as a feature collection. To return the result as a Pandas DataFrame, set the `to_pandas` parameter to `True`.

The following code extracts mean elevation values along 100 segments of the transect:

```
reducer = 'mean'
transect = geemap.extract_transect(
    image, line, n_segments=100, reducer=reducer, to_pandas=True
)
transect
```

The output DataFrame should look like the following:

index	mean	distance
0	99.243094	0.000000
1	91.486990	2783.303405
2	87.383866	5566.606810
...	...	...
99	31.881802	275547.037093

To create a line chart for the elevation profile, use the line_chart() function:

```
geemap.line_chart(
    data=transect,
    x='distance',
    y='mean',
    markers=True,
    x_label='Distance (m)',
    y_label='Elevation (m)',
    height=400,
)
```

The resulting line chart should look like Fig. 6.19.

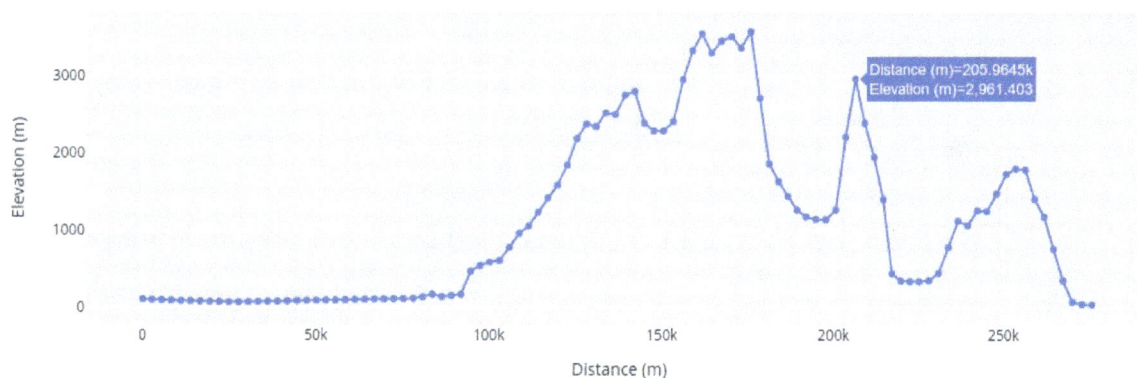

Figure 6.19: A line chart of elevation values along a transect.

Note that the line chart is interactive. You can pan and zoom the chart to explore the elevation profile in more details. Lastly, you can also export the elevation profile as a CSV file:

```
transect.to_csv('transect.csv')
```

Interactive region reduction

Besides extracting pixels values by providing an input feature as introduced above, you can also extract pixel values interactively by drawing a point or a region on the map. The following example shows how to extract NDVI timeseries interactive from the MODIS MOD13A2 16-day vegetation data[132].

First, filter the image collection to select images acquired between 2015 and 2019. Then, select the NDVI band and convert the collection to a multi-band image using the collection.toBands() method. Finally, specify the visualization parameters and add the image to the map:

```
Map = geemap.Map()
```

---

[132]MODIS MOD13A2 16-day vegetation data: tiny.geemap.org/mhyy

```python
collection = (
    ee.ImageCollection('MODIS/061/MOD13A2')
    .filterDate('2015-01-01', '2019-12-31')
    .select('NDVI')
)

image = collection.toBands()

ndvi_vis = {
    'min': 0.0,
    'max': 9000.0,
    'palette': 'ndvi',
}

Map.addLayer(image, {}, 'MODIS NDVI Time-series')
Map.addLayer(image.select(0), ndvi_vis, 'First image')

Map
```

Note that the first image of the collection is added to the map with the specified visualization parameters. This gives you a quick idea of the spatial pattern of NDVI values at the global scale. To check the image acquisition dates of all images, use the `image_dates()` function:

```python
dates = geemap.image_dates(collection).getInfo()
dates
```

To find out how many images are in the collection we can pass the list of dates to the `len` function:

```python
len(dates)
```

In total, there are 115 images in this collection.

Before using the plotting tool to extract pixel values interactively, you can specify the plotting options and set the reducer to use. For example, set the `add_marker_cluster` parameter to `True` for the `Map.set_plot_options()` function if you want to place a marker on the map for the clicked location. Since the interactive plotting tool allows drawing a region of interest, you need to set the reducer for aggregating the pixel values within the region. The default reducer is `ee.Reducer.mean()`.

```python
Map.set_plot_options(add_marker_cluster=True)
Map.roi_reducer = ee.Reducer.mean()
Map
```

Once the plotting options are set, you can activate the plotting tool and start clicking on the map to extract pixel values, which will be plotted in a line chart displayed in the lower right corner of the map (see Fig. 6.20). You can also draw a region on the map to extract pixel values within the selected region. Simply click on the rectangle icon in the drawing toolbar to activate the drawing tool. Then, draw a region on the map and click on the map to extract pixel values within the region. The aggregated pixel values based on the specified reducer will be plotted in the line chart.

To export the plotted data as a CSV file or shapefile, use the `Map.extract_values_to_points()` function:

```python
Map.extract_values_to_points('ndvi.csv')
```

6.8 Mapping available image count

When performing spatial analysis using satellite imagery, it is important to know how many images are available for a given region of interest. The `image_count()` function can be used to map the number of available images at each pixel location for a given region of interest.

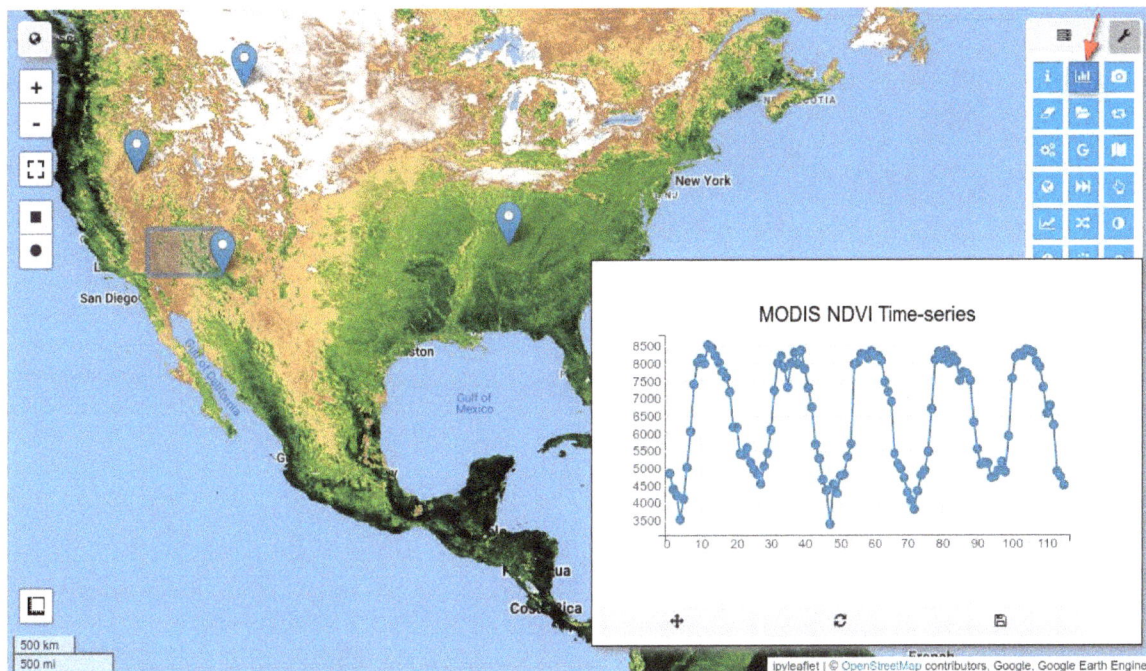

Figure 6.20: Plotting MODIS NDVI timeseries interactively.

The following example maps the number of available Landsat 8 images globally for the year 2021:

```
collection = ee.ImageCollection("LANDSAT/LC08/C02/T1_L2")
image = geemap.image_count(
    collection, region=None, start_date='2021-01-01', end_date='2022-01-01', clip=False
)
```

Specify the `region` parameter to filter the image collection for a given region of interest. Specify the `start_date` and `end_date` parameters to limit the time period of the image collection. Set `clip=True` to clip the output image to the region of interest. The following code adds the resulting image with a corresponding colorbar to the map and overlays a country boundary layer on the image. This process is very computationally intensive. It might take a few minutes to render the layer, so please be patient.

```
Map = geemap.Map()
vis = {'min': 0, 'max': 60, 'palette': 'coolwarm'}
Map.addLayer(image, vis, 'Image Count')
Map.add_colorbar(vis, label='Landsat 8 Image Count')

countries = ee.FeatureCollection(geemap.examples.get_ee_path('countries'))
style = {"color": "00000088", "width": 1, "fillColor": "00000000"}
Map.addLayer(countries.style(**style), {}, "Countries")
Map
```

The resulting map should look like Fig. 6.21.

6.9 Cloud-free composites

When working on a large geographic region, it is often necessary to create a cloud-free composite for a given time period. Earth Engine provides the `ee.Algorithms.Landsat.simpleComposite()` method for creating a simple cloud-free Landsat composite with only one line of code. This method utilizes the available Landsat scenes at each location, converts them to TOA reflectance, applies the simple

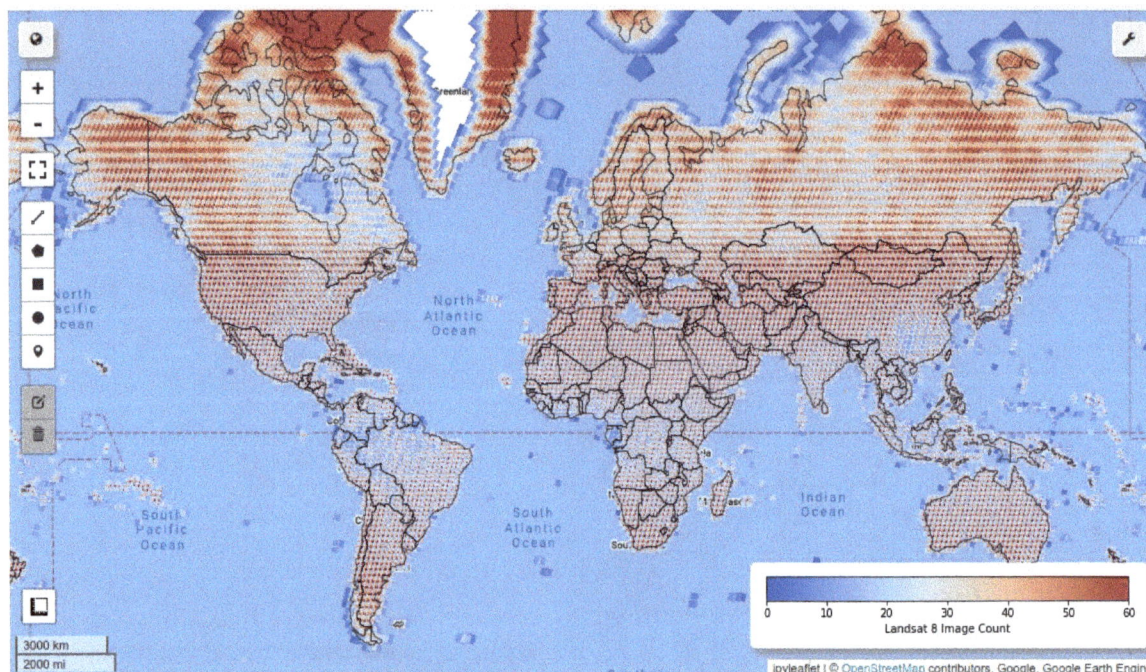

Figure 6.21: The number of available Landsat 8 imagery at each pixel location.

cloud score algorithm, and takes the median of the all available pixels at each location. The following code creates a cloud-free Landsat 8 composite at the global scale for the year 2021 using the default parameters for the `simpleComposite()` method:

```python
Map = geemap.Map()

collection = ee.ImageCollection('LANDSAT/LC08/C02/T1').filterDate(
    '2021-01-01', '2022-01-01'
)

composite = ee.Algorithms.Landsat.simpleComposite(collection)

vis_params = {'bands': ['B5', 'B4', 'B3'], 'max': 128}

Map.setCenter(-122.3578, 37.7726, 10)
Map.addLayer(composite, vis_params, 'TOA composite')
Map
```

The resulting map should look like Fig. 6.22.

If the resulting composite is not satisfactory (e.g., some pixels might still be cloudy), you can adjust the parameters for the `simpleComposite()` method. For example, you can set the `cloudScoreRange` parameter to a smaller range to reduce the number of cloudy pixels in the composite. The cloud score range represents the size of the range of cloud scores to accept per pixel. The default value is 10. You can also adjust the `percentile` parameter rather than using the default percentile of 50, i.e., taking the median value of all available pixels at each location. The following code creates a custom cloud-free composite with a cloud score range of 5 and a percentile of 30, which should result in a less cloudy composite than the previous composite with default parameters:

```python
customComposite = ee.Algorithms.Landsat.simpleComposite(
    **{'collection': collection, 'percentile': 30, 'cloudScoreRange': 5}
)
```

Figure 6.22: The Landsat 8 cloud-free composite for the year 2021.

```
Map.addLayer(customComposite, vis_params, 'Custom TOA composite')
Map.setCenter(-105.4317, 52.5536, 11)
```

Pan and zoom the map and change the layer opacities to see the difference between the two composites. To inspect the cloud-free composite in more detail, you can create a linked map to compare the image with different band combinations. First, customize the visualization parameters and labels for the 2*2 linked map:

```
vis_params = [
    {'bands': ['B4', 'B3', 'B2'], 'min': 0, 'max': 128},
    {'bands': ['B5', 'B4', 'B3'], 'min': 0, 'max': 128},
    {'bands': ['B7', 'B6', 'B4'], 'min': 0, 'max': 128},
    {'bands': ['B6', 'B5', 'B2'], 'min': 0, 'max': 128},
]

labels = [
    'Natural Color (4, 3, 2)',
    'Color Infrared (5, 4, 3)',
    'Short-Wave Infrared (7, 6 4)',
    'Agriculture (6, 5, 2)',
]
```

Then, use the `linked_maps()` function to create the linked map with specified parameters, such as `center`, `zoom`, `vis_params`, and `labels`:

```
geemap.linked_maps(
    rows=2,
    cols=2,
    height="300px",
    center=[37.7726, -122.1578],
    zoom=9,
    ee_objects=[composite],
```

```
    vis_params=vis_params,
    labels=labels,
    label_position="topright",
)
```

Pan and zoom the map to see the difference between the different band combinations (see Fig. 6.23).
Note that the linked map may not work properly in Google Colab or JupyterLab.

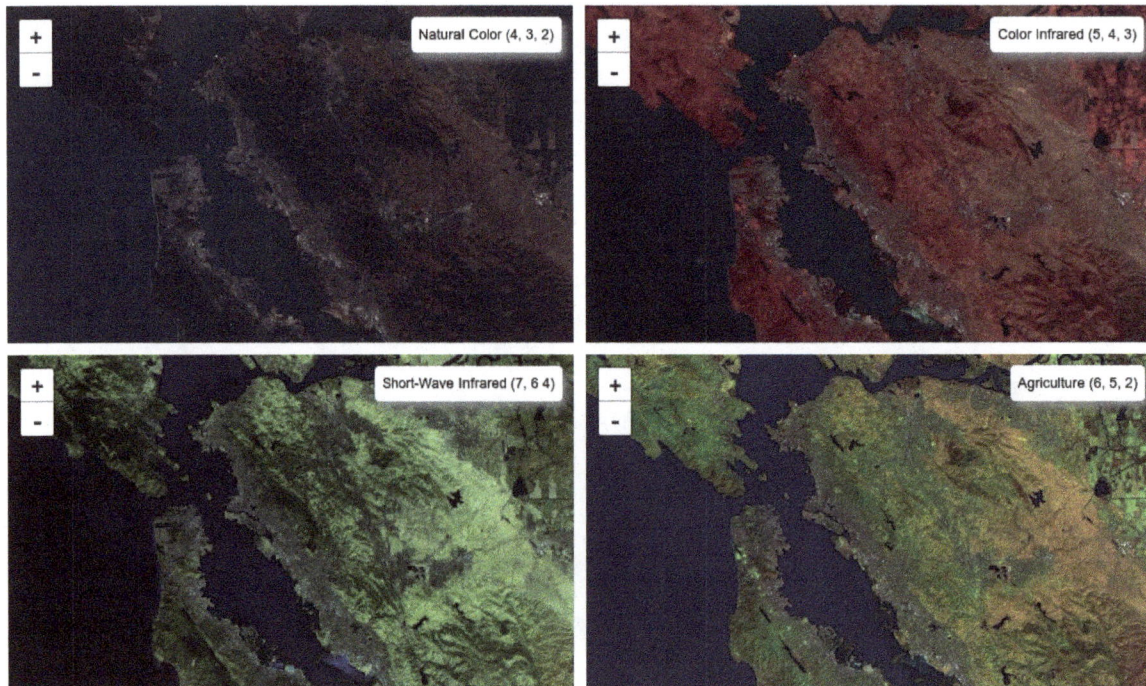

Figure 6.23: A linked map for comparing the Landsat 8 cloud-free composite with different band combinations.

6.10 Quality mosaicking

When performing time series analysis, it is often necessary to create a quality mosaicked image for a
given time period. Earth Engine provides the ee.ImageCollection.qualityMosaic() method for creating a quality mosaicked image with only one line of code. It composites all the images in a collection
using a quality band as a per-pixel ordering function. The resulting image is a single image with the
same number of bands as the input images. Each pixel represents the best quality pixel (i.e., the highest
value based on the quality band) from the input images.

For example, if you have a collection of Landsat images, and you want to identify the greenest pixel
(i.e., the highest NDVI) in each location, you can use the qualityMosaic() method to create a composite
image of the greenest pixel. The following example identifies the greenest pixel in each location for the
year 2020 using the qualityMosaic() method with Landsat 8 imagery.

First, create an interactive map and add the US FeatureCollection to the map:

```
Map = geemap.Map(center=[40, -100], zoom=4)
countries = ee.FeatureCollection(geemap.examples.get_ee_path('countries'))
roi = countries.filter(ee.Filter.eq('ISO_A3', 'USA'))
Map.addLayer(roi, {}, 'roi')
Map
```

Next, filter the Landsat 8 TOA Reflectance collection by the date range and the ROI:

```
start_date = '2020-01-01'
end_date = '2021-01-01'
collection = (
    ee.ImageCollection('LANDSAT/LC08/C01/T1_TOA')
    .filterBounds(roi)
    .filterDate(start_date, end_date)
)
```

Create a simple median composite for comparing with the quality mosaicked image later:

```
median = collection.median()
vis_rgb = {
    'bands': ['B4', 'B3', 'B2'],
    'min': 0,
    'max': 0.4,
}
Map.addLayer(median, vis_rgb, 'Median')
Map
```

The resulting map should look like Fig. 6.24.

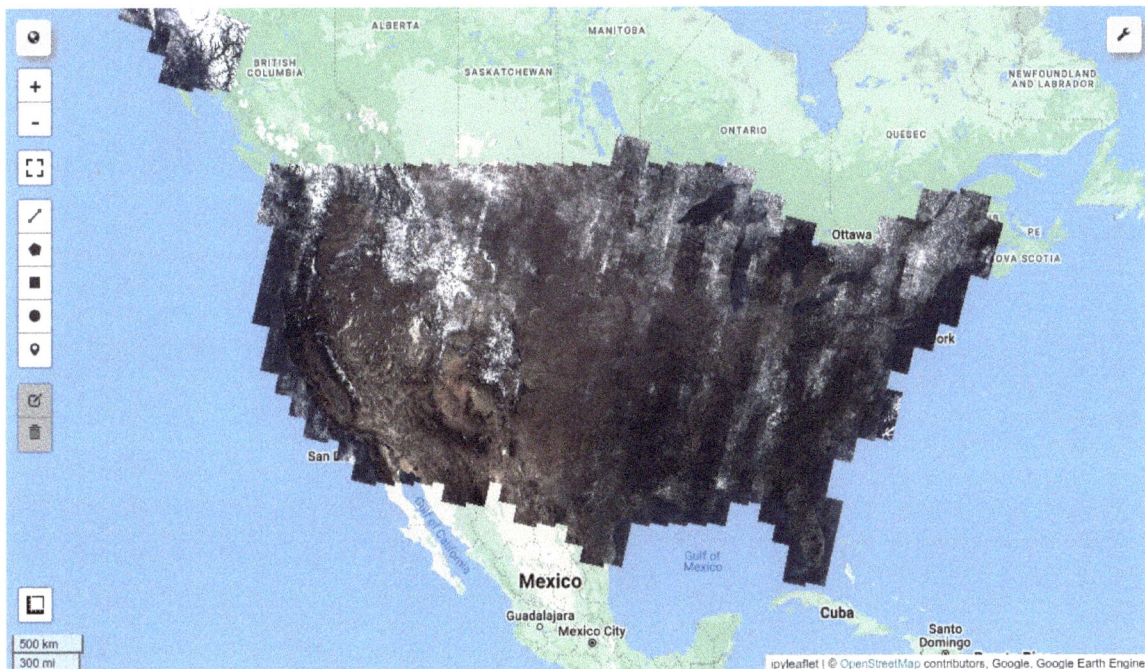

Figure 6.24: A median composite of Landsat 8 TOA reflectance images for the US for the year 2020.

Note that the median composite still contains some cloudy pixels, indicating that more than half of the available pixels at certain locations are cloudy. Therefore, using the median composite for time series analysis might not necessarily be the best choice. In this case, you can use the `qualityMosaic()` method to create a quality mosaicked image. First, create a quality band by calculating the NDVI for each image in the collection:

```
def add_ndvi(image):
    ndvi = image.normalizedDifference(['B5', 'B4']).rename('NDVI')
    return image.addBands(ndvi)
```

Next, define an `add_time()` function to add one or more time bands to each image in the collection. For example, you can add the month and day of the year bands to each image in the collection:

```python
def add_time(image):
    date = ee.Date(image.date())

    img_date = ee.Number.parse(date.format('YYYYMMdd'))
    image = image.addBands(ee.Image(img_date).rename('date').toInt())

    img_month = ee.Number.parse(date.format('M'))
    image = image.addBands(ee.Image(img_month).rename('month').toInt())

    img_doy = ee.Number.parse(date.format('D'))
    image = image.addBands(ee.Image(img_doy).rename('doy').toInt())

    return image
```

Apply the `add_ndvi()` and `add_time()` functions to each image in the collection by using the `map()` method:

```python
images = collection.map(add_ndvi).map(add_time)
```

With the NDVI and time bands added to each image in the collection, you can now create a quality mosaicked image using the `qualityMosaic()` method. The `qualityMosaic()` method takes a quality band name as an input parameter. In this example, the quality band is the NDVI band. The `qualityMosaic()` method will use the NDVI band to identify the greenest pixel in each location:

```python
greenest = images.qualityMosaic('NDVI')
```

The resulting image has the same number of bands as the input images:

```python
greenest.bandNames()
```

Now, you can add the NDVI band from the quality mosaicked image and visualize it (Fig. 6.25):

```python
ndvi = greenest.select('NDVI')
vis_ndvi = {'min': 0, 'max': 1, 'palette': 'ndvi'}
Map.addLayer(ndvi, vis_ndvi, 'NDVI')
Map.add_colorbar(vis_ndvi, label='NDVI', layer_name='NDVI')
Map
```

Since the quality mosaicked image has the same number of bands as the input images, you can add visualize it with different band combinations, such as the natural color composite (see Fig. 6.26):

```python
Map.addLayer(greenest, vis_rgb, 'Greenest pixel')
```

Compared with the median composite, the quality mosaicked image has fewer cloudy pixels. The reason for this is that if a pixel has the highest NDVI value in a location, it is less likely to be a cloudy pixel. Cloudy pixels tend to have negative NDVI values. In addition to displaying a spectral band composite, you can also display a time band composite. For example, you can identify the month when each pixel has the highest NDVI value:

```python
vis_month = {'palette': ['red', 'blue'], 'min': 1, 'max': 12}
Map.addLayer(greenest.select('month'), vis_month, 'Greenest month')
Map.add_colorbar(vis_month, label='Month', layer_name='Greenest month')
```

You can also identify the day of the year when each pixel has the highest NDVI value:

```python
vis_doy = {'palette': ['brown', 'green'], 'min': 1, 'max': 365}
Map.addLayer(greenest.select('doy'), vis_doy, 'Greenest doy')
Map.add_colorbar(vis_doy, label='Day of year', layer_name='Greenest doy')
```

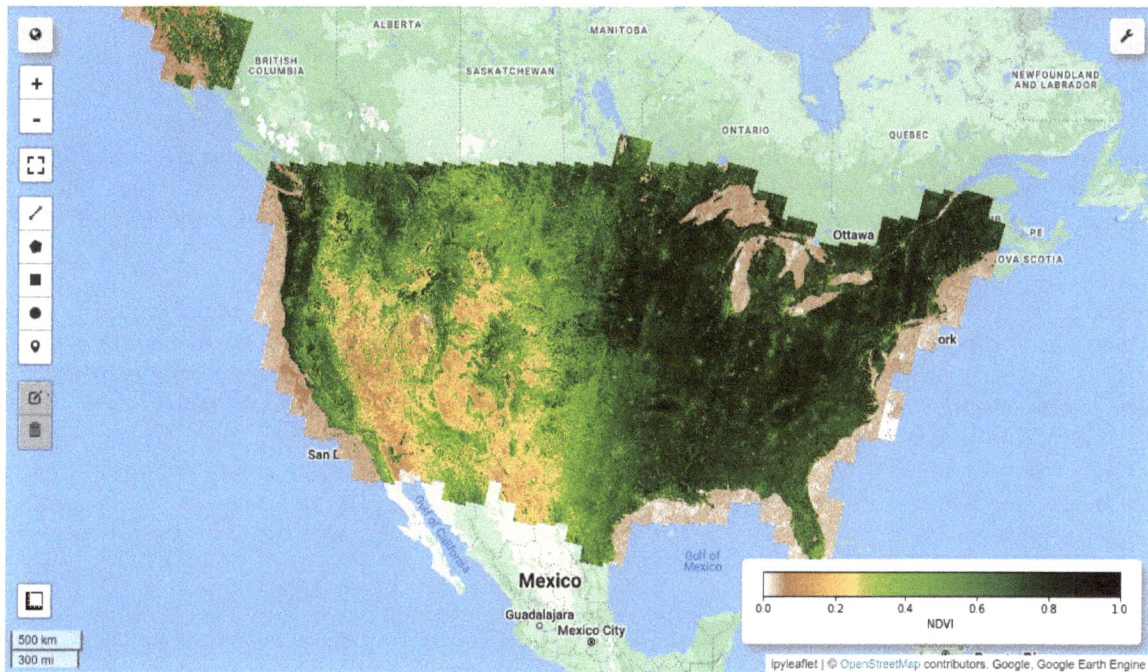

Figure 6.25: The greenest pixel in each location for the year 2020 in the US.

Figure 6.26: A quality mosaicked image of the greenest pixel in each location for the year 2020 in the US.

6.11   Interactive charts

Chart Overview

The Earth Engine JavaScript API provides `ui.Chart` functions for creating interactive charts from data tables and Earth Engine objects. However, the Python API does not provide these functions. Luckily, geemap has some similar functions for creating interactive charts from data tables and Earth Engine objects.

Geemap has two modules for creating interactive charts: geemap.chart[133] and geemap.plot[134]. The `geemap.chart` module provides functions for creating interactive charts from Earth Engine objects, such as the `feature_histogram()`, `feature_byFeature()`, and `feature_byProperty()` functions. The `geemap.plot` module provides functions for creating interactive charts from data tables, such as the `bar_chart()`, `line_chart()`, `pie_chart()` and `histogram()` functions. These two modules are still under active development. Therefore, the functions in these modules might change in the future. Check out the geemap API reference for the latest information.

Data table charts

The geemap.plot[134] module provides functions for creating interactive charts from data tables, such as CSV files and Pandas dataframes. You can create interactive charts with only one line of code. In this section, you will learn how to create bar charts, line charts, and pie charts from data tables.

First, let's load the countries GeoJSON file and display the attribute table (see Fig. 6.27).

```
data = geemap.examples.get_path('countries.geojson')
df = geemap.geojson_to_df(data)
df.head()
```

	type	fid	NAME	POP_EST	POP_RANK	GDP_MD_EST	INCOME_GRP	ISO_A2	ISO_A3	CONTINENT	SUBREGION
0	Feature	1	Fiji	920938	11	8374.0	4. Lower middle income	FJ	FJI	Oceania	Melanesia
1	Feature	2	Tanzania	53950935	16	150600.0	5. Low income	TZ	TZA	Africa	Eastern Africa
2	Feature	3	W. Sahara	603253	11	906.5	5. Low income	EH	ESH	Africa	Northern Africa
3	Feature	4	Canada	35623680	15	1674000.0	1. High income: OECD	CA	CAN	North America	Northern America
4	Feature	5	United States of America	326625791	17	18560000.0	1. High income: OECD	US	USA	North America	Northern America

Figure 6.27: Displaying the attributes of the first five countries in the data table.

We are interested in the POP_EST column, which contains the population estimates for each country. You can create a bar chart from the POP_EST column using the `bar_chart()` function:

```
geemap.bar_chart(
    data=df,
    x='NAME',
    y='POP_EST',
    x_label='Country',
    y_label='Population',
    descending=True,
    max_rows=30,
    title='World Population',
    height=500,
    layout_args={'title_x': 0.5, 'title_y': 0.85},
)
```

---

[133]geemap.chart: geemap.org/chart
[134]geemap.plot: geemap.org/plot

Note that the data parameter accepts a Pandas dataframe or a path to a CSV file. The x parameter
specifies the column name for the x-axis. The y parameter specifies the column name for the y-axis. The
x_label parameter specifies the label for the x-axis. The y_label parameter specifies the label for the
y-axis. The descending parameter specifies whether to sort the x column in descending order. There are
176 countries in the data table. It might be overwhelming to display all of them in the chart, therefore,
you can use the max_rows parameter to limit the number of rows to display. The title parameter
specifies the chart title. The height parameter specifies the chart height. The layout_args parameter
specifies arguments that can be passed to the update_layout() function in Plotly. For example, you can
use title_x and title_y to specify the title position (see Fig. 6.28).

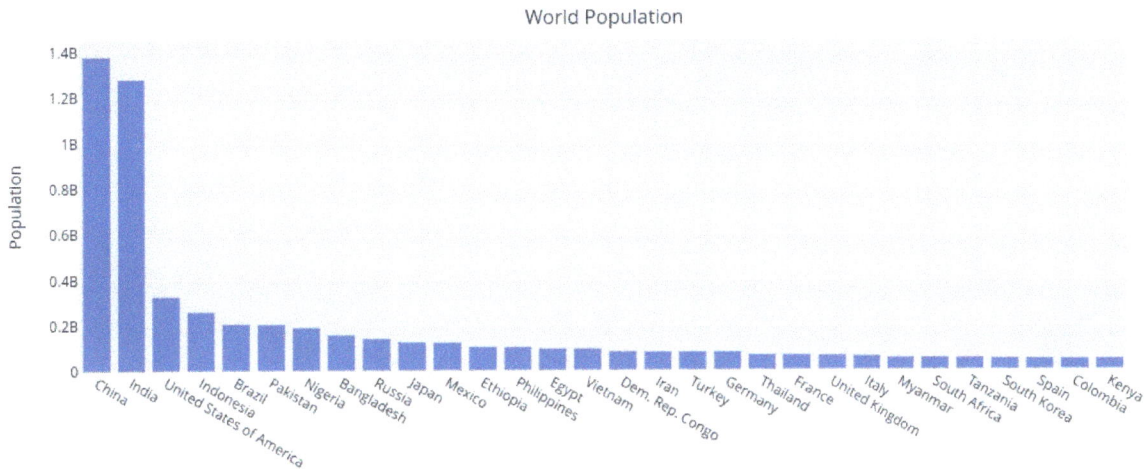

Figure 6.28: A bar chart showing the population of the top 30 countries in the world.

To create a pie chart, use the pie_chart() function. The pie_chart() function accepts similar parameters as the bar_chart() function. The only difference is that the pie_chart() function does not use the x
and x_label parameters. Instead, it uses the names and values parameters to specify the column names
for the names and values.

```
geemap.pie_chart(
    data=df,
    names='NAME',
    values='POP_EST',
    max_rows=30,
    height=600,
    title='World Population',
    legend_title='Country',
    layout_args={'title_x': 0.47, 'title_y': 0.87},
)
```

The resulting pie chart is shown in Fig. 6.29.

As can be seen from the pie chart above, China and India have the largest populations in the world.
The top 30 countries account for 77.3% of the world's population. The remaining 22.7% of the world's
population is distributed among the remaining 146 countries.

Next, let's create a line chart using a life expectancy data table. The data table contains the life expectancy for each country in the world from 1950 to 2010. It can be a bit overwhelming to display all
the countries in the chart. You can filter the data table to display countries in a specific region (e.g.,
Oceania):

```
data = geemap.examples.get_path('life_exp.csv')
df = geemap.csv_to_df(data)
```

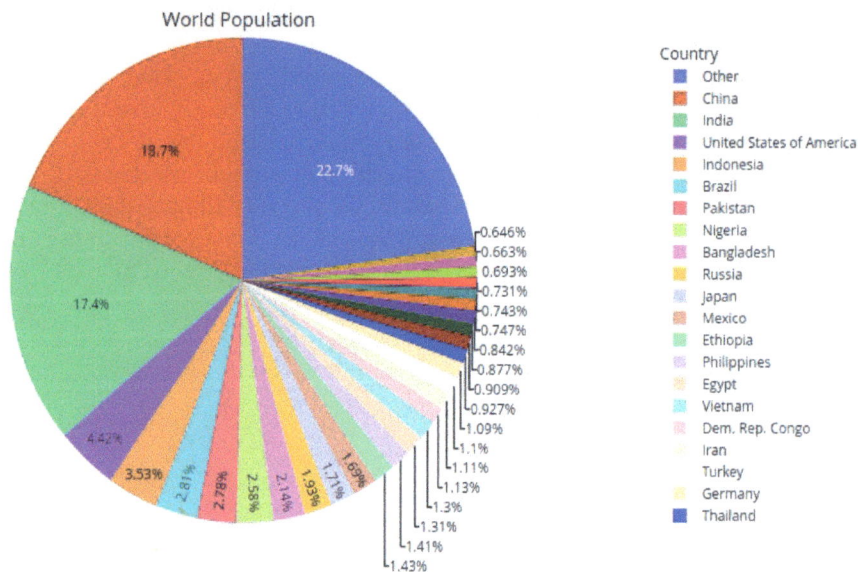

Figure 6.29: A pie chart showing the population of the top 30 countries in the world.

```
df = df[df['continent'] == 'Oceania']
df.head()
```

The resulting data table is shown in Fig. 6.30.

	country	continent	year	lifeExp	pop	gdpPercap	iso_alpha	iso_num
60	Australia	Oceania	1952	69.12	8691212	10039.59564	AUS	36
61	Australia	Oceania	1957	70.33	9712569	10949.64959	AUS	36
62	Australia	Oceania	1962	70.93	10794968	12217.22686	AUS	36
63	Australia	Oceania	1967	71.10	11872264	14526.12465	AUS	36
64	Australia	Oceania	1972	71.93	13177000	16788.62948	AUS	36

Figure 6.30: Displaying the first five rows of the data table of life expectancy in Oceania.

With the filtered data table, you can create a line chart using the `line_chart()` function:

```
geemap.line_chart(
    df,
    x='year',
    y='lifeExp',
    color='country',
    x_label='Year',
    y_label='Life expectancy',
    legend_title='Country',
    height=400,
    markers=True,
)
```

The resulting line chart is shown in Fig. 6.31.

Earth Engine object charts

The previous section showed you how to create charts from data tables. In this section, you will learn how to create charts directly from Earth Engine objects, such as the `ee.FeatureCollection` object. Support for creating charts from `ee.Image` and `ee.ImageCollection` objects will be added in the future.

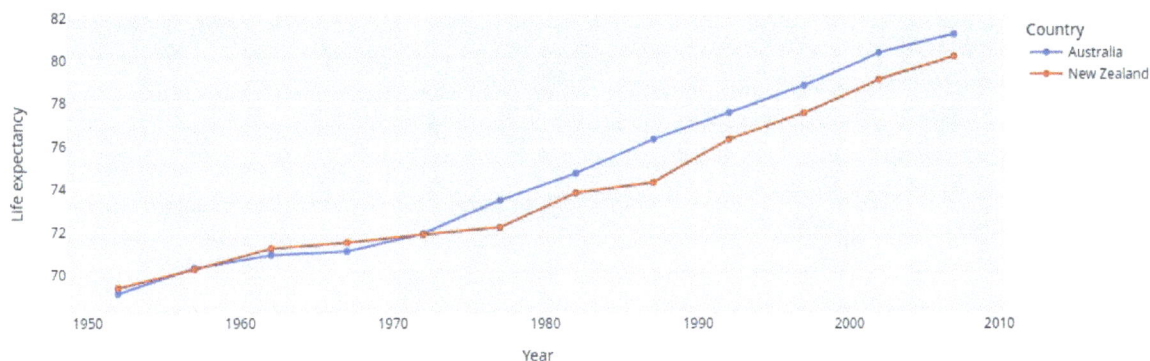

Figure 6.31: A line chart showing the life expectancy of Australia and New Zealand.

Please check the geemap API documentation for the latest updates. All the charting functions are in the `geemap.chart` module. First, import the `geemap.chart` module:

```
import geemap.chart as chart
```

The following examples use an `ee.FeatureCollection` object containing three ecoregion features (i.e., desert, forest, and grassland) with climate variables such as `tmean` (30-year average of monthly mean temperature) and `ppt` (3-year average of monthly total precipitation). This sample dataset was derived from the PRISM Long-Term Average Climate Dataset[135].

First, load the sample dataset to the map and use the inspector tool to inspect the feature properties (see Fig. 6.32):

```
Map = geemap.Map(center=[40, -100], zoom=4)
collection = ee.FeatureCollection('projects/google/charts_feature_example')
Map.addLayer(collection, {}, "Ecoregions")
Map
```

Chart by feature

The sample dataset contains three features with a variety of properties, such as the `label`, `ppt`, `tmean`, `tmin`, and `tmax` properties. To create a chart of features, you need to specify the feature property to use for the x-axis and the feature property to use for the y-axis. The following code selects the `label` and `tmean` properties for the x- and y-axes, respectively:

```
features = collection.select('[0-9][0-9]_tmean|label')
df = geemap.ee_to_df(features, sort_columns=True)
df
```

The resulting data table is shown in Fig. 6.33.

Note that the `sort_column` parameter passed to the `ee_to_df()` function is set to `True` to sort the column names in ascending order. Otherwise, the column names might not be in alphabetical order. Next, customize the chart options, such as the `labels`, `colors`, `title`, `x_label`, and `y_label` options:

```
xProperty = "label"
yProperties = df.columns[:12]

labels = [
    'Jan',
```

---

[135]PRISM Long-Term Average Climate Dataset: tiny.geemap.org/uigt

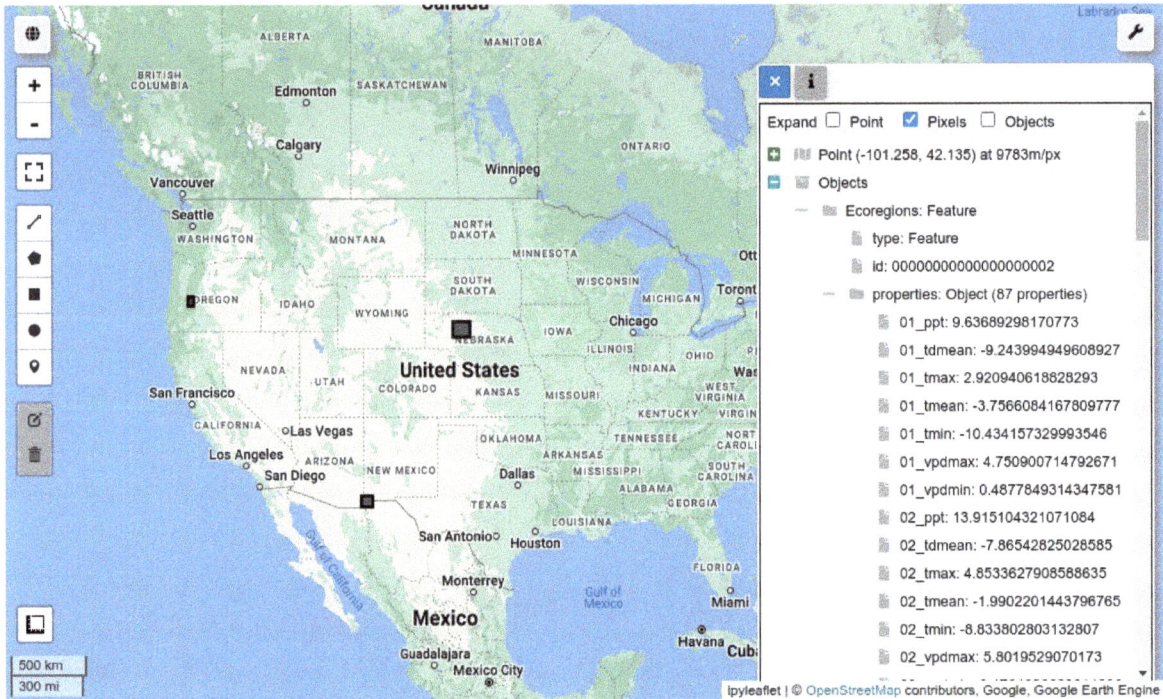

Figure 6.32: A feature collection containing three ecoregion features.

	01_tmean	02_tmean	03_tmean	04_tmean	05_tmean	06_tmean	07_tmean	08_tmean	09_tmean	10_tmean	11_tmean	12_tmean	label
0	5.791036	7.645011	10.454658	14.251640	19.032815	23.790506	25.066577	23.845259	21.454795	15.997988	9.849321	5.641386	Desert
1	2.792467	3.609074	5.032932	7.120137	10.395376	13.776134	17.850177	17.919983	15.206573	10.081709	4.784706	2.317886	Forest
2	-3.756608	-1.990220	2.570146	7.721306	13.643875	19.033558	22.753059	21.848346	16.401770	9.263021	2.021918	-3.426706	Grassland

Figure 6.33: A data table of the average monthly temperature by ecoregion in the United States.

```
        'Feb',
        'Mar',
        'Apr',
        'May',
        'Jun',
        'Jul',
        'Aug',
        'Sep',
        'Oct',
        'Nov',
        'Dec',
    ]
    colors = [
        '#604791',
        '#1d6b99',
        '#39a8a7',
        '#0f8755',
        '#76b349',
        '#f0af07',
        '#e37d05',
        '#cf513e',
        '#96356f',
        '#724173',
        '#9c4f97',
        '#696969',
```

```
    ]
title = "Average Monthly Temperature by Ecoregion"
xlabel = "Ecoregion"
ylabel = "Temperature"
```

Create a dictionary of options for the chart:

```
options = {
    "labels": labels,
    "colors": colors,
    "title": title,
    "xlabel": xlabel,
    "ylabel": ylabel,
    "legend_location": "top-left",
    "height": "500px",
}
```

Create the chart using the `chart.feature_byFeature()` function (see Fig. 6.34):

```
chart.feature_byFeature(features, xProperty, yProperties, **options)
```

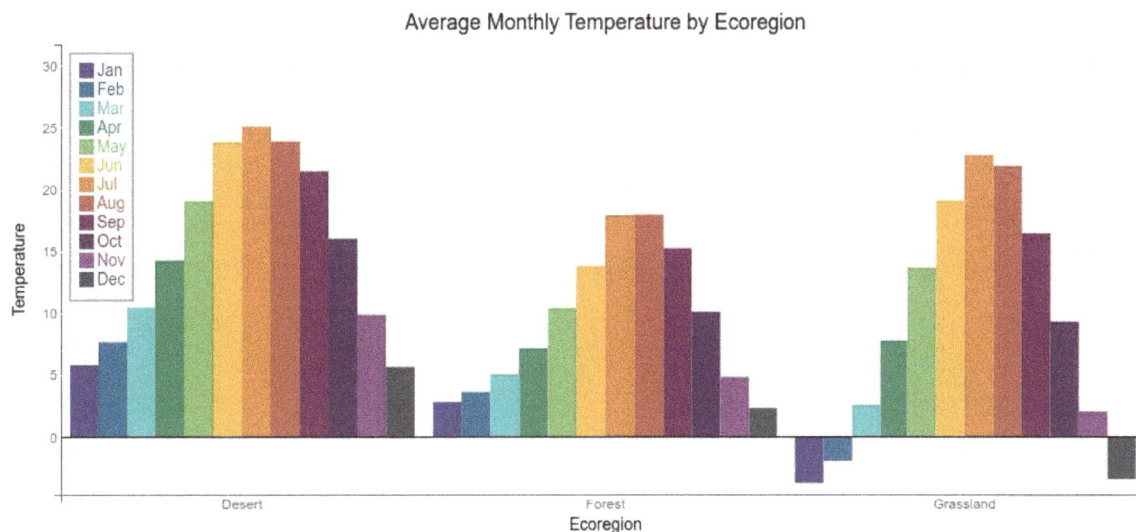

Figure 6.34: A bar chart of the average monthly temperature by ecoregion in the United States.

Chart by property

The `chart.feature_byFeature()` function creates a chart by feature. To create a chart by property, use the `chart.feature_byProperty()` function. Using the same sample dataset, let's select the `label` and `ppt` properties for the x- and y-axes, respectively:

```
features = collection.select('[0-9][0-9]_ppt|label')
df = geemap.ee_to_df(features, sort_columns=True)
df
```

The resulting data table is shown in Fig. 6.35.

Customize the chart options, such as `xProperties`, and `seriesProperty`. Note that `xProperties` accepts a (`property`, `label`) dictionary specifying labels for properties to be used as values on the x-axis:

```
keys = df.columns[:12]
```

	01_ppt	02_ppt	03_ppt	04_ppt	05_ppt	06_ppt	07_ppt	08_ppt	09_ppt	10_ppt	11_ppt	12_ppt	label
0	27.954341	21.858469	17.579124	8.252543	8.372216	14.802123	80.389409	79.326886	37.412247	32.728694	22.809099	35.557367	Desert
1	235.373540	181.531780	181.917962	150.759019	112.392139	72.709877	23.409216	25.395566	54.064534	121.976155	261.008798	273.243438	Forest
2	9.636893	13.915104	29.761295	57.736112	84.276450	86.136860	76.181755	60.189630	45.833161	38.134907	20.085923	11.198115	Grassland

Figure 6.35: A data table of the average monthly total precipitation by ecoregion in the United States.

```
values = [
    'Jan',
    'Feb',
    'Mar',
    'Apr',
    'May',
    'Jun',
    'Jul',
    'Aug',
    'Sep',
    'Oct',
    'Nov',
    'Dec',
]
xProperties = dict(zip(keys, values))
seriesProperty = "label"
```

Create a dictionary of chart options:

```
options = {
    'title': "Average Ecoregion Precipitation by Month",
    'colors': ['#f0af07', '#0f8755', '#76b349'],
    'xlabel': "Month",
    'ylabel': "Precipitation (mm)",
    'legend_location': "top-left",
    "height": "500px",
}
```

Lastly, create the chart using the `chart_by_property()` function:

```
chart.feature_byProperty(features, xProperties, seriesProperty, **options)
```

The resulting chart is shown in Fig. 6.36.

Feature histograms

To create a histogram using a range of values from the property in an `ee.FeatureCollection`, use the `chart.feature_histogram()` function. The following code generates a dataset with 5000 points randomly sampled from the PRISM Long-Term Average Climate Dataset[135]. In this example, we select the July precipitation property `07_ppt` for the histogram:

```
source = ee.ImageCollection('OREGONSTATE/PRISM/Norm81m').toBands()
region = ee.Geometry.Rectangle(-123.41, 40.43, -116.38, 45.14)
samples = source.sample(region, 5000)
prop = '07_ppt'
```

Create a dictionary of chart options:

```
options = {
    "title": 'July Precipitation Distribution for NW USA',
    "xlabel": 'Precipitation (mm)',
    "ylabel": 'Pixel count',
```

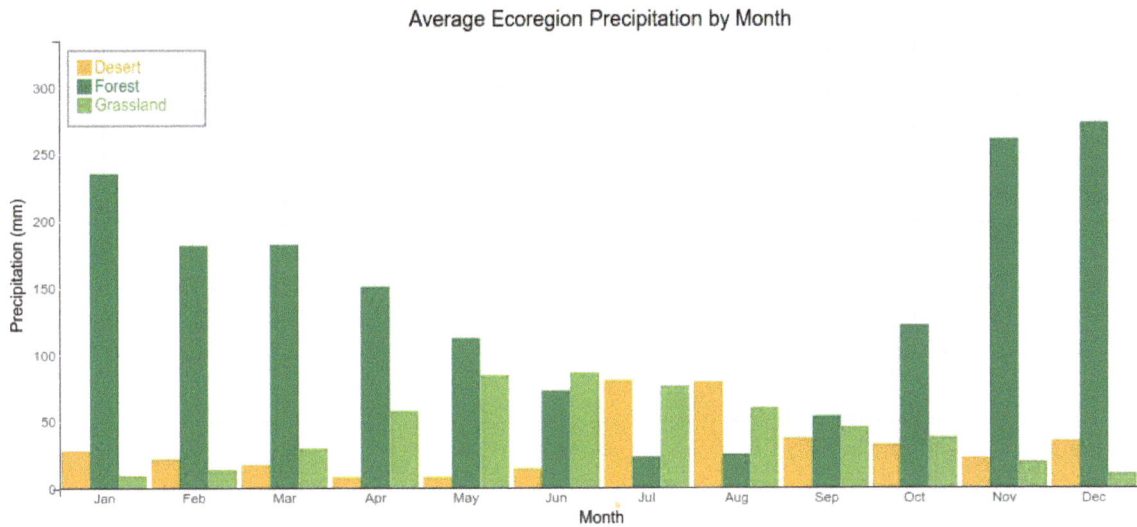

Figure 6.36: A bar chart of the average monthly total precipitation by ecoregion in the United States.

```
    "colors": ['#1d6b99'],
}
```

Generate a histogram with default parameters:

```
chart.feature_histogram(samples, prop, **options)
```

The resulting histogram is shown in Fig. 6.37.

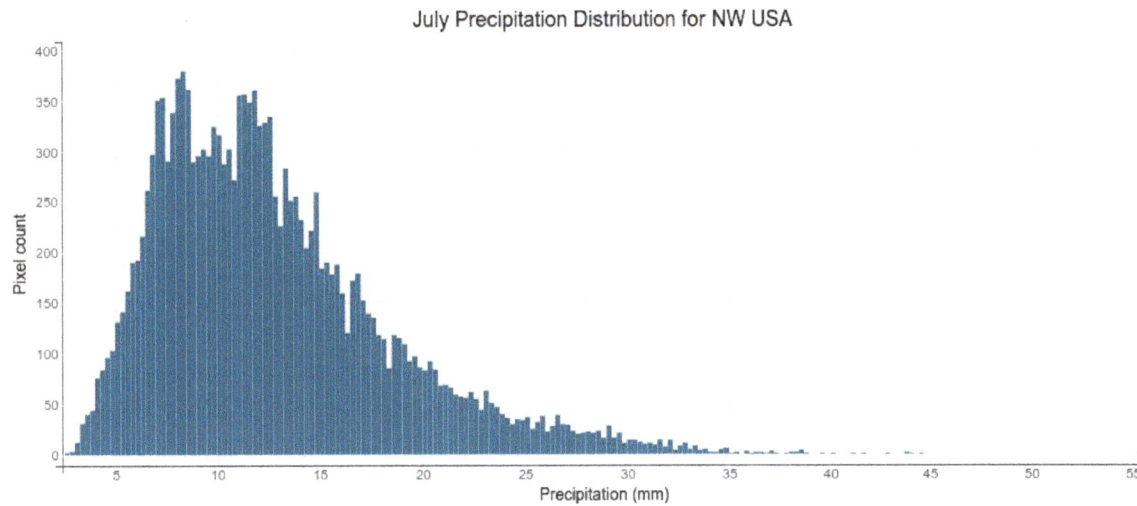

Figure 6.37: A histogram of the July precipitation distribution for the northwest United States.

To customize the number of bins, use the maxBuckets parameter:

```
chart.feature_histogram(samples, prop, maxBuckets=30, **options)
```

The resulting histogram is shown in Fig. 6.38.

July Precipitation Distribution for NW USA

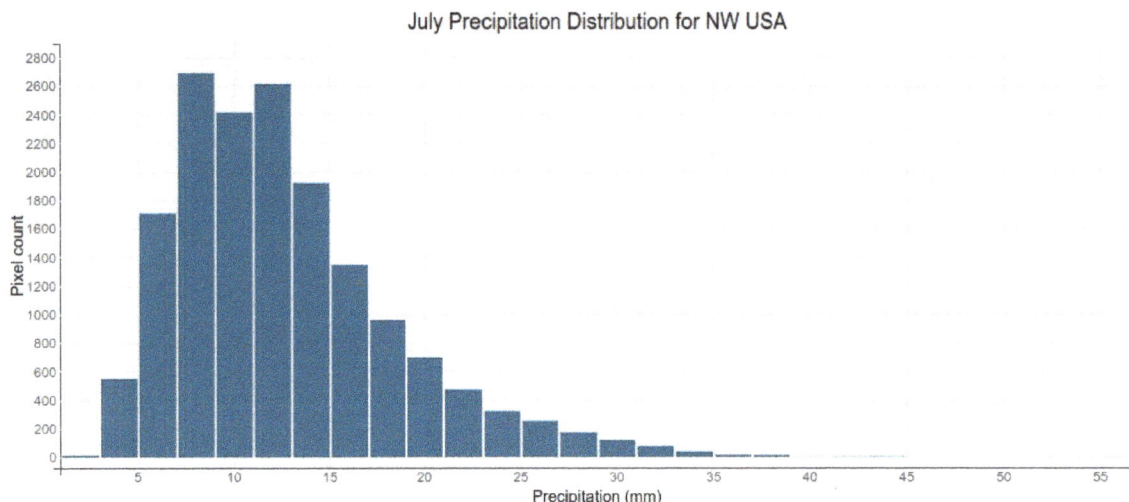

Figure 6.38: A histogram of the July precipitation distribution for the northwest United States. The number of bins is set to 30.

To customize the bin width, use the `minBucketWidth` parameter:

```
chart.feature_histogram(samples, prop, minBucketWidth=0.5, **options)
```

You can set both the `maxBuckets` and `minBucketWidth` parameters to customize the histogram:

```
chart.feature_histogram(samples, prop, minBucketWidth=3, maxBuckets=30, **options)
```

6.12 Unsupervised classification

Unsupervised classification is a machine learning technique that groups similar pixels into classes based on their spectral characteristics. Earth Engine provides the `ee.Clusterer` class for performing unsupervised classification. The supported clustering algorithms include: `wekaKmeans`, `wekaXmeans`, `wekaCascadeKMeans`, `wekaCobweb`, and `wekaVQ`. The general workflow for performing unsupervised classification in Earth Engine is as follows:

1. Prepare a multiband image for classification.

2. Generate training samples from the image.

3. Initialize a clusterer and adjust the parameters as needed.

4. Train the clusterer using the training samples.

5. Apply the clusterer to the image.

6. Label the clusters as needed.

7. Export the classified image.

The following example demonstrates how to perform unsupervised classification on a Landsat 9 image. First, filter the Landsat 9 image collection by a region of interest and date range. Select the least cloudy image and select the seven spectral bands. Note that a scaling factor of 0.0000275 and a bias of -0.2 are applied to the image to convert the DN values to reflectance values:

```python
Map = geemap.Map()

point = ee.Geometry.Point([-88.0664, 41.9411])

image = (
    ee.ImageCollection('LANDSAT/LC09/C02/T1_L2')
    .filterBounds(point)
    .filterDate('2022-01-01', '2022-12-31')
    .sort('CLOUD_COVER')
    .first()
    .select('SR_B[1-7]')
)

region = image.geometry()
image = image.multiply(0.0000275).add(-0.2).set(image.toDictionary())
vis_params = {'min': 0, 'max': 0.3, 'bands': ['SR_B5', 'SR_B4', 'SR_B3']}

Map.centerObject(region, 8)
Map.addLayer(image, vis_params, "Landsat-9")
Map
```

Next, use the `get_info()` function to inspect the image metadata and inspect it in a tree structure:

```python
geemap.get_info(image)
```

You can also get an image property using the `get()` method. For example, to get the image acquisition date:

```python
image.get('DATE_ACQUIRED').getInfo()
```

To check the image cloud cover:

```python
image.get('CLOUD_COVER').getInfo()
```

With the image ready, you can now generate training samples. You can specify a region where the training samples will be generated. There are several ways to create a region for generating training samples:

- Draw a shape (e.g., rectangle) on the map and the use `region = Map.user_roi`
- Define a geometry, such as `region = ee.Geometry.Rectangle([xmin, ymin, xmax, ymax])`
- Create a buffer around a point, such as `region = ee.Geometry.Point([x, y]).buffer(v)`
- If you don't define a region, it will use the image footprint by default

The following code generates 5000 training samples from the image and add them to the map. Note that the `region` parameter is not specified, so the image footprint will be used as the region (see Fig. 6.39):

```python
training = image.sample(
    **{
        # "region": region,
        'scale': 30,
        'numPixels': 5000,
        'seed': 0,
        'geometries': True,  # Set this to False to ignore geometries
    }
)

Map.addLayer(training, {}, 'Training samples')
Map
```

To inspect the attribute table of training data, use the `ee_to_df()` function on the first few features of

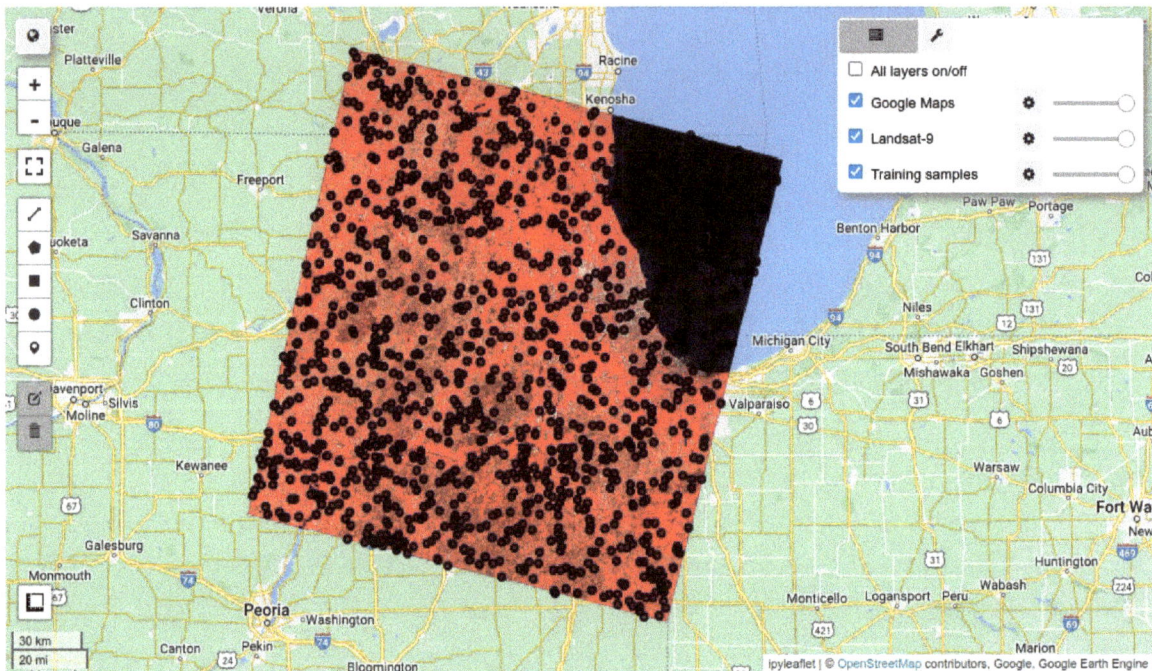

Figure 6.39: Training data generated from a Landsat-9 image.

the collection:

```
geemap.ee_to_df(training.limit(5))
```

The training data is ready. Next, you need to initialize a clusterer and train it using the training data. The following code initializes a wekaKmeans clusterer and trains it using the training data by specifying the number of clusters (e.g., 5):

```
n_clusters = 5
clusterer = ee.Clusterer.wekaKMeans(n_clusters).train(training)
```

With the clusterer trained, you can now apply it to the image. The following code applies the clusterer to the image and adds the classified image to the map with a random color palette:

```
result = image.cluster(clusterer)
Map.addLayer(result.randomVisualizer(), {}, 'clusters')
Map
```

Note the value and color of each cluster are randomly assigned. Use the Inspector tool to inspect the value of each cluster and label them as needed. Define a legend dictionary with pairs of cluster labels and colors, which can be used to create a legend for the classified image:

```
legend_dict = {
    'Open Water': '#466b9f',
    'Developed, High Intensity': '#ab0000',
    'Developed, Low Intensity': '#d99282',
    'Forest': '#1c5f2c',
    'Cropland': '#ab6c28'

}

palette = list(legend_dict.values())
```

```
Map.addLayer(
    result, {'min': 0, 'max': 4, 'palette': palette}, 'Labelled clusters'
)
Map.add_legend(title='Land Cover Type',legend_dict=legend_dict , position='bottomright')
Map
```

The unsupervised classification result is shown in Fig. 6.40.

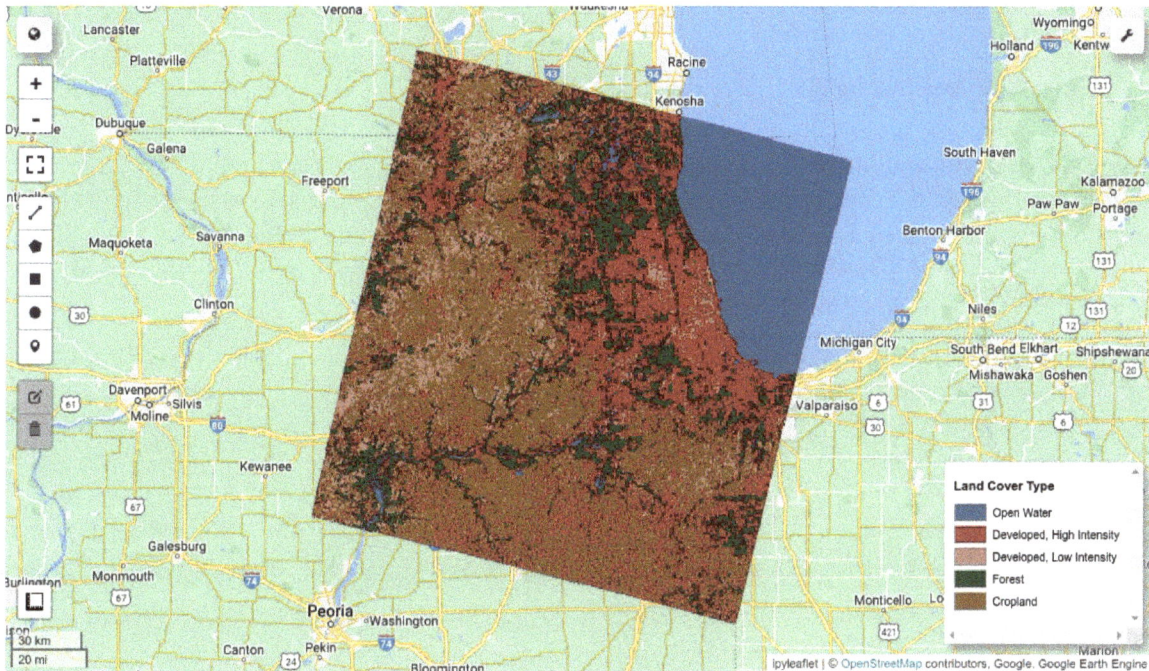

Figure 6.40: Unsupervised classification result of a Landsat-9 image.

Finally, you can export the classified image to your computer. Specify the image region, scale, and output file path as needed:

```
geemap.download_ee_image(image, filename='unsupervised.tif', region=region, scale=90)
```

6.13   Supervised classification

Supervised classification is a machine learning technique that uses labeled training data to classify an image. The training data is used to train a classifier, which is then applied to the image to generate a classified image. Earth Engine provides the ee.Classifier class for performing supervised classification. The supported supervised classification algorithms include: Classification and Regression Trees (CART)[136] , Support Vector Machine (SVM)[137] , Random Forest[138] , Naive Bayes[139] , and Gradient Tree Boost[140]. The general workflow for supervised classification is as follows:

1. Prepare an image for classification.

---

[136]Classification and Regression Trees (CART): tiny.geemap.org/hkbd

[137]Support Vector Machine (SVM): tiny.geemap.org/acbk

[138]Random Forest: tiny.geemap.org/otdk

[139]Naive Bayes: tiny.geemap.org/qclj

[140]Gradient Tree Boost: tiny.geemap.org/ybkk

2. Collect training data. Each training sample should have a class label and a set of properties storing numeric values for the predictors.

3. Initialize a classifier and set its parameters as needed.

4. Train the classifier using the training data.

5. Apply the classifier to the image.

6. Perform accuracy assessment.

7. Export the classified image.

In this section, you will learn how to perform supervised classification using the CART algorithm. You can easily adapt the code to other supervised classification algorithms, such as SVM and Random Forest. We will use labeled training data from the USGS National Land Cover Database (NLCD)[141] dataset and train a CART classifier using the training data. The trained classifier will then be applied to the Landsat-9 image to generate a classified image.

First, filter the Landsat 8 image collection[142] to select a cloud-free image acquired in 2019 for your region of interest:

```python
Map = geemap.Map()
point = ee.Geometry.Point([-122.4439, 37.7538])

image = (
    ee.ImageCollection('LANDSAT/LC08/C02/T1_L2')
    .filterBounds(point)
    .filterDate('2019-01-01', '2020-01-01')
    .sort('CLOUD_COVER')
    .first()
    .select('SR_B[1-7]')
)

image = image.multiply(0.0000275).add(-0.2).set(image.toDictionary())
vis_params = {'min': 0, 'max': 0.3, 'bands': ['SR_B5', 'SR_B4', 'SR_B3']}

Map.centerObject(point, 8)
Map.addLayer(image, vis_params, "Landsat-8")
Map
```

The least cloud image is shown in Fig. 6.41.

Use the `get_info()` function to check the image properties:

```python
geemap.get_info(image)
```

To get a specific image property, use the `get()` method with the property name as the argument. For example, to retrieve the image acquisition date:

```python
image.get('DATE_ACQUIRED').getInfo()
```

To check the cloud cover of the image:

```python
image.get('CLOUD_COVER').getInfo()
```

---

[141]USGS National Land Cover Database (NLCD): tiny.geemap.org/sifh

[142]Landsat 8 image collection: tiny.geemap.org/myuh

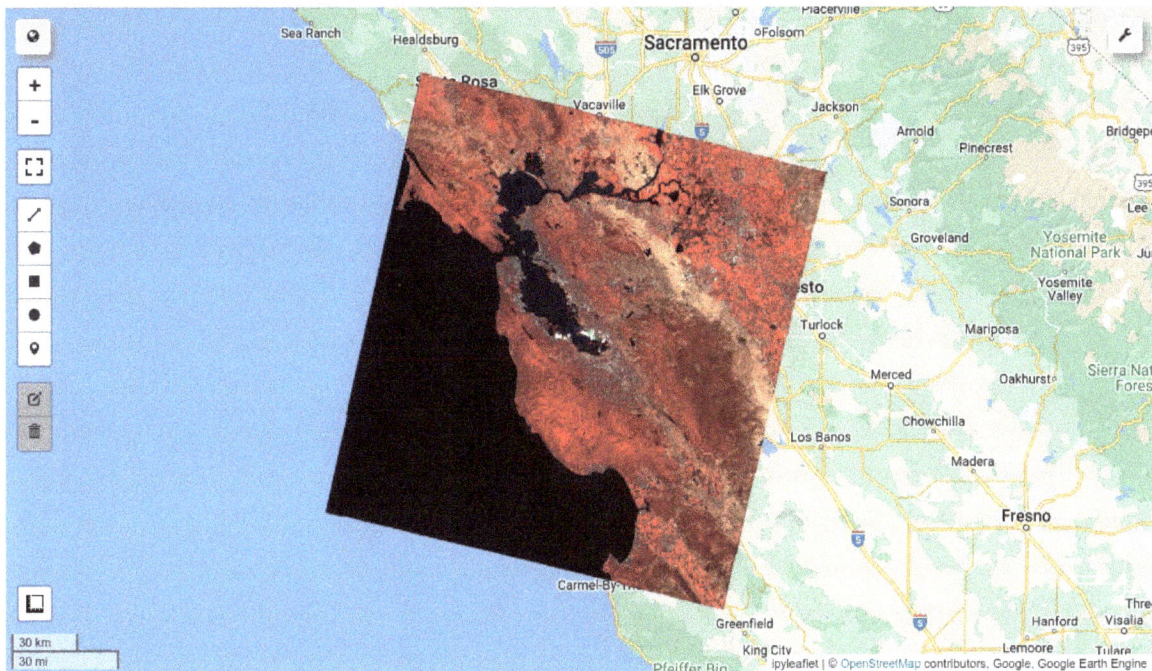

Figure 6.41: A Landsat 8 image for supervised classification.

Next, create a training dataset from the NLCD dataset, which is a 30-m resolution dataset covering the conterminous United States. The NLCD dataset contains 21 land cover classes. A detailed description of each NLCD land cover type can be found at `https://bit.ly/3EyvacV`. The following code filters the NLCD dataset to select the land cover image of interest and clips the dataset to the region of interest (the footprint of the selected Landsat image in the previous step):

```
nlcd = ee.Image('USGS/NLCD_RELEASES/2019_REL/NLCD/2019')
landcover = nlcd.select('landcover').clip(image.geometry())
Map.addLayer(landcover, {}, 'NLCD Landcover')
Map
```

With the land cover image ready, we can now sample the image to collect training data with land cover labels. Similar to the unsupervised classification introduced above, you can specify a region of interest, scale, and number of points to sample. The following code samples 5000 points from the land cover image:

```
points = landcover.sample(
    **{
        'region': image.geometry(),
        'scale': 30,
        'numPixels': 5000,
        'seed': 0,
        'geometries': True,
    }
)

Map.addLayer(points, {}, 'training', False)
```

Note that the resulting number of training samples may be less than the specified number of points. This is because the sampling algorithm will discard pixels with no data values. To check the number of training samples:

```
print(points.size().getInfo())
```

Revise the number of points to sample in the previous step if needed.

Next, we will add the spectral bands of the Landsat image to the training data. Note that the training data created from the previous step already contains the land cover labels (i.e., the `landcover` property). The following code adds the seven spectral bands of the Landsat image to the training data:

```
bands = ['SR_B1', 'SR_B2', 'SR_B3', 'SR_B4', 'SR_B5', 'SR_B6', 'SR_B7']
label = 'landcover'
features = image.select(bands).sampleRegions(
    **{'collection': points, 'properties': [label], 'scale': 30}
)
```

Dislay the attribute table of the training data using the ee_to_df() function:

```
geemap.ee_to_df(features.limit(5))
```

The attribute table should look like Fig. 6.42.

	landcover	SR_B1	SR_B2	SR_B3	SR_B4	SR_B5	SR_B6	SR_B7
0	71	0.047363	0.066833	0.106625	0.150295	0.237965	0.372523	0.251000
1	71	0.024097	0.031963	0.049012	0.066613	0.154942	0.182663	0.116745
2	21	0.015325	0.017333	0.029185	0.021100	0.196110	0.108880	0.047967
3	11	0.013290	0.020000	0.031605	0.011997	-0.001945	0.000310	0.000612
4	52	0.019918	0.030697	0.056767	0.077062	0.175513	0.182965	0.118148

Figure 6.42: The attribute table of the training data.

The training dataset is ready. You can now train a classifier using the training data. The following code initializes a CART classifier[143] and trains it using the training data:

```
params = {

    'features': features,
    'classProperty': label,
    'inputProperties': bands,

}
classifier = ee.Classifier.smileCart(maxNodes=None).train(**params)
```

The `features` parameter specifies the training data. The `classProperty` parameter specifies the property name of the training data that contains the class labels. The `inputProperties` parameter specifies the property names of the training data that contain the predictor values.

All Earth Engine classifiers have a `train()` function to train the classifier using the training data. The CART classifier has a `maxNodes` parameter to specify the maximum number of nodes in the tree. The default value is `None`, which means that the tree will be grown until all leaves are pure or until all leaves contain less than 5 training samples.

Since the classifier has been trained, you can now apply it to the Landsat image to generate a classified image. Make sure you use the same spectral bands as the training data. The following code applies the trained classifier to the selected Landsat image and adds the classified image with a random color palette to the map:

```
classified = image.select(bands).classify(classifier).rename('landcover')
Map.addLayer(classified.randomVisualizer(), {}, 'Classified')
Map
```

---

[143]CART classifier: tiny.geemap.org/orul

To compare the classified image with the referenced NLCD land cover image, it is better to use the same color palette. To set the color palette of an Earth Engine image with a predefined palette, set the `bandname_class_values` and `bandname_class_palette` properties of the image. For example, the NLCD land cover has the `landcover_class_values` and `landcover_class_palette` properties. When the land cover band is added to the map, the color palette will be automatically applied so that users don't have to specify the color palette manually. To check the color palette of the NLCD land cover image:

```
geemap.get_info(nlcd)
```

We can use the same approach to set the color palette of the classified image. Note that in the previous step, we already renamed the classified image band to `landcover`. Therefore, we can use the `landcover_class_values` and `landcover_class_palette` properties to set the color palette of the classified image. The following code sets the color palette of the classified image:

```
class_values = nlcd.get('landcover_class_values')
class_palette = nlcd.get('landcover_class_palette')
classified = classified.set({
    'landcover_class_values': class_values,
    'landcover_class_palette': class_palette
})
```

The classified image should now have the same color palette as the NLCD land cover image. Add the classified image and associated legend to the map (see Fig. 6.43):

```
Map.addLayer(classified, {}, 'Land cover')
Map.add_legend(title="Land cover type", builtin_legend='NLCD')
Map
```

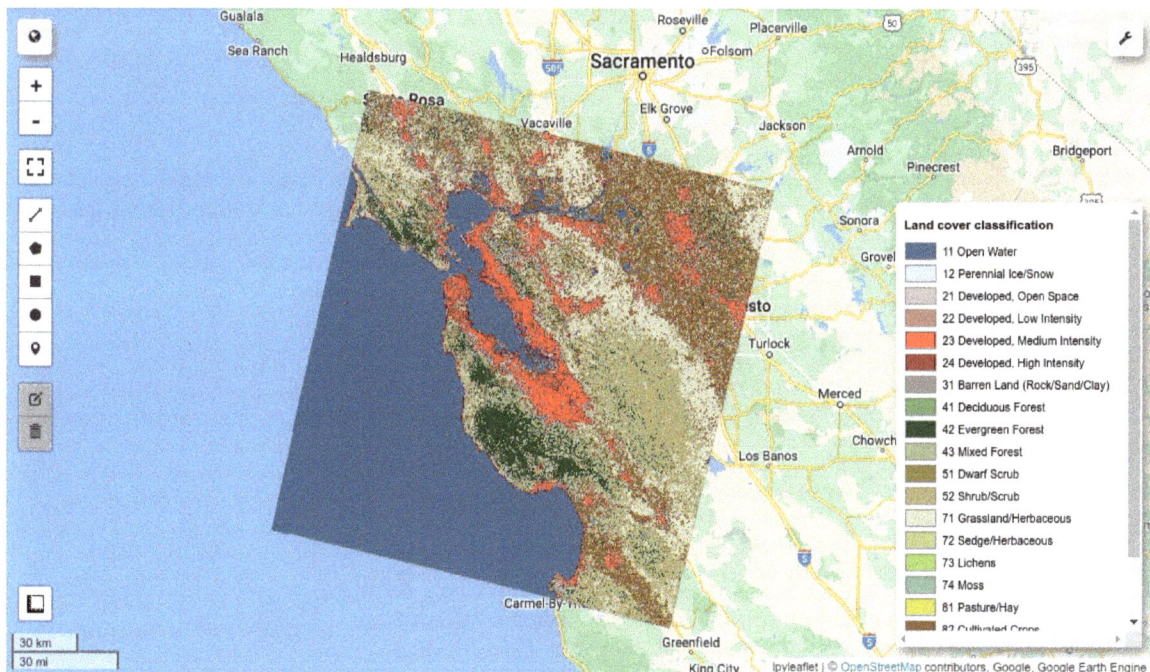

Figure 6.43: The classified image and its associated legend.

Use the layer control widget to change the opacity of the classified image and compare it visually with the NLCD land cover image. You might need to use the full-screen mode as the legend may block the view of layer control widget.

Finally, you can export the classified image to your computer:

```
geemap.download_ee_image(
    landcover,
    filename='supervised.tif',
    region=image.geometry(),
    scale=30
    )
```

6.14   Accuracy assessment

After performing image classification, you may want to assess the accuracy of the classification. Earth Engine provides several functions for assessing the accuracy of a classification. In this section, we will classify a Sentinel-2 image using random forest[144] and assess the accuracy of the classification.

First, filter the Sentinel-2 image collection and select an image for your region of interest:

```
Map = geemap.Map()
point = ee.Geometry.Point([-122.4439, 37.7538])

img = (
    ee.ImageCollection('COPERNICUS/S2_SR')
    .filterBounds(point)
    .filterDate('2020-01-01', '2021-01-01')
    .sort('CLOUDY_PIXEL_PERCENTAGE')
    .first()
    .select('B.*')
)

vis_params = {'min': 100, 'max': 3500, 'bands': ['B11', 'B8', 'B3']}

Map.centerObject(point, 9)
Map.addLayer(img, vis_params, "Sentinel-2")
Map
```

The ESA 10-m WorldCover[145] can be used to create labeled training data. First, we need to remap the land cover class values to a 0-based sequential series so that we can create a confusion matrix later:

```
lc = ee.Image('ESA/WorldCover/v100/2020')
classValues = [10, 20, 30, 40, 50, 60, 70, 80, 90, 95, 100]
remapValues = ee.List.sequence(0, 10)
label = 'lc'
lc = lc.remap(classValues, remapValues).rename(label).toByte()
```

Next, add the ESA land cover as a band of the Sentinel-2 reflectance image and sample 100 pixels at a 10m scale from each land cover class within the region of interest:

```
sample = img.addBands(lc).stratifiedSample(**{
  'numPoints': 100,
  'classBand': label,
  'region': img.geometry(),
  'scale': 10,
  'geometries': True
})
```

Add a random value field to the sample and use it to approximately split 80% of the features into a training set and 20% into a validation set:

```
sample = sample.randomColumn()
trainingSample = sample.filter('random <= 0.8')
validationSample = sample.filter('random > 0.8')
```

With the training data ready, we can train a random forest classifier using the training data.

The following code trains a random forest classifier with 10 trees:

```
trainedClassifier = ee.Classifier.smileRandomForest(numberOfTrees=10).train(**{
    'features': trainingSample,
    'classProperty': label,
    'inputProperties': img.bandNames()
})
```

To get information about the trained classifier:

```
print('Results of trained classifier', trainedClassifier.explain().getInfo())
```

To get a confusion matrix and overall accuracy for the training sample:

```
trainAccuracy = trainedClassifier.confusionMatrix()
trainAccuracy.getInfo()

[[81, 0, 0, 0, 1, 0, 0, 0, 0],
 [0, 83, 1, 0, 0, 0, 0, 0, 0],
 [1, 1, 73, 2, 0, 0, 0, 0, 0],
 [0, 0, 1, 77, 0, 0, 0, 0, 0],
 [0, 0, 0, 1, 81, 1, 0, 0, 0],
 [0, 1, 2, 3, 2, 70, 0, 0, 2],
 [0, 0, 0, 0, 0, 0, 0, 0, 0],
 [0, 0, 0, 0, 0, 0, 0, 71, 0],
 [1, 0, 0, 1, 0, 4, 0, 0, 71]]
```

The horizontal axis of the confusion matrix corresponds to the input classes, and the vertical axis corresponds to the output classes. The rows and columns start at class 0 and increase sequentially up to the maximum class value, so some rows or columns might be empty if the input classes aren't 0-based or sequential. That's the reason why we remapped the ESA land cover class values to a 0-based sequential series earlier. Note that your confusion matrix may look slightly different from the one shown above as the training data is randomly sampled.

The overall accuracy essentially tells us what proportion of al the reference sites was mapped correctly. The overall accuracy is usually expressed as a percent, with 100% accuracy being a perfect classification where all reference sites were classified correctly.

```
trainAccuracy.accuracy().getInfo()
```

The Kappa Coefficient is generated from a statistical test to evaluate the accuracy of a classification. Kappa essentially evaluates how well the classification performed as compared to just randomly assigning values, i.e., did the classification do better than random? The Kappa Coefficient can range from -1 to 1. A value of 0 indicated that the classification is no better than a random classification. A negative number indicates the classification is significantly worse than random. A value close to 1 indicates that the classification is significantly better than random.

```
trainAccuracy.kappa().getInfo()
```

To get a confusion matrix and overall accuracy for the validation sample:

```
validationSample = validationSample.classify(trainedClassifier)
validationAccuracy = validationSample.errorMatrix(label, 'classification')
validationAccuracy.getInfo()
```

```
[[13, 1, 3, 0, 1, 0, 0, 0, 0],
 [0, 11, 2, 0, 1, 0, 0, 0, 2],
 [1, 0, 12, 7, 1, 2, 0, 0, 0],
 [0, 3, 6, 9, 3, 1, 0, 0, 0],
 [2, 0, 3, 0, 10, 2, 0, 0, 0],
 [1, 1, 1, 3, 7, 6, 0, 0, 1],
 [0, 0, 0, 0, 0, 0, 0, 0, 0],
 [0, 0, 0, 0, 0, 0, 0, 29, 0],
 [2, 2, 2, 0, 2, 2, 0, 0, 13]]
```

To compute the overall accuracy for the validation sample:

```
validationAccuracy.accuracy().getInfo()
```

Producer's Accuracy, also known as the Sensitivity or Recall, is a measure of how well a classifier correctly identifies positive instances. It is the ratio of true positive classifications to the total number of actual positive instances. This metric is used to evaluate the performance of a classifier when the focus is on minimizing false negatives (i.e., instances that are actually positive but are classified as negative).

```
validationAccuracy.producersAccuracy().getInfo()
```

On the other hand, Consumer's Accuracy, also known as Precision, is a measure of how well a classifier correctly identifies negative instances. It is the ratio of true negative classifications to the total number of actual negative instances. This metric is used to evaluate the performance of a classifier when the focus is on minimizing false positives (i.e., instances that are actually negative but are classified as positive).

```
validationAccuracy.consumersAccuracy().getInfo()
```

The confusion matrices can be saved as a csv file:

```
import csv

with open("training.csv", "w", newline="") as f:
    writer = csv.writer(f)
    writer.writerows(trainAccuracy.getInfo())

with open("validation.csv", "w", newline="") as f:
    writer = csv.writer(f)
    writer.writerows(validationAccuracy.getInfo())
```

If the validation accuracy is acceptable, the trained classifier can then be applied to the entire image:

```
imgClassified = img.classify(trainedClassifier)
```

Lastly, add the resulting data layers (e.g., Sentinel-2 image, classified image, training and validation samples) to the map (see Fig. 6.44). Use the layer control widget the change the opacity of the classified image and compare it to the ESA WorldCover.

```
classVis = {
  'min': 0,
  'max': 10,
  'palette': ['006400' ,'ffbb22', 'ffff4c', 'f096ff', 'fa0000', 'b4b4b4',
              'f0f0f0', '0064c8', '0096a0', '00cf75', 'fae6a0']
}
Map.addLayer(lc, classVis, 'ESA Land Cover', False)
Map.addLayer(imgClassified, classVis, 'Classified')
Map.addLayer(trainingSample, {'color': 'black'}, 'Training sample')
Map.addLayer(validationSample, {'color': 'white'}, 'Validation sample')
Map.add_legend(title='Land Cover Type', builtin_legend='ESA_WorldCover')
```

```
Map.centerObject(img)
Map
```

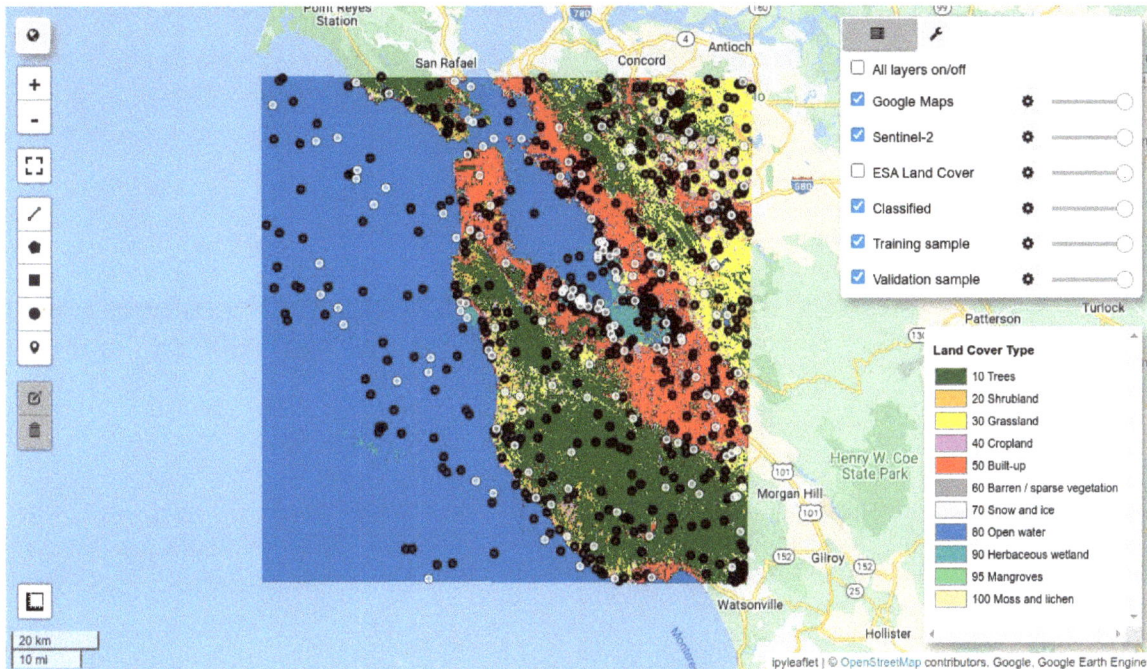

Figure 6.44: Supervised classification of a Sentinel-2 imagery using a random forest classifier.

6.15   Using locally trained machine learning models

The previous sections demonstrated how to use machine learning models that are trained on Earth Engine. However, it is also possible to train machine learning models locally using the scikit-learn[146] library, which is a popular machine learning library for Python. It provides a wide range of machine learning algorithms and tools for data analysis and model training. Geemap provides functions for converting locally trained machine learning models to Earth Engine classifiers, which can then be used to classify images on Earth Engine. The following example illustrates how to train a random forest (or any other ensemble tree estimator) locally using scikit-learn and convert the estimator into a string representation that Earth Engine can interpret, and apply the machine learning model with Earth Engine.

First, import the necessary libraries:

```
import pandas as pd
from geemap import ml
from sklearn import ensemble
```

Train a model locally using scikit-learn

To train a machine learning model, we first need to prepare the training data. A sample training dataset[147] is available on the geemap GitHub repository. Let's read the training data into a Pandas

---

[146]scikit-learn: scikit-learn.org

DataFrame (see Fig. 6.45):

```
url = "https://raw.githubusercontent.com/gee-community/geemap/master/examples/data/rf_example.csv"
df = pd.read_csv(url)
df
```

	B2	B3	B4	B5	B6	B7	landcover
0	0.139846	0.114738	0.109982	0.119542	0.125795	0.105720	0
1	0.130316	0.109207	0.107499	0.140210	0.132006	0.106497	0
2	0.146690	0.135766	0.146550	0.225686	0.218105	0.167111	0
3	0.119413	0.106924	0.105196	0.144868	0.158775	0.122056	0
4	0.155492	0.139932	0.137486	0.151377	0.153771	0.133134	0
...	...	...	...	...	...	...	...

Figure 6.45: Training data for a random forest classifier.

The training dataset was generated from a Landsat 8 TOA imagery. It contains six features (i.e., band names) and a class label (i.e., landcover). The class label is the last column in the dataset. The first six columns are the features. Specify the feature names and the class label name:

```
feature_names = ['B2', 'B3', 'B4', 'B5', 'B6', 'B7']
label = "landcover"
```

Get the features and labels from the training dataset and create a random forest classifier using scikit-learn. The classifier is trained using the training data and the specified number of trees:

```
X = df[feature_names]
y = df[label]
n_trees = 10
rf = ensemble.RandomForestClassifier(n_trees).fit(X, y)
```

Convert a sklearn classifier object to a list of strings

The Earth Engine API cannot use scikit-learn machine learning models directly. Therefore, we need to convert the trained scikit-learn model into a string representation that Earth Engine can interpret. To do so, use the `ml.rf_to_strings()` function:

```
trees = ml.rf_to_strings(rf, feature_names)
```

The number of trees we converted should be equal to the number of trees we defined for the model. Make sure the number of trees is correct:

```
print(len(trees))
```

Check the first tree in the list:

```
print(trees[0])
```

```
1) root 62 9999 9999 (1.153966895141798)
  2) B5 <= 0.064203 24 0.0000 2 *
  3) B5 > 0.064203 62 0.6554 2
    6) B2 <= 0.116759 18 0.0000 1 *
    7) B2 > 0.116759 20 0.0000 0 *
```

Note that random forest classifiers are composed of multiple decision trees. Each tree is represented as a string. The first line of the string represents the root node of the tree. The root node is followed by the child nodes. One space indents the child nodes. The first line of the child node represents the

---

[147]sample training dataset: tiny.geemap.org/sqah

left child node, and the second line represents the right child node. Random forest is often treated as a black box. Interpreting each individual tree in the random forest is not very intuitive.

Convert sklearn classifier to GEE classifier

At this point, you can take the list of strings and save them locally to avoid training again. However, we want to use the model with Earth Engine, so we need to create an ee.Classifier and persist the data on Earth Engine for best results. First, create an ee.Classifier to use with Earth Engine objects from the trees by converting the list of strings to a ee.Classifier using ml.strings_to_classifier():

```
ee_classifier = ml.strings_to_classifier(trees)
ee_classifier.getInfo()
```

Classify image using GEE classifier

The Earth Engine classifier is now ready, and we can use it to classify an image. The following code creates a Landsat 8 image composite of 2018 and classifies it using the Earth Engine classifier:

```
# Make a cloud-free Landsat 8 TOA composite (from raw imagery).
l8 = ee.ImageCollection('LANDSAT/LC08/C01/T1')

image = ee.Algorithms.Landsat.simpleComposite(
    collection=l8.filterDate('2018-01-01', '2018-12-31'), asFloat=True
)
```

Note the asFloat=True argument. This is necessary to ensure that the image is converted to floating point values, the same as the training data. With both the image and classifier ready, we can classify the image using the classify() method:

```
classified = image.select(feature_names).classify(ee_classifier)
```

Finally, display the Landsat 8 image and the classified image on the map:

```
Map = geemap.Map(center=(37.75, -122.25), zoom=11)

Map.addLayer(
    image,
    {"bands": ['B7', 'B5', 'B3'], "min": 0.05, "max": 0.55, "gamma": 1.5},
    'image',
)
Map.addLayer(
    classified,
    {"min": 0, "max": 2, "palette": ['red', 'green', 'blue']},
    'classification',
)
Map
```

The classified image should look like Fig. 6.46.

Use the layer control to change the opacity of the classified image and compare it with the original image.

Note that this workflow has several limitations, particularly due to how much data you can pass from the client to the server and how large of a model ee can actually handle. Earth Engine can only handle 40MB of data passed to the server, so if you have a lot of large decision tree strings, then this will not work.

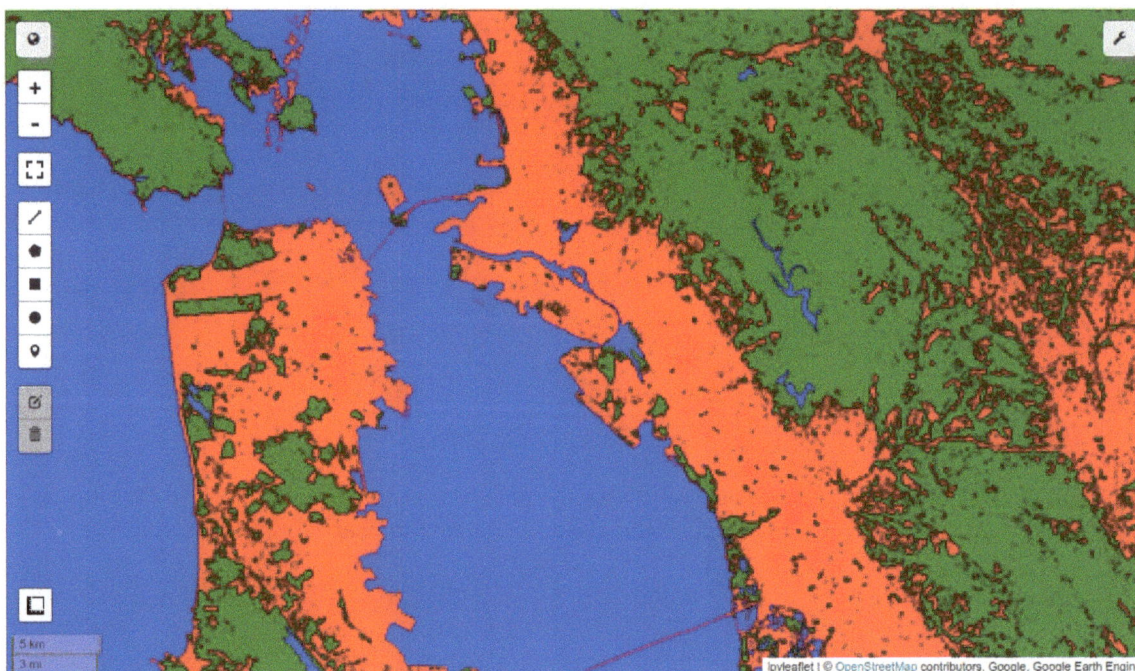

Figure 6.46: Random forest classification of a Landsat 8 image using a locally trained model.

Save trees to the cloud

Now we have the strings in a format that Earth Engine can use, we may want to save it for later use. There is a function to export a list of tree strings to a feature collection. First, specify the asset id where to save trees:

```
user_id = geemap.ee_user_id()
asset_id = user_id + "/random_forest_strings_test"
asset_id
```

The `ee_user_id()` function can be used to get your Earth Engine user id. Then, use `ml.export_trees_to_fc()` to convert the tree strings to a feature collection and export it to Earth Engine:

```
ml.export_trees_to_fc(trees, asset_id)
```

This export task might take a few minutes to complete, so wait a few minutes before moving on. You can also check the status of the export task using the Earth Engine JavaScript Code Editor. Once the export task is complete, you can load the asset as an `ee.FeatureCollection`. Then convert it to an `ee.Classifier` and use it to classify an image:

```
rf_fc = ee.FeatureCollection(asset_id)
another_classifier = ml.fc_to_classifier(rf_fc)
classified = image.select(feature_names).classify(another_classifier)
```

Save trees locally

In addition to saving the trees to the cloud, you can also save them locally as a CSV file using the `ml.trees_to_csv()` function. The CSV file can then be used to be converted to an `ee.Classifier` using the `ml.csv_to_classifier()` function. In this way, you don't need to train the model again.

```
out_csv = "trees.csv"
ml.trees_to_csv(trees, out_csv)
another_classifier = ml.csv_to_classifier(out_csv)
classified = image.select(feature_names).classify(another_classifier)
```

6.16   Sankey diagrams

The sankee[148] Python package allows users to visualize land cover change with interactive Sankey diagrams. Users can use existing land cover datasets or any classified images as input to create Sankey diagrams. Some existing land cover datasets supported by sankee include:

- NOAA Coastal Change Analysis Program (C-CAP) Land Cover[149]

- Canada Forested Ecosystem Land Cover[150]

- Copernicus CORINE Land Cover[151]

- Copernicus Global Land Cover[152]

- MODIS Global Land Cover[153]

- USFS Landscape Change Monitoring System (LCMS)[154]

- USGS National Land Cover Database[155]

The following example shows how to create a Sankey diagram for visualizing land cover change in the Mount St. Helens region using the USFS Landscape Change Monitoring System (LCMS)[154] dataset:

```
import sankee

sankee.datasets.LCMS_LC.sankify(
    years=[1990, 2000, 2010, 2020],
    region=ee.Geometry.Point([-122.192688, 46.25917]).buffer(2000),
    max_classes=3,
    title="Mount St. Helens Recovery",
)
```

The years parameter specifies the years for which to compute the land cover change. The region parameter specifies the region for which to compute the land cover change. The max_classes parameter specifies the maximum number of land cover classes to include in the Sankey diagram. The title parameter specifies the title of the Sankey diagram. Pass these parameters to the sankify() function to create a Sankey diagram. The output should look like the following Fig. 6.47.

Geemap has integrated the Sankee package, allowing users to visualize land cover change with interactive Sankey diagrams without coding. Simply create an interactive map and activate the Sankey plotting tool. A dialog box will appear to allow users to select the dataset and parameters for the Sankey diagram. Select "User-drawn ROI" from the **Region** dropdown list or select a pre-defined ROI (e.g., Las

---

[148]sankee: github.com/aazuspan/sankee

[149]NOAA Coastal Change Analysis Program (C-CAP) Land Cover: tiny.geemap.org/smrt

[150]Canada Forested Ecosystem Land Cover: tiny.geemap.org/gthz

[151]Copernicus CORINE Land Cover: tiny.geemap.org/dwbg

[152]Copernicus Global Land Cover: tiny.geemap.org/npkc

[153]MODIS Global Land Cover: tiny.geemap.org/ghtx

[154]USFS Landscape Change Monitoring System (LCMS): tiny.geemap.org/pauh

[155]USGS National Land Cover Database: tiny.geemap.org/ingx

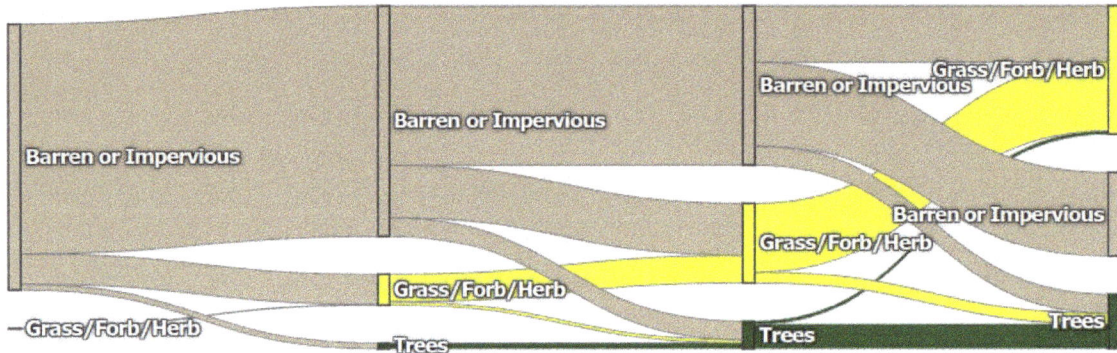

Figure 6.47: A Sankey diagram showing land cover change in Mount St. Helens in Skamania County, Washington.

Vegas). Choose an existing dataset from the **Dataset** dropdown list, such as the NLCD - National Land Cover Database. Adjust other parameters as needed and click the **Apply** button. The Sankey diagram will be created and displayed in the lower-right corner of the map. Click on the toolbar icons to resize the Sankey diagram (see Fig. 6.48).

```
Map = geemap.Map(height=650)
Map
```

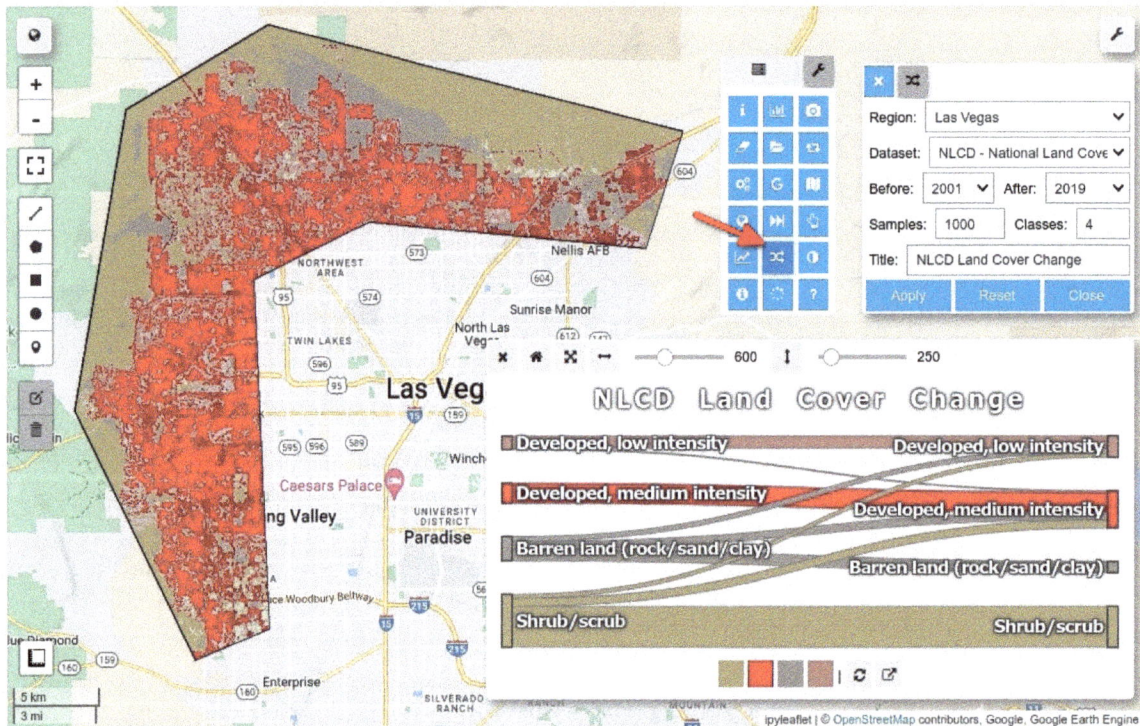

Figure 6.48: A Sankey diagram showing land cover change in Las Vegas, NV.

6.17 Summary

In this chapter, we learned how to use geemap to perform geospatial analysis with Earth Engine, such as computing zonal statistics, extracting pixel values, creating interactive charts, classifying images with machine learning algorithms (e.g., random forest), and visualizing land cover change with Sankey diagrams. In the next chapter, we will learn how to export data and analysis results from Earth Engine.

7. Exporting Earth Engine Data

7.1 Introduction

In the last chapter, we have covered how to analyze Earth Engine data. It is often desirable to export analysis results from Earth Engine. You can export images, tables, videos, and map tiles from Earth Engine. The exported data can be saved to a local drive on your computer, to your Google Drive account, to Google Cloud Storage or to a new Earth Engine asset.

In this chapter, we will focus on exporting various data from Earth Engine, including images, image collections, videos, maps, and image thumbnails. We will also learn how to use the high-volume endpoint for exporting thousands of image chips in just a few minutes. At the end of this chapter, you will be able to export data from Earth Engine to a variety of formats.

7.2 Technical requirements

To follow along with this chapter, you will need to have geemap and several optional dependencies installed. If you have already followed the Section 1.5 - *Installing geemap* section, then you should already have a conda environment with all the necessary packages installed. Otherwise, you can create a new conda environment and install pygis[30] with the following commands, which will automatically install geemap and all the required dependencies:

```
conda create -n gee python
conda activate gee
conda install -c conda-forge mamba
mamba install -c conda-forge pygis
```

Next, launch JupyterLab by typing the following commands in your terminal or Anaconda prompt:

```
jupyter lab
```

Alternatively, you can use geemap with a Google Colab cloud environment without installing anything on your local computer. Click 07_data_export.ipynb[156] to launch the notebook in Google Colab.

Once in Colab, you can uncomment the following line and run the cell to install pygis, which includes geemap and all the necessary dependencies:

```
# %pip install pygis
```

The installation process may take 2-3 minutes. Once pygis has been installed successfully, click the **RESTART RUNTIME** button that appears at the end of the installation log or go to the **Runtime** menu and select **Restart runtime**. After that, you can start coding.

To begin, import the necessary libraries that will be used in this chapter:

```
import ee
import geemap
```

---

[156]07_data_export.ipynb: tiny.geemap.org/ch07

Initialize the Earth Engine Python API:

```
geemap.ee_initialize()
```

If this is your first time running the code above, you will need to authenticate Earth Engine first. Follow the instructions in Section 1.7 - *Earth Engine authentication* to authenticate Earth Engine.

7.3   Exporting images

You can export images directly to a local drive, to a Google Drive, to Google Cloud Storage, to a new Earth Engine asset, or to a NumPy array. Let's use a Landsat imagery covering California as an example.

```
Map = geemap.Map()

image = ee.Image('LANDSAT/LC08/C02/T1_TOA/LC08_044034_20140318').select(
    ['B5', 'B4', 'B3']
)

vis_params = {'min': 0, 'max': 0.5, 'gamma': [0.95, 1.1, 1]}

Map.centerObject(image, 8)
Map.addLayer(image, vis_params, 'Landsat')
Map
```

You can export the entire image or a subset of the image by specifying a region of interest. Let's specify a bounding box as a region of interest that covers the San Francisco Bay Area (see Fig. 7.1).

```
region = ee.Geometry.BBox(-122.5955, 37.5339, -122.0982, 37.8252)
fc = ee.FeatureCollection(region)
style = {'color': 'ffff00ff', 'fillColor': '00000000'}
Map.addLayer(fc.style(**style), {}, 'ROI')
Map
```

To local drive

Small images can be exported directly to a local drive using the `ee_export_image()` function. Under the hood, the function uses the `ee.Image.getDownloadURL()` method to get the download URL of the image. The maximum request size is 32 MB, and the maximum grid dimension is 10000. You can specify parameters to control the export, such as the output `filename`, spatial resolution (`scale`), coordinate system (`crs`), and region of interest (`region`). The following example exports the Landsat imagery to a local drive:

```
geemap.ee_export_image(image, filename="landsat.tif", scale=30, region=region)
```

Within a few seconds, you should see a file named `landsat.tif` in your current working directory. If only the `scale` parameter is specified while the `crs` and `crs_transform` parameters are not specified, Earth Engine will calculate a `crs_tranform` parameter for you. However, this may result in an image that is shifted relative to another image with the same pixel size. Therefore, the exported images might not be aligned with the projection's origin. Let's retrieve the projection information of the image so that we can use the information to specify the `crs` and `crs_transform` parameters:

```
projection = image.select(0).projection().getInfo()
projection

{'type': 'Projection',
 'crs': 'EPSG:32610',
 'transform': [30, 0, 460785, 0, -30, 4264215]}
```

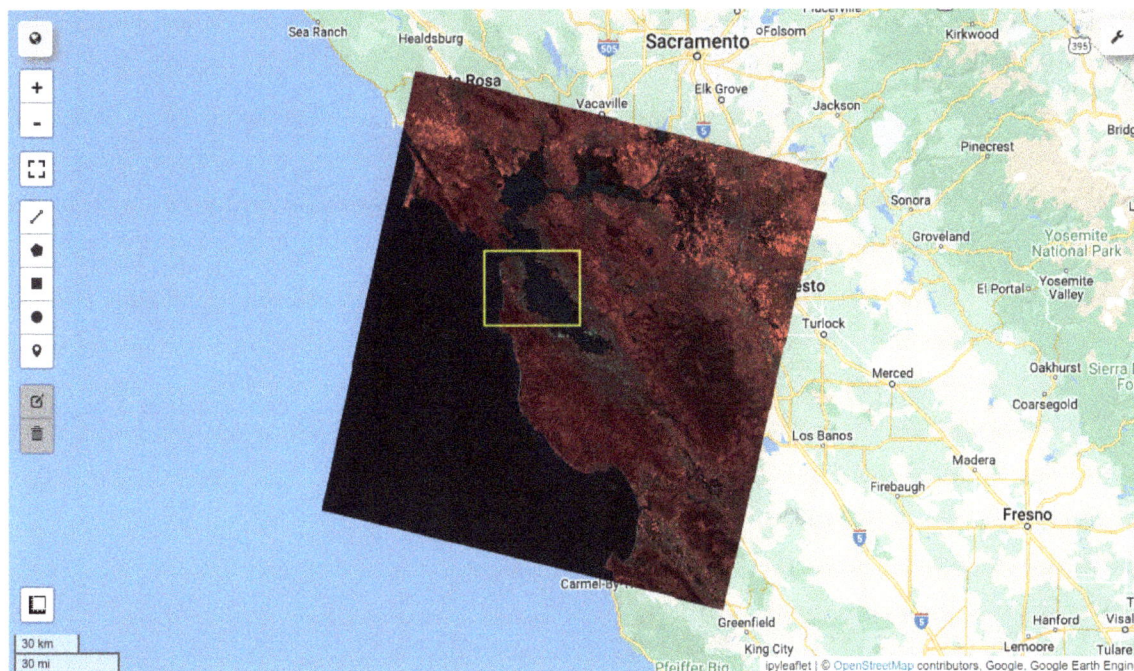

Figure 7.1: A Landsat imagery covering the San Francisco Bay Area.

The `crs` parameter is the coordinate reference system of the image. The `crs_transform` parameter is a list of parameters from an affine transformation matrix in row-major order [xScale, xShearing, xTranslation, yShearing, yScale, yTranslation]. An image's origin is defined by the xTranslation and yTranslation values, and the image's pixel size is defined by the xScale and yScale values.

```
crs = projection['crs']
crs_transform = projection['transform']
```

To align the exported image's pixels with a specific image, make sure to specify the `crs` and `crs_transform` parameters for full control of the grid when exporting the image.

```
geemap.ee_export_image(
    image,
    filename="landsat_crs.tif",
    crs=crs,
    crs_transform=crs_transform,
    region=region,
)
```

For images larger than 32 MB, the `ee_export_image()` function will throw an error. In that case, you can use the `download_ee_image()` function to export a large image to a local drive. This function requires the geedim[157] package, which can automatically split and download a large image as separate tiles, then re-assemble into a single GeoTIFF. Install the package using `pip install geedim` or `conda install -c conda-forge geedim`.

The following example downloads a Landsat image larger than 32 MB to a local drive:

```
geemap.download_ee_image(image, filename='landsat_full.tif', scale=60)
```

---

[157]geedim: github.com/dugalh/geedim

It should be noted that the `download_ee_image()` function is not very stable, and it might not work for all images. If you encounter an error, you can try to download the image to your Google Drive using the `ee_export_image_to_drive()` function. See the next section for more information.

Alternatively, you can divide a large image into smaller tiles and download each tile separately. The `fishnet()` function can create a fishnet of rectangular grids based on an input vector dataset. Specify the number of rows and columns in the fishnet, or the horizontal and vertical intervals (`h_interval` and `v_interval`) in degrees. The `delta` parameter is the buffer distance of the grid size to make sure that the fishnet fully covers the image. It should be a value between 0 and 1. The following example creates a fishnet of 4x4 rectangular grids based on the bounding box of the Landsat image and adds it to the map:

```python
fishnet = geemap.fishnet(image.geometry(), rows=4, cols=4, delta=0.5)
style = {'color': 'ffff00ff', 'fillColor': '00000000'}
Map.addLayer(fishnet.style(**style), {}, 'Fishnet')
Map
```

The resulting fishnet is shown in Fig. 7.2.

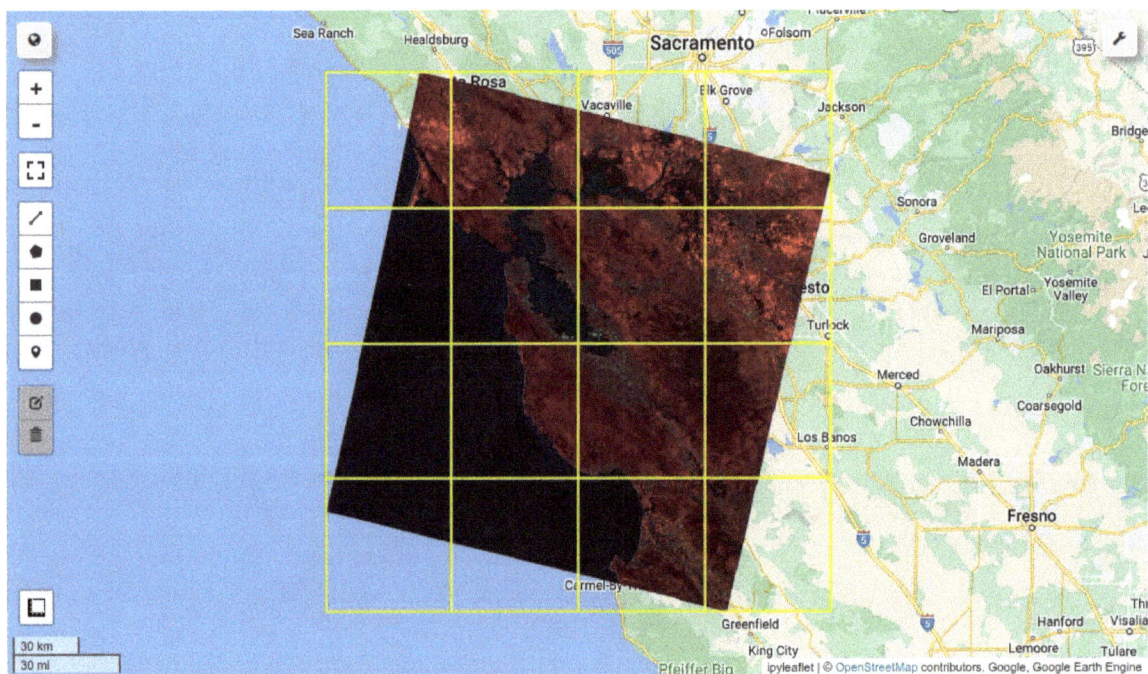

Figure 7.2: Creating a fishnet covering the Landsat image.

The fishnet created above can then be used to divide the image into smaller tiles and download each tile automatically to your local drive. Simply specify the output directory and scale of the tiles.

```python
out_dir = 'Downloads'
geemap.download_ee_image_tiles(
    image, fishnet, out_dir, prefix="landsat_", crs="EPSG:3857", scale=30
)
```

To Google Drive

As mentioned above, the `ee_export_image()` function can only export images smaller than 32 MB. For larger images, you can export them to your Google Drive account using the `ee_export_image_to_` `drive()` function. The default export format is GeoTIFF. The `description` parameter specifies the task description, which is also being used as the output filename without the file extension. Use the `folder` parameter to specify the Google Drive folder where the exported image will be stored. Other parameters are similar to the `ee_export_image()` function, such as `region`, `scale`, `crs`, and `crsTransform`. Note that the `ee_export_image_to_drive()` function only creates a task to export an Image to your Google Drive account. You need to check the status of the task in the JavaScript Code Editor Task Manager to see if the export is complete. If the export task fails, check the error message and revise the parameters accordingly.

The following example exports the Landsat image to your Google Drive account with a 30-m resolution:

```
geemap.ee_export_image_to_drive(
    image, description='landsat', folder='export', region=region, scale=30
)
```

Open the Earth Engine JavaScript Code Editor[18] and click the **Tasks** tab to check the status of the export task (see Fig. 7.3). If the export is successful, you should see a file named `landsat.tif` in your Google Drive folder. If the export fails, you can try to reduce the image size by increasing the `scale` parameter or by reducing the size of the region of interest.

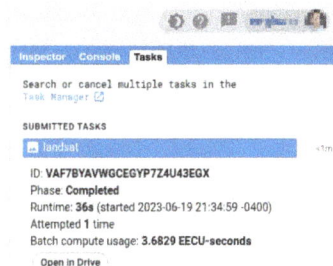

Figure 7.3: Earth Engine Task Manager.

To Asset

Besides exporting images to Google Drive, you can also export images to Earth Engine Assets using the `ee_export_image_to_asset` function. This is particularly useful if you need the exported images for further analysis in Earth Engine. Earth Engine assets in your Earth Engine account can be directly used by the Earth Engine APIs for computation and analysis just like any datasets in the Earth Engine Data Catalog. However, the Earth Engine APIs cannot use images in your Google Drive account unless you ingest them into your Earth Engine account.

To export an image to an Earth Engine asset, you need to specify parameter destination asset ID (`assetId`) of the asset where you want to store the image, e.g., `landsat_sfo`. Note that no file extension is allowed. Earth Engine assets do not have a file extension like GeoTIFF files. If the `assetId` only contains letters, numbers, and underscores, the asset will be stored in the root folder of your Earth Engine account. If the `assetId` contains a forward slash (/), the asset will be stored in the folder specified by the path. For example, if the `assetId` is `data/landsat_sfo`, the asset will be stored in the `data` folder with the name `landsat_sfo`. Other optional parameters include `region`, `scale`, `crs`, `crsTransform`, `maxPixels`, among others. Note that the `ee_export_image_to_asset` only creates a task to export an Image to an Earth Engine asset. You need to check the status of the task in the JavaScript Code Editor to see if the export is complete.

```
assetId = 'landsat_sfo'
geemap.ee_export_image_to_asset(
    image, description='landsat', assetId=assetId, region=region, scale=30
```

```
    )
```

To Cloud Storage

If your Google Drive has limited storage, you may consider exporting images to Google Cloud Storage using the ee_export_image_to_cloud_storage function. Google Cloud Storage is a paid service, so you will need to set up a project with billing enabled. After that, you can create a storage bucket. Similar to the folder parameter in the ee_export_image_to_drive function, you need to specify the bucket where the exported image will be stored. Other parameters are similar to the ee_export_image_to_drive function, such as region, scale, crs, crsTransform and maxPixels. Note that the ee_export_image_to_cloud_storage function only creates a task to export an Image to Google Cloud Storage. You need to check the status of the task in the JavaScript Code Editor to see if the export is complete.

```
bucket = 'your-bucket'
geemap.ee_export_image_to_cloud_storage(
    image, description='landsat', bucket=None, region=region, scale=30
)
```

To NumPy array

To extract a rectangular region of pixels from an image into a NumPy array per band, use the ee_to_numpy function. This is only suitable for extracting a small rectangular region of pixels from an image. The maximum number of pixels that can be extracted is 262,144, which is roughly the size of a 512x512 image. Specify a bounding box in the form of an ee.Geometry object. If you encounter an error saying that the number of pixels exceeds the maximum, you can try to reduce the bounding box size.

```
region = ee.Geometry.BBox(-122.5003, 37.7233, -122.3410, 37.8026)
rgb_img = geemap.ee_to_numpy(image, region=region)
```

Check the shape of the NumPy array. It is 3-band image with 298 rows and 471 columns, consisting of 140,358 pixels.

```
print(rgb_img.shape)
```

```
(298, 471, 3)
```

To display the NumPy array as an image, use the plt.imshow function from the matplotlib module.

```
import matplotlib.pyplot as plt

rgb_img_test = (255 * ((rgb_img[:, :, 0:3]) + 0.2)).astype('uint8')
plt.imshow(rgb_img_test)
plt.show()
```

The resulting image should look like Fig. 7.4.

Figure 7.4: Extract a rectangular region of pixels from an image into a NumPy array.

7.4 Exporting image collections

The functions introduced above for exporting images can only export one image at a time. To export an ImageCollection, use the ee_export_image_collection_to_drive(), ee_export_image_collection_to_drive(), or ee_export_image_collection_to_asset() function. Under the hood, the function loops through the images in the collection and exports each image individually using the export image functions introduced above.

The following code creates a NAIP image collection with 4-band images to be used in the subsequent examples:

```
point = ee.Geometry.Point(-99.2222, 46.7816)
collection = (
    ee.ImageCollection('USDA/NAIP/DOQQ')
    .filterBounds(point)
    .filterDate('2008-01-01', '2018-01-01')
    .filter(ee.Filter.listContains("system:band_names", "N"))
)
```

Check the number of images in the collection:

```
collection.aggregate_array('system:index').getInfo()
```

The output should look like this:

```
['m_4609915_sw_14_1_20090818',
 'm_4609915_sw_14_1_20100629',
 'm_4609915_sw_14_1_20120714',
 'm_4609915_sw_14_1_20140901',
 'm_4609915_sw_14_1_20150926',
 'm_4609915_sw_14_h_20160704',
 'm_4609915_sw_14_h_20170703']
```

To local drive

To export an ImageCollection object to your local drive, use the ee_export_image_collection() function. The out_dir parameter specifies the output directory where the images will be stored. The optional parameters include scale, crs, crs_tranform, region, dimensions, format, etc. The default format is ZIPPED_GEO_TIFF, GeoTIFF file(s) wrapped in a zip file. It can also be GEO_TIFF. The following example downloads the NAIP image collection to the user's Downloads folder:

```
out_dir = 'Downloads'
geemap.ee_export_image_collection(collection, out_dir=out_dir, scale=10)
```

To Google Drive

To export an ImageCollection object to your Google Drive, use the ee_export_image_collection_to_drive() function. The folder parameter specifies the folder where the images will be stored. Other optional parameters include scale, crs, crs_transform, region, dimensions, etc. The following example exports the NAIP image collection to your Google Drive:

```
geemap.ee_export_image_collection_to_drive(collection, folder='export', scale=10)
```

To Assets

To export an `ImageCollection` object to Earth Engine assets, use the `ee_export_image_collection_to_asset()` function. By default, the system index of each image in the collection is used as the asset ID. You can provide a list of asset IDs using the `assetIds` parameter. The other optional parameters include `scale`, `crs`, `crs_transform`, `region`, `dimensions`, etc. The following example exports the NAIP image collection to Earth Engine assets:

```
geemap.ee_export_image_collection_to_asset(collection, scale=10)
```

Open the Earth Engine JavaScript Code Editor Task Manager to see if the export is complete (see Fig. 7.5).

Figure 7.5: Exporting an ImageCollection to Earth Engine assets.

7.5 Exporting videos

To export an `ImageCollection` as a video, use the `ee_export_video_to_drive()` function. First, let's create a Landsat `ImageCollection`, which contains Landsat 5 images acquired between 1991 and 2011:

```
collection = (
    ee.ImageCollection('LANDSAT/LT05/C01/T1_TOA')
    .filter(ee.Filter.eq('WRS_PATH', 44))
    .filter(ee.Filter.eq('WRS_ROW', 34))
    .filter(ee.Filter.lt('CLOUD_COVER', 30))
    .filterDate('1991-01-01', '2011-12-30')
    .select(['B4', 'B3', 'B2'])
    .map(lambda img: img.multiply(512).uint8())
)
region = ee.Geometry.Rectangle([-122.7286, 37.6325, -122.0241, 37.9592])
```

Next, use the `ee_export_video_to_drive()` function to export the `ImageCollection` object as a video to Google Drive. The `folder` parameter specifies the output directory where the video will be stored. You can configure other parameters, such as `framesPerSecond`, `dimensions`, `region`, `scale`, `crs`, `crsTransform`, etc. to customize the export.

```
geemap.ee_export_video_to_drive(
    collection, folder='export', framesPerSecond=12, dimensions=720, region=region
)
```

7.6 Exporting image thumbnails

To download an image thumbnail, use the `get_image_thumbnail()` method, which should be much faster than the `ee_export_image()` function due to the much smaller file size being downloaded. The following example demonstrates how to download an image thumbnail.

First, create a Landsat `ImageCollection` object with 10 images with the least cloud coverage:

```
roi = ee.Geometry.Point([-122.44, 37.75])
collection = (
    ee.ImageCollection('LANDSAT/LC08/C02/T1_TOA')
    .filterBounds(roi)
    .sort("CLOUD_COVER")
    .limit(10)
)

image = collection.first()
```

Create an interactive map and add the first image to the map:

```
Map = geemap.Map()

vis_params = {
    'bands': ['B5', 'B4', 'B3'],
    'min': 0,
    'max': 0.3,
    'gamma': [0.95, 1.1, 1],
}

Map.addLayer(image, vis_params, "LANDSAT 8")
Map.setCenter(-122.44, 37.75, 8)
Map
```

The `region` parameter specifies the region of interest for the thumbnail. If the `region` parameter is not specified, the thumbnail will be the entire image. The `dimensions` parameter takes a number or pair of numbers in the format WIDTHxHEIGHT. If only one number is passed, it is used as the maximum, and the other dimension is computed by proportional scaling. The following example downloads the thumbnail for the first image in the collection:

```
out_img = 'landsat.jpg'
region = ee.Geometry.BBox(-122.5955, 37.5339, -122.0982, 37.8252)
geemap.get_image_thumbnail(image, out_img, vis_params, dimensions=1000, region=region)
```

Let's show the image using the `show_image()` function:

```
geemap.show_image(out_img)
```

The image thumbnail is shown in Fig. 7.6.

To download image thumbnails of all images in the collection, use the get_image_collection_thumbnails() function:

```
out_dir = 'Downloads'
geemap.get_image_collection_thumbnails(
    collection,
    out_dir,
    vis_params,
    dimensions=1000,
    region=region,
)
```

Figure 7.6: A thumbnail of a Landsat imagery.

7.7 Exporting feature collections

You can export an Earth Engine `FeatureCollection` object to a local drive, to a Google Drive, to Google Cloud Storage, or to a new Earth Engine asset. The format of the exported dataset can be in the `CSV`, `SHP` (shapefile), `GeoJSON`, `KML`, `KMZ`, or `TFRecord` formats.

Let's use the Large-Scale International Boundary (LSIB[158]) dataset as an example and select European countries as the region of interest (see Fig. 7.7).

```
Map = geemap.Map()
fc = ee.FeatureCollection('USDOS/LSIB_SIMPLE/2017').filter(
    ee.Filter.eq('wld_rgn', 'Europe')
)

Map.addLayer(fc, {}, "Europe")
Map.centerObject(fc, 3)
Map
```

The map of the European countries is shown in Fig. 7.7.

To local drive

To export a `FeatureCollection` object to a local drive, use the `ee_export_vector()` function. Make sure to specify the output file format in the `filename` parameter. Alternatively, you can use one of the following functions to export a `FeatureCollection` object to a specific file format:

- `ee_to_shp()`

---

[158]LSIB: tiny.geemap.org/qkuu

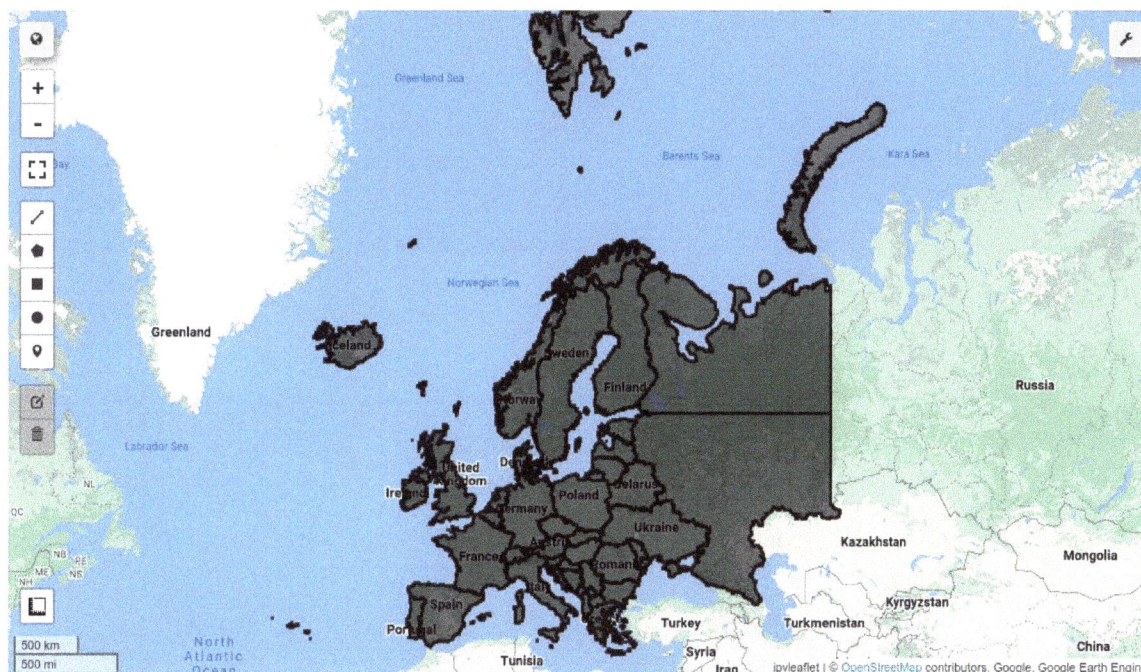

Figure 7.7: European countries in the Large Scale International Boundary (LSIB) dataset.

- `ee_to_geojson()`
- `ee_to_csv()`
- `ee_to_gdf()`
- `ee_to_df()`

To export a `FeatureCollection` object to a shapefile, use the `ee_to_shp()` function. The features in the `FeatureCollection` object must have the same geometry type and projection. Column names have a limit of 10 characters, and longer column names will be truncated to 10 characters. Specify the output `filename` and a list of attributes to export in the `selectors` parameter. If `selectors` is not specified, all attributes will be exported. The following example exports the `FeatureCollection` object to a shapefile:

```
geemap.ee_to_shp(fc, filename='europe.shp', selectors=None)
```

Alternatively, you can export a `FeatureCollection` object to a shapefile using the `ee_export_vector()` function. Make sure to add the `.shp` suffix to the end of the `filename` parameter:

```
geemap.ee_export_vector(fc, filename='europe2.shp')
```

To export a `FeatureCollection` object to a GeoJSON file, use the `ee_to_geojson()` function. Specify the output `filename` where the GeoJSON file will be saved. If `filename` is not specified, the function will return the GeoJSON object as a dictionary.

The following example exports the `FeatureCollection` object to a GeoJSON file in the current directory:

```
geemap.ee_to_geojson(fc, filename='europe.geojson')
```

To export a `FeatureCollection` object to a CSV file, use the `ee_to_csv()` function. Note that the output CSV file is essentially the attribute table of the vector dataset without geometry information. Specify the output `filename` and a list of attributes to export in the `selectors` parameter. If `selectors` is not specified, all attributes will be exported.

The following example exports the `FeatureCollection` object to a CSV file in the current directory:

```
geemap.ee_to_csv(fc, filename='europe.csv')
```

If you need the exported data for further processing, the `ee_to_gdf()` and `ee_to_df()` functions can be used to convert the `FeatureCollection` object to a GeoPandas `GeoDataFrame` or Pandas `DataFrame`, respectively.

To convert a `FeatureCollection` object to a GeoPandas `GeoDataFrame`, use the `ee_to_gdf()` function:

```
gdf = geemap.ee_to_gdf(fc)
gdf
```

The output should look like Fig. 7.8.

	geometry	abbreviati	country_co	country_na	wld_rgn
0	MULTIPOLYGON ((((31.57599 52.46329, 31.58122 52...	Rus.	RS	Russia	Europe
1	MULTIPOLYGON ((((29.68999 45.20052, 29.69058 45...	Ukr.	UP	Ukraine	Europe
2	POLYGON ((22.67614 44.21546, 22.68241 44.20975...	Rom.	RO	Romania	Europe
3	POLYGON ((28.21134 45.46691, 28.25694 45.51638...	Mol.	MD	Moldova	Europe
4	MULTIPOLYGON ((((62.08986 81.51786, 62.09978 81...	Rus.	RS	Russia	Europe
5	MULTIPOLYGON ((((22.00997 60.14789, 22.01161 60...	Fin.	FI	Finland	Europe
6	MULTIPOLYGON ((((11.75161 65.01125, 11.75656 65...	Nor.	NO	Norway	Europe

Figure 7.8: Converting a FeatureCollection to a GeoPandas GeoDataFrame.

To convert a `FeatureCollection` object to a Pandas `DataFrame`, use the `ee_to_df()` function:

```
df = geemap.ee_to_df(fc)
df
```

The output should look like Fig. 7.9.

	wld_rgn	country_na	abbreviati	country_co
0	Europe	Russia	Rus.	RS
1	Europe	Ukraine	Ukr.	UP
2	Europe	Romania	Rom.	RO
3	Europe	Moldova	Mol.	MD
4	Europe	Russia	Rus.	RS
5	Europe	Finland	Fin.	FI

Figure 7.9: Converting a 'FeatureCollection' to a Pandas 'DataFrame'.

To Google Drive

To export a `FeatureCollection` object to your Google Drive, use the `ee_export_vector_to_drive()` function. The file format can be CSV, SHP, GeoJSON, KML, KMZ, or TFRecord. Note that the function only creates a task to export a `FeatureCollection` object to Google Drive. You need to check the task status in the JavaScript Code Editor Task Manager to see if the export is complete.

```
geemap.ee_export_vector_to_drive(
    fc, description="europe", fileFormat='SHP', folder="export"
)
```

To Asset

To export a `FeatureCollection` object to an Earth Engine asset, use the `ee_export_vector_to_asset()` function. The file format can be CSV, SHP, GeoJSON, KML, KMZ, or TFRecord. Specify the `assetId` of the asset where the data will be saved. If the `assetId` parameter only contains letters, numbers, and underscores, the asset will be stored in the root folder of your Earth Engine assets. If the `assetId` contains a forward slash (/), the asset will be stored in the folder specified by the path.

For example, if the `assetId` is `data/europe`, the asset will be stored in the `data` folder with the name `europe`. Note that the function only creates a task to export a `FeatureCollection` object to an Earth Engine asset. You need to check the task status in the JavaScript Code Editor Task Manager to see if the export is complete. The following example exports the `FeatureCollection` object to an Earth Engine asset:

```
geemap.ee_export_vector_to_asset(fc, description='Exporting Europe', assetId='europe')
```

7.8 Exporting maps

To export Earth Engine layers along with the interactive map as a web page, use the `Map.to_html()` method. This can be useful sharing maps with others. The HTML file can be easily hosted on a web server.

The following example creates an interactive map and adds the SRTM DEM to it:

```
Map = geemap.Map()
image = ee.Image('USGS/SRTMGL1_003')
vis_params = {
    'min': 0,
    'max': 4000,
    'palette': ['006633', 'E5FFCC', '662A00', 'D8D8D8', 'F5F5F5'],
}
Map.addLayer(image, vis_params, 'SRTM DEM', True)
Map
```

Use the `Map.to_html()` function to export the map as an HTML file. Specify the output `filename` where the HTML file will be saved. If `filename` is not specified, the function will return the HTML as a string. Other parameters such as `title`, `width`, and `height` can be customized as needed. Note that the Earth Engine map tiles expire after a few hours. Once the tiles expire, the Earth Engine layers will not be displayed on the map. You will need to re-run the following code to export the map again. It is also important to note that exporting the map to HTML will lose the bidirectional functionality. The geemap toolbar will not appear in the exported map.

```
Map.to_html(
    filename="mymap.html", title="Earth Engine Map", width='100%', height='800px'
)
```

7.9 Using the high-volume endpoint

The functions introduced above for exporting vector and raster data use the standard Earth Engine API endpoint. Each user is limited to a small number of concurrently running tasks with the standard endpoint. In contrast, the high-volume API endpoint[159] is designed to handle more requests in parallel than the standard endpoint, with the tradeoff of higher average latency and reduced caching. This endpoint is best suited for making automated requests, for example, using the `getThumbUrl()` and `getDownloadURL()` functions. The high-volume endpoint does not affect the number of concurrent tasks you can run.

The following example illustrates how to use the high-volume endpoint to export hundreds of smaller images in parallel. The source code was adopted from the Medium post - Fast(er) Downloads[160] by Noel Gorelick. Due to the limitation of the multiprocessing[161] package, the source code can only be run in the top-level. It could not be implemented as a geemap. Therefore, the source code is a bit longer than other examples.

First, import the libraries:

```
import ee
import geemap
import logging
import multiprocessing
import os
import requests
import shutil
from retry import retry
```

Initialize Earth Engine by passing an `opt_url` parameter to use the high-volume endpoint:

```
ee.Initialize(opt_url='https://earthengine-highvolume.googleapis.com')
```

Define the Region of Interest (ROI) for downloading images. You can use the drawing tools on the map to draw an ROI, then you can use the `Map.user_roi` property to retrieve the geometry. Alternatively, you can define the ROI as an `ee.Geometry` object as shown below:

```
region = Map.user_roi

if region is None:
    region = ee.Geometry.Polygon(
        [
            [
                [-122.513695, 37.707998],
                [-122.513695, 37.804359],
                [-122.371902, 37.804359],
                [-122.371902, 37.707998],
                [-122.513695, 37.707998],
            ]
        ],
        None,
        False,
    )
```

Next, specify the image source to download. The following example uses the 1-m NAIP imagery[162] covering the continental U.S. It mosaics all image tiles covering the ROI into a single image:

```
image = (
    ee.ImageCollection('USDA/NAIP/DOQQ')
    .filterBounds(region)
    .filterDate('2020', '2021')
    .mosaic()
    .clip(region)
    .select('N', 'R', 'G')
)
```

---

[159]high-volume API endpoint: tiny.geemap.org/hjqe

[160]Fast(er) Downloads: tiny.geemap.org/sjpd

[161]multiprocessing: tiny.geemap.org/rjmk

[162]NAIP imagery: tiny.geemap.org/msiq

Create an interactive map and add the image to it (see Fig. 7.10).

```
Map = geemap.Map()
Map.addLayer(image, {}, "Image")
Map.addLayer(region, {}, "ROI", False)
Map.centerObject(region, 12)
Map
```

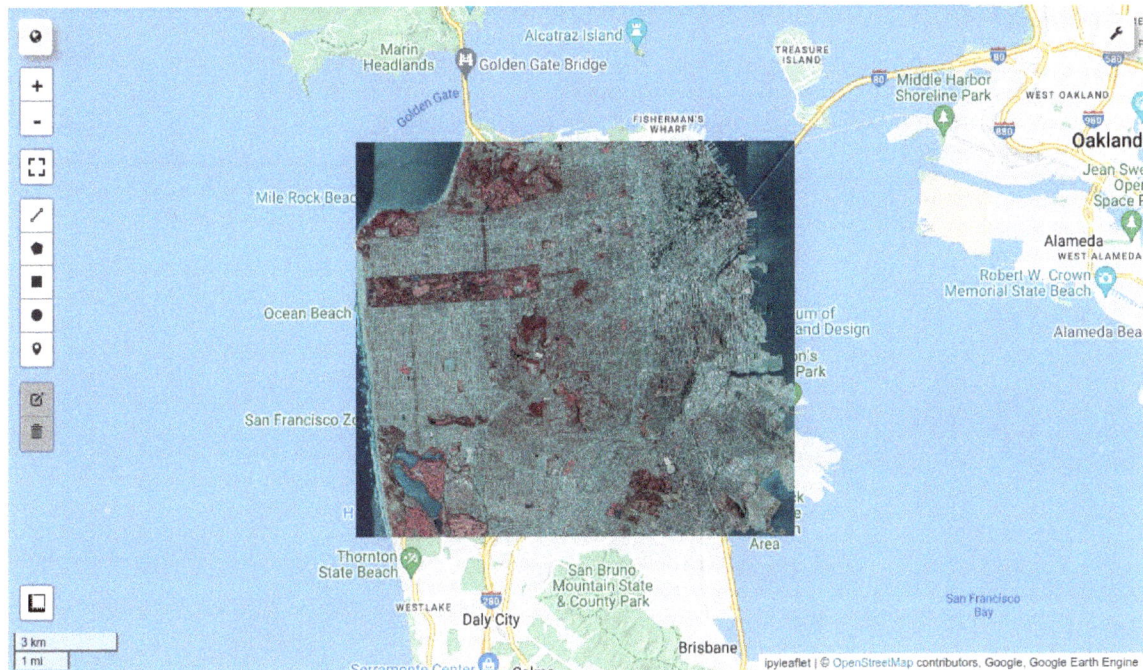

Figure 7.10: A NAIP imagery covering the San Francisco Bay Area.

Define a dictionary of parameters for the image export. The `count` parameter specifies the number of images to download. The `dimensions` parameter specifies the image size. The `format` parameter specifies the image format, which can be `png`, `jpg`, `GEO_TIFF`, etc.

```
out_dir = 'Downloads'
params = {
    'count': 1000,  # How many image chips to export
    'buffer': 127,  # The buffer distance (m) around each point
    'scale': 100,  # The scale to do stratified sampling
    'seed': 1,  # A randomization seed to use for subsampling.
    'dimensions': '256x256',  # The dimension of each image chip
    'format': "png",  # The output image format, can be png, jpg, ZIPPED_GEO_TIFF, GEO_TIFF, NPY
    'prefix': 'tile_',  # The filename prefix
    'processes': 25,  # How many processes to used for parallel processing
    'out_dir': out_dir,  # The output directory. Default to the current working directly
}
```

Next, define a `getRequests()` function for generating 1000 points using the stratified random sampling, which requires a `Class` band. It is the name of the band containing the classes to use for stratification. If unspecified, the first band of the input image is used. Therefore, we have to add a new band with a constant value (e.g., 1) to the image. The result of the `getRequests()` function returns a list of dictionaries containing points. You can save the generated points to a Map attribute (e.g., `Map.data`) to use them later

```
def getRequests():
```

```python
    img = ee.Image(1).rename("Class").addBands(image)
    points = img.stratifiedSample(
        numPoints=params['count'],
        region=region,
        scale=params['scale'],
        seed=params['seed'],
        geometries=True,
    )
    Map.data = points
    return points.aggregate_array('.geo').getInfo()
```

The getResult() function then takes one of those points and generates an image centered on that
location, which is then downloaded as a PNG image and saved to a file. This function uses the
getThumbURL() function to select the pixels, however, you could also use the getDownloadURL() function
if you wanted the output to be in GeoTIFF or NumPy format. Note that both functions are limited to
32MB of data per request.

```python
@retry(tries=10, delay=1, backoff=2)
def getResult(index, point):
    point = ee.Geometry.Point(point['coordinates'])
    region = point.buffer(params['buffer']).bounds()

    if params['format'] in ['png', 'jpg']:
        url = image.getThumbURL(
            {
                'region': region,
                'dimensions': params['dimensions'],
                'format': params['format'],
            }
        )
    else:
        url = image.getDownloadURL(
            {
                'region': region,
                'dimensions': params['dimensions'],
                'format': params['format'],
            }
        )

    if params['format'] == "GEO_TIFF":
        ext = 'tif'
    else:
        ext = params['format']

    r = requests.get(url, stream=True)
    if r.status_code != 200:
        r.raise_for_status()

    out_dir = os.path.abspath(params['out_dir'])
    basename = str(index).zfill(len(str(params['count'])))
    filename = f"{out_dir}/{params['prefix']}{basename}.{ext}"
    with open(filename, 'wb') as out_file:
        shutil.copyfileobj(r.raw, out_file)
    print("Done: ", basename)
```

All the parameters and functions defined above, you can now initialize the downloading process as
follows. Note that the parallel processing below might not work on Windows machines.

```python
%%time
logging.basicConfig()
items = getRequests()

pool = multiprocessing.Pool(params['processes'])
```

```
pool.starmap(getResult, enumerate(items))

pool.close()
```

Using 25 parallel processes, it took about one minute to download and save 1,000 images. Note that earlier we saved the generated points to the Map.data attribute, which can be retrieved and plotted on the map (see Fig. 7.11).

```
Map.addLayer(Map.data, {}, "Sample points")
Map
```

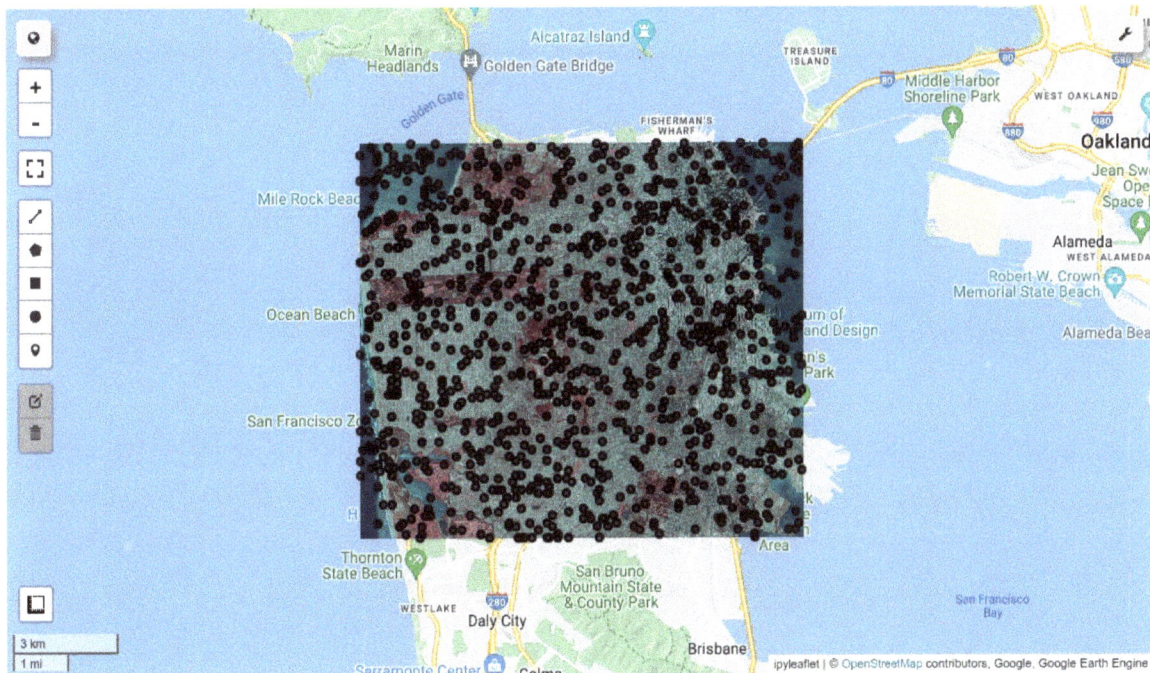

Figure 7.11: 1000 random points generated from the NAIP imagery.

To export the randomly generated points corresponding to the downloaded image tiles, use the ee_to_shp() function:

```
geemap.ee_to_shp(Map.data, filename='points.shp')
```

7.10 Summary

Exporting data from Earth Engine is a common task for all Earth Engine users. In this chapter, we learned how to export both vector and raster data from Earth Engine, such as images, image collections, image thumbnails, feature collections, videos, and maps. In the next chapter, we will learn how to make publication-quality maps with the cartoee module.

8. Making Maps with Cartoee

8.1 Introduction

Unlike desktop GIS with rich functionality for creating high-quality maps to be used in journal publications and project reports, cloud computing platforms (Earth Engine, Planetary Computer, etc.) largely lack interactive functionality for creating publication-quality maps with essential map elements (e.g., legend, scale bar, north arrow).

This chapter introduces the geemap `cartoee` module that can be used to create publication-quality maps from Earth Engine datasets and processes results without having to download data. The `cartoee` module does this by requesting PNG images from Earth Engine datasets and processing the results (which are usually good enough for visualization) while `cartopy` is used to create the plots. Utility functions are available to create plot aesthetics such as grid lines or color bars. The cartoee[163] package was originally developed by Kel Markert[164] , and it has been integrated into the geemap package.

In this chapter, we will explore how to plot Earth Engine vector and raster data on the map, and how to add legends, scale bars, and north arrows to the map.

8.2 Technical requirements

To follow this chapter, you will need to install geemap and several optional dependencies. If you have followed Section 1.5 - *Installing geemap*, then you should already have a conda environment with these packages installed. Otherwise, you need to create a new conda environment and install pygis[30] with the following commands, which will automatically install geemap and all the necessary dependencies:

```
conda create -n gee python
conda activate gee
conda install -c conda-forge mamba
mamba install -c conda-forge geemap pygis
mamba install -c conda-forge cartopy
```

Next, launch JupyterLab by typing the following command in the terminal or Anaconda prompt:

```
jupyter lab
```

Alternatively, you can use geemap with a Google Colab cloud environment without installing anything on your computer. Click 08_cartoee.ipynb[165] to launch the notebook in Google Colab.

Next, press **Ctrl + /** to uncomment the following line and press **Shift + Enter** to install pygis[30] , which includes geemap and all the necessary dependencies:

```
# %pip install pygis
```

```
# %pip install cartopy
```

---

[163]cartoee: github.com/KMarkert/cartoee
[164]Kel Markert: github.com/KMarkert
[165]08_cartoee.ipynb: tiny.geemap.org/ch08

The installation will take 2-3 minutes. After pygis has been installed successfully, click the **RESTART RUNTIME** button appearing at the end of the installation log or click on the menu **Runtime -> Restart runtime**. Then you can start coding.

Import the libraries that will be used in this chapter:

```
import ee
import geemap
```

Initialize the Earth Engine Python API:

```
geemap.ee_initialize()
```

If this is your first time running the code above, you will need to authenticate Earth Engine first. Follow the instructions in Section 1.7 - *Earth Engine authentication* to authenticate Earth Engine.

```
from geemap import cartoee
import matplotlib.pyplot as plt
```

8.3 Plotting single-band images

In this first example, we will explore how to plot a single-band Earth Engine image on a map. We will use the SRTM Global Digital Elevation Model (DEM)[166] dataset, which is a 30-meter resolution elevation model of the Earth's surface. The DEM dataset is available in the Earth Engine Data Catalog as an Earth Engine Image with the asset ID `NASA/NASADEM_HGT/001`. The DEM dataset is a single-band image with elevation values in meters. Let's specify the region of interest (ROI) and visualization parameters for the DEM:

```
srtm = ee.Image("CGIAR/SRTM90_V4")

# define bounding box [east, south, west, north] to request data
region = [180, -60, -180, 85]
vis = {'min': 0, 'max': 3000}
```

Next, create a matplotlib figure and plot the DEM on the map through the `cartoee.get_map()` function. Optionally, you can also add a color bar and grid lines to the map. Lastly, use the `plt.show()` function to display the map:

```
fig = plt.figure(figsize=(15, 9))

# use cartoee to get a map
ax = cartoee.get_map(srtm, region=region, vis_params=vis)

# add a color bar to the map using the visualization params we passed to the map
cartoee.add_colorbar(ax, vis, loc="bottom", label="Elevation (m)", orientation="horizontal")

# add grid lines to the map at a specified interval
cartoee.add_gridlines(ax, interval=[60, 30], linestyle=":")

# add coastlines using the cartopy api
ax.coastlines(color="red")

plt.show()
```

The resulting map is shown in Fig. 8.1.

This is a decent map with a minimal amount of code, but we can also easily use matplotlib colormaps

---

[166]SRTM Global Digital Elevation Model (DEM): tiny.geemap.org/rtra

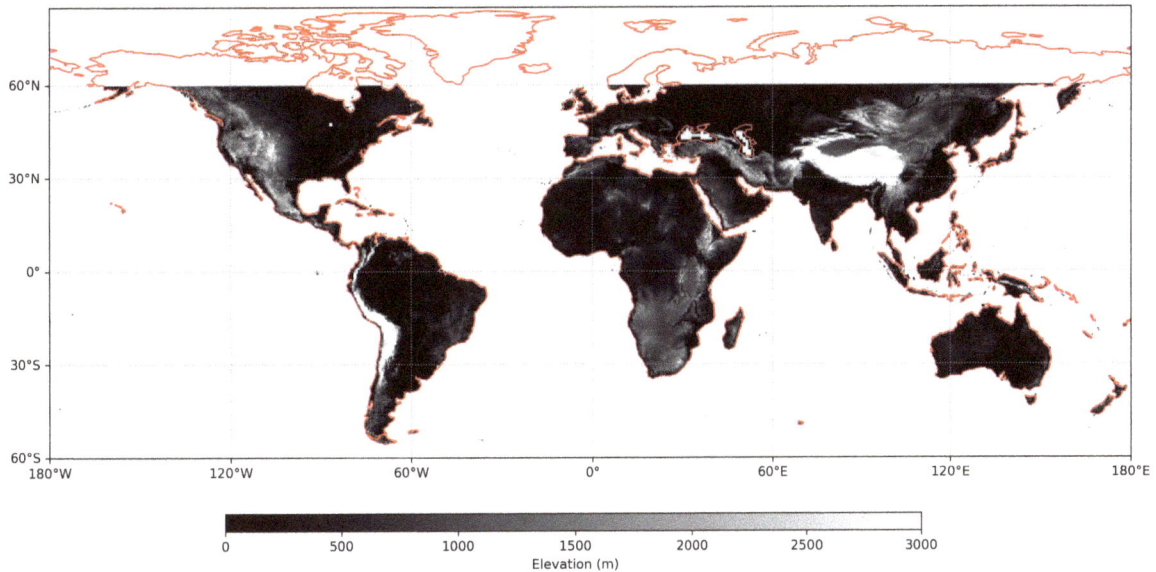

Figure 8.1: A grayscale map of the SRTM global elevation data created with cartoee.

to visualize Earth Engine datasets to add more color. Here we add a `cmap` keyword to the `cartoee.get_map()` and `cartoee.add_colorbar()` functions:

```
fig = plt.figure(figsize=(15, 7))

cmap = "terrain"

ax = cartoee.get_map(srtm, region=region, vis_params=vis, cmap=cmap)
cartoee.add_colorbar(
    ax, vis, cmap=cmap, loc="right", label="Elevation (m)", orientation="vertical"
)

cartoee.add_gridlines(ax, interval=[60, 30], linestyle="--")
ax.coastlines(color="red")
ax.set_title(label='Global Elevation Map', fontsize=15)

plt.show()
```

The resulting map is shown in Fig. 8.2.

Note that we set `orientation="vertical"` and `loc="right"` for the colorbar to make the layout more compact. We also added a title to the map using the `ax.set_title()` function.

To save the map as an image file, use the `cartoee.savefig()` function:

```
cartoee.savefig(fig, fname="srtm.jpg", dpi=300, bbox_inches='tight')
```

Set the `bbox_inches='tight'` keyword to remove the white space around the map. Customize the map `dpi` parameter to adjust the resolution of the saved image as needed.

8.4 Plotting multi-band images

Cartoee can also plot multi-band images. Here is an example of plotting the false-color composite of a Landsat scene (see Fig. 8.3).

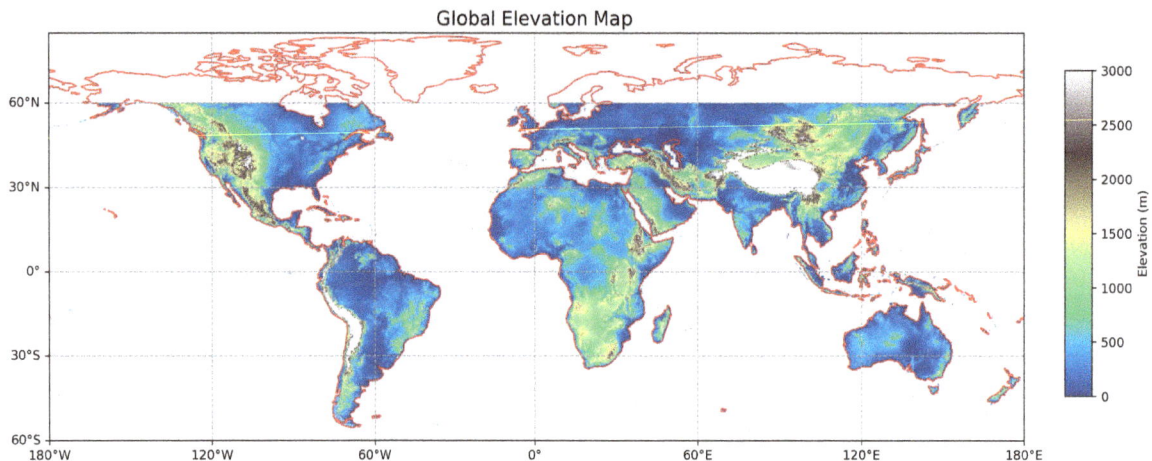

Figure 8.2: A color map of the SRTM global elevation data created with cartoee.

```
image = ee.Image('LANDSAT/LC08/C01/T1_SR/LC08_044034_20140318')
vis = {"bands": ['B5', 'B4', 'B3'], "min": 0, "max": 5000, "gamma": 1.3}

fig = plt.figure(figsize=(15, 10))

ax = cartoee.get_map(image, vis_params=vis)
cartoee.pad_view(ax)
cartoee.add_gridlines(ax, interval=0.5, xtick_rotation=0, linestyle=":")
ax.coastlines(color="yellow")

plt.show()
```

By default, if the `region` parameter of the `cartoee.get_map()` function is not specified, the whole extent of the image will be plotted as seen in the previous Landsat example. We can also zoom to a specific region of an image by defining the `region` parameter in the format of [east, south, west, north] to plot. Here is an example of plotting a zoomed-in region of the Landsat scene around the San Francisco Bay Area:

```
fig = plt.figure(figsize=(15, 10))

region = [-121.8025, 37.3458, -122.6265, 37.9178]
ax = cartoee.get_map(image, vis_params=vis, region=region)
cartoee.add_gridlines(ax, interval=0.15, xtick_rotation=0, linestyle=":")
ax.coastlines(color="yellow")

plt.show()
```

8.5 Adding north arrows and scale bars

To add a north arrow to the map, use the `cartoee.add_north_arrow()` function. There are two functions for adding a scale bar: `cartoee.add_scale_bar()` and `cartoee.add_scale_bar_lite()`. In the following example, we will use the `cartoee.add_scale_bar_lite()` function. The `cartoee.add_scale_bar()` function is more customizable, which will be covered in the next example. Both the north arrow and color bar are customizable. You can change the location, size, color, and more.

Here is an example of adding a north arrow and a simple scale bar to the zoomed-in map of the San Francisco Bay Area (see Fig. 8.4).

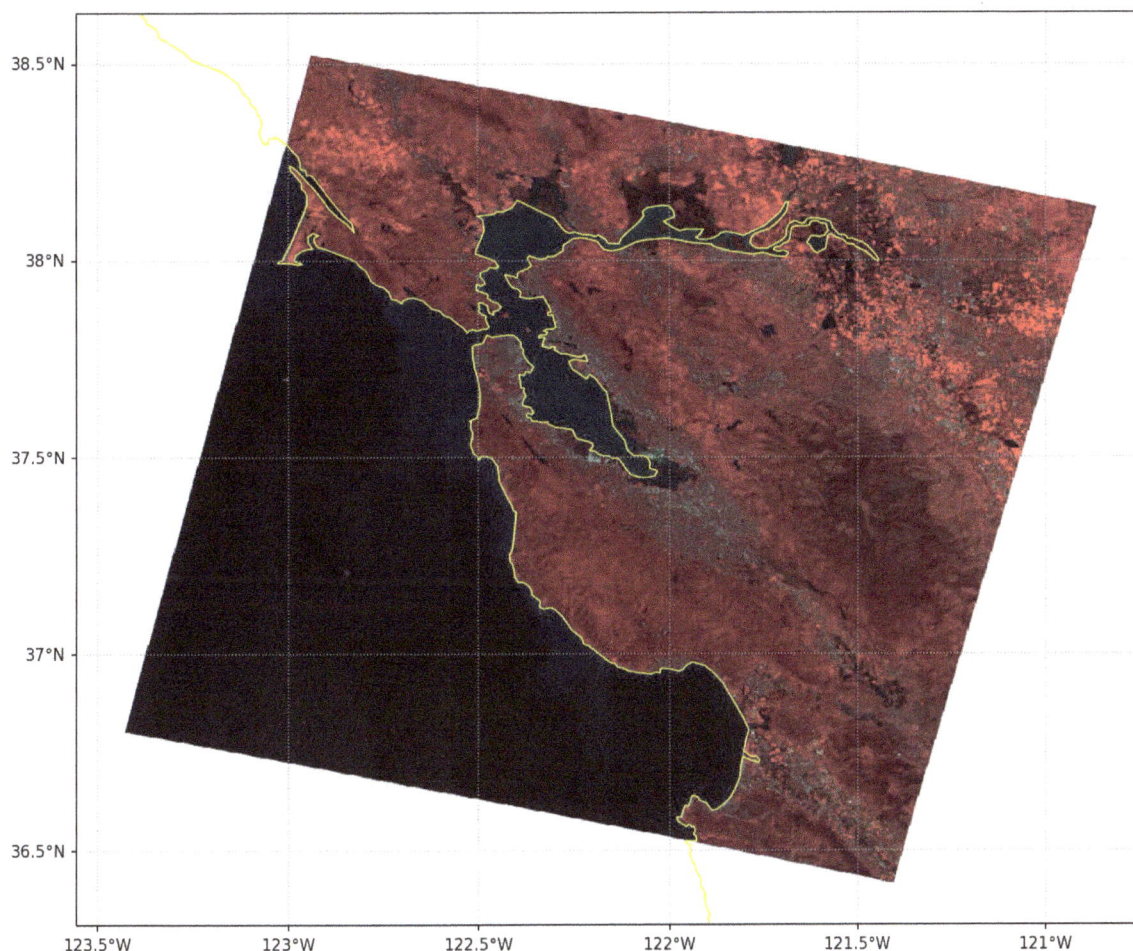

Figure 8.3: False-color composite of a Landsat scene.

```
fig = plt.figure(figsize=(15, 10))

region = [-121.8025, 37.3458, -122.6265, 37.9178]
ax = cartoee.get_map(image, vis_params=vis, region=region)
cartoee.add_gridlines(ax, interval=0.15, xtick_rotation=0, linestyle=":")
ax.coastlines(color="yellow")

cartoee.add_north_arrow(
    ax, text="N", xy=(0.05, 0.25), text_color="white", arrow_color="white", fontsize=20
)
cartoee.add_scale_bar_lite(
    ax, length=10, xy=(0.1, 0.05), fontsize=20, color="white", unit="km"
)
ax.set_title(label='Landsat False Color Composite (Band 5/4/3)', fontsize=15)

plt.show()
```

The `cartoee.add_scale_bar_lite()` function adds a simple scale bar as a straight line with a distance label on it. To add a better-looking scale bar, use the `cartoee.add_scale_bar()` function:

```
fig = plt.figure(figsize=(15, 10))
```

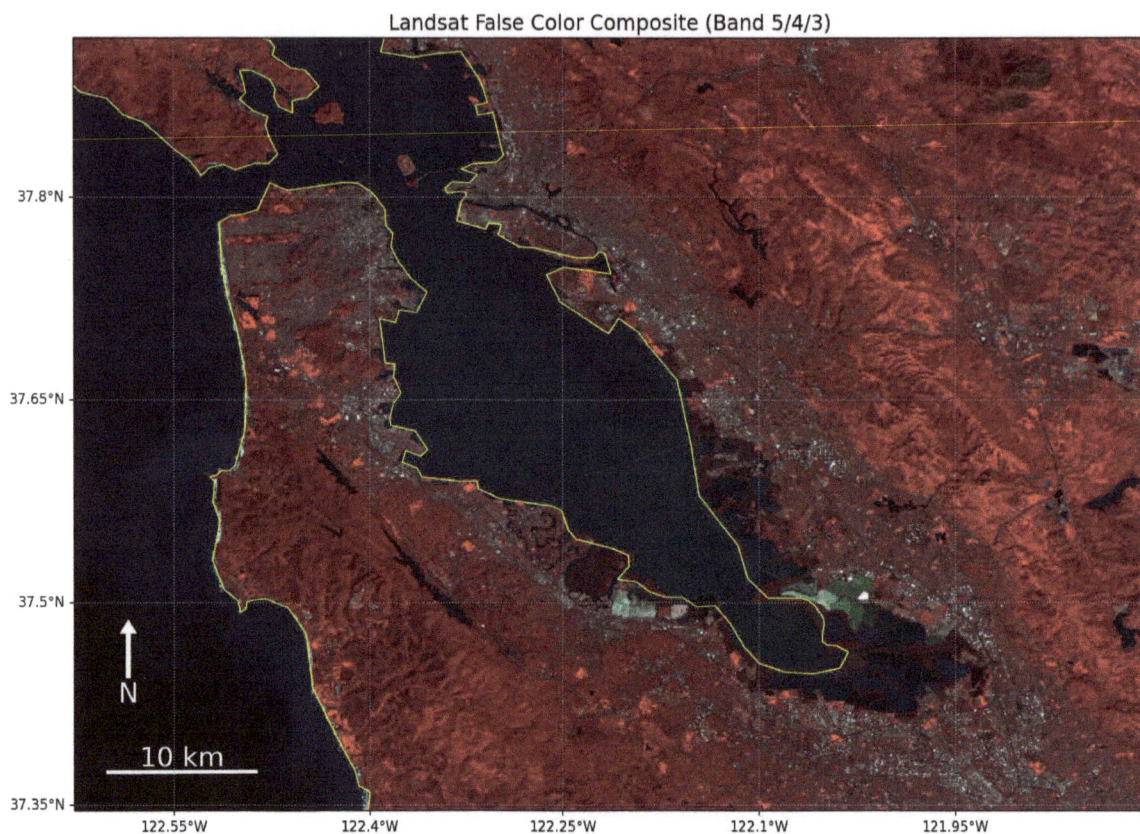

Figure 8.4: False-color composite of a Landsat scene with a north arrow and scale bar.

```python
region = [-121.8025, 37.3458, -122.6265, 37.9178]
ax = cartoee.get_map(image, vis_params=vis, region=region)
cartoee.add_gridlines(ax, interval=0.15, xtick_rotation=0, linestyle=":")
ax.coastlines(color="yellow")

# add north arrow
north_arrow_dict = {
    "text": "N",
    "xy": (0.05, 0.30),
    "arrow_length": 0.10,
    "text_color": "white",
    "arrow_color": "white",
    "fontsize": 20,
    "width": 5,
    "headwidth": 15,
    "ha": "center",
    "va": "center",
}
cartoee.add_north_arrow(ax, **north_arrow_dict)

# add scale bar
scale_bar_dict = {
    'metric_distance': 4,
    'unit': "km",
    'at_x': (0.03, 0.15),
    'at_y': (0.08, 0.11),
    'max_stripes': 4,
    'ytick_label_margins': 0.25,
```

```
    'fontsize': 8,
    'font_weight': "bold",
    'rotation': 0,
    'zorder': 999,
    'paddings': {"xmin": 0.05, "xmax": 0.05, "ymin": 1.5, "ymax": 0.5},
}

cartoee.add_scale_bar(ax, **scale_bar_dict)
ax.set_title(label='Landsat False Color Composite (Band 5/4/3)', fontsize=15)

plt.show()
```

The improved scale bar is shown in Fig. 8.5.

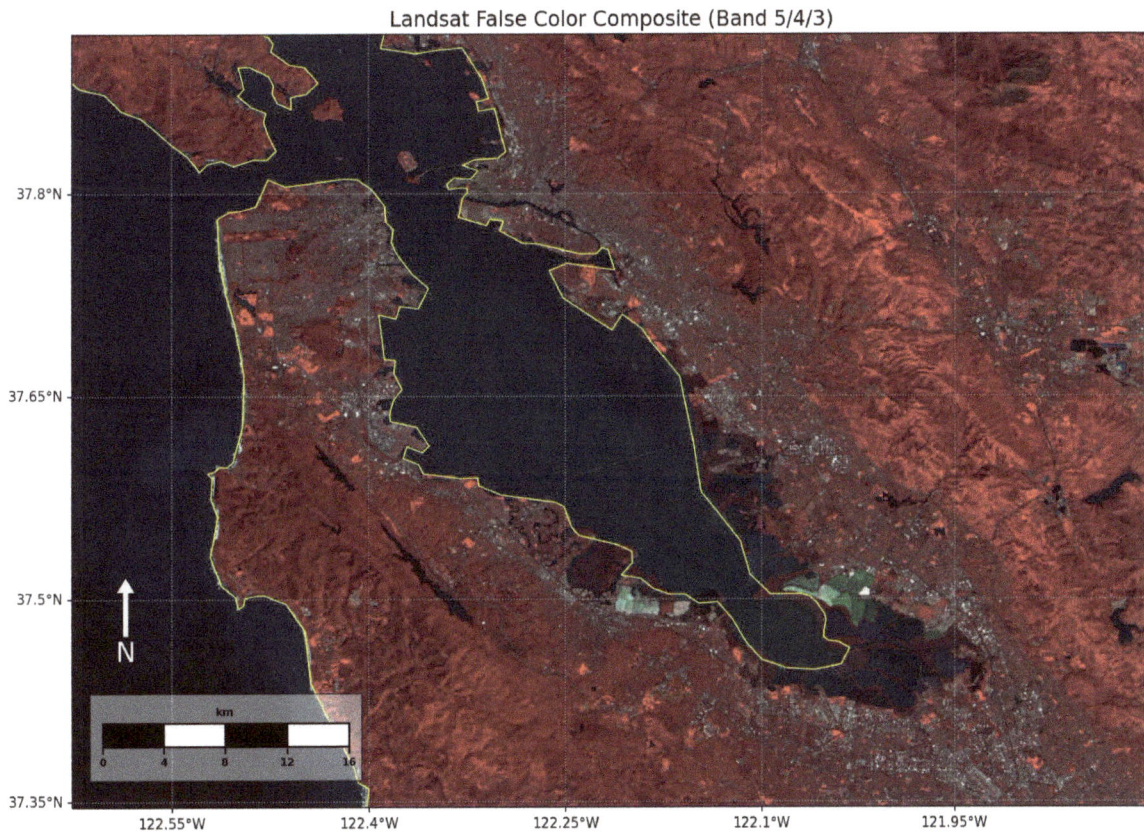

Figure 8.5: False-color composite of a Landsat scene with an improved scale bar.

To save the map as an image file, use the `fig.savefig()` function:

```
cartoee.savefig(fig, fname="landsat.jpg")
```

8.6 Adding legends

Another important element of a map is a legend. The `cartoee.add_legend()` function can be used to add a legend to a map. In this example, we will create a legend based on the Global Power Plant Database[74] , which is a comprehensive, open source database of power plants around the world. The database contains information on the location, capacity, and fuel type of power plants. There are numerous fuel types in the database, but we will only use the following 10 fuel types in this example.

Let's filter the power plant database to only include the 10 fuel types:

```
fuels = [
    'Coal',
    'Oil',
    'Gas',
    'Hydro',
    'Nuclear',
    'Solar',
    'Waste',
    'Wind',
    'Geothermal',
    'Biomass',
]

fc = ee.FeatureCollection("WRI/GPPD/power_plants").filter(
    ee.Filter.inList('fuel1', fuels)
)

colors = [
    '000000',
    '593704',
    'BC80BD',
    '0565A6',
    'E31A1C',
    'FF7F00',
    '6A3D9A',
    '5CA2D1',
    'FDBF6F',
    '229A00',
]

styled_fc = geemap.ee_vector_style(fc, column="fuel1", labels=fuels, color=colors, pointSize=1)
```

The geemap.ee_vector_style() function can be used to style a vector layer. In this example, we use the column parameter to specify the column name that contains the fuel type information. The labels parameter is used to specify the fuel types that we want to include in the legend. The color parameter is used to specify the colors for each fuel type. The pointSize parameter is used to specify the size of the points in the vector layer. With the styled vector layer, we can now add the layer and its corresponding legend to an interactive map (see Fig. 8.6).

```
Map = geemap.Map()
Map.addLayer(styled_fc, {}, 'Power Plants')
Map.add_legend(title="Power Plant Fuel Type", labels=fuels, colors=colors)
Map
```

To export the map with a legend, we need to create a legend manually. The legend is created by creating a list of Line2D objects where each Line2D object represents a fuel type in the legend. The Line2D object is created by specifying the marker style, marker color, marker size, and label. The Line2D objects are then added to a list and the list is then passed to the cartoee.add_legend() function:

```
from matplotlib.lines import Line2D

legend = []

for index, fuel in enumerate(fuels):
    item = Line2D(
                [],
                [],
                marker="o",
                color='#' + colors[index],
                label=fuel,
                markerfacecolor='#' + colors[index],
```

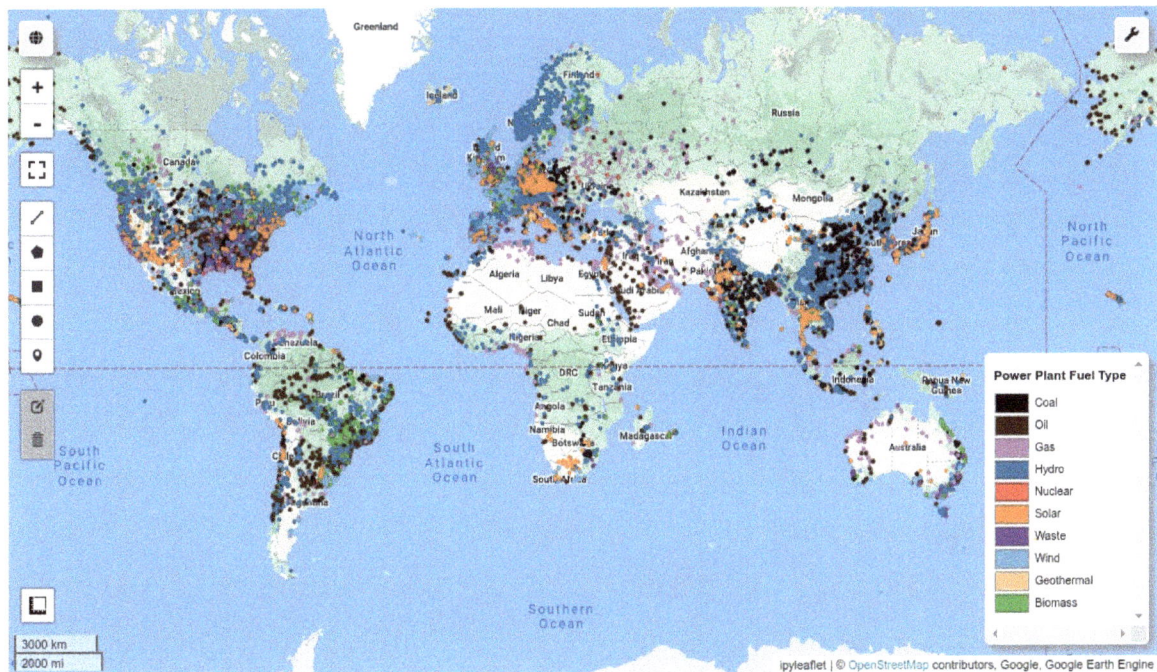

Figure 8.6: Visualizing the global power plant database on an interactive map with a legend.

```
                    markersize=5,
                    ls="",
                )
    legend.append(item)

fig = plt.figure(figsize=(15, 10))

bbox = [180, -88, -180, 88]
ax = cartoee.get_map(styled_fc, region=bbox)
ax.set_title(label='Global Power Plant Database', fontsize=15)
cartoee.add_gridlines(ax, interval=30)
cartoee.add_legend(ax, legend_elements=legend, font_size=10, title='Fule Type', title_fontize=12, loc='lower
    left')
ax.coastlines(color="black")

plt.show()
```

The resulting map with the legend is shown in Fig. 8.7.

```
cartoee.savefig(fig, 'ch08_power_plants.jpg', dpi=150)
```

8.7 Adding basemaps

Besides plotting the image itself, `cartoee` can also add basemaps to the map. The `cartoee.get_map()` function has a `basemap` parameter that can be set to one of the following values: ROADMAP, SATELLITE, TERRAIN, and HYBRID. In this example, we will use the ROADMAP basemap (see Fig. 8.8).

```
image = ee.Image('LANDSAT/LC08/C01/T1_SR/LC08_044034_20140318')
vis = {"bands": ['B5', 'B4', 'B3'], "min": 0, "max": 5000, "gamma": 1.3}

fig = plt.figure(figsize=(15, 10))
ax = cartoee.get_map(image, vis_params=vis, basemap='ROADMAP', zoom_level=8)
```

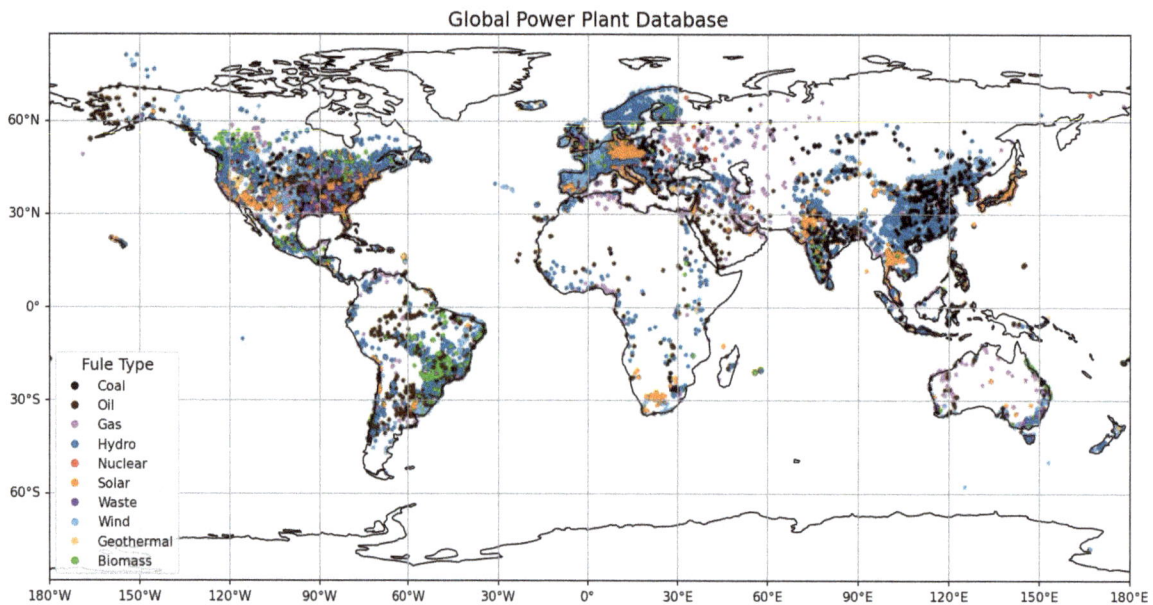

Figure 8.7: Map of the global power plant database.

```
cartoee.pad_view(ax)
cartoee.add_gridlines(ax, interval=0.5, xtick_rotation=0, linestyle=":")
ax.coastlines(color="yellow")

plt.show()
```

Depending on the image size, you might need to adjust the zoom_level parameter of the cartoee.get_ map() function to get the best view of the image. The default value is 2. The higher the zoom level, the more detailed the basemap will be, and the smaller the labels will be.

Cartopy also provides a number of basemaps that can be used with cartoee, such as Stamen and OSM. First, we need to import the cartopy.io.img_tiles module:

```
import cartopy.io.img_tiles as cimgt
```

The available basemaps can be found in the Cartopy documentation - img_tiles.html[167]. In this example, we will use the Stamen terrain-background basemap (see Fig. 8.9).

```
basemap = cimgt.Stamen('terrain-background')

fig = plt.figure(figsize=(15, 10))

ax = cartoee.get_map(image, vis_params=vis, basemap=basemap, zoom_level=8)
cartoee.pad_view(ax)
cartoee.add_gridlines(ax, interval=0.5, xtick_rotation=0, linestyle=":")
ax.coastlines(color="yellow")

plt.show()
```

To use the OpenStreetMap basemap, use the OSM class:

```
basemap = cimgt.OSM()
```

---

[167]img_tiles.html: tiny.geemap.org/bqzw

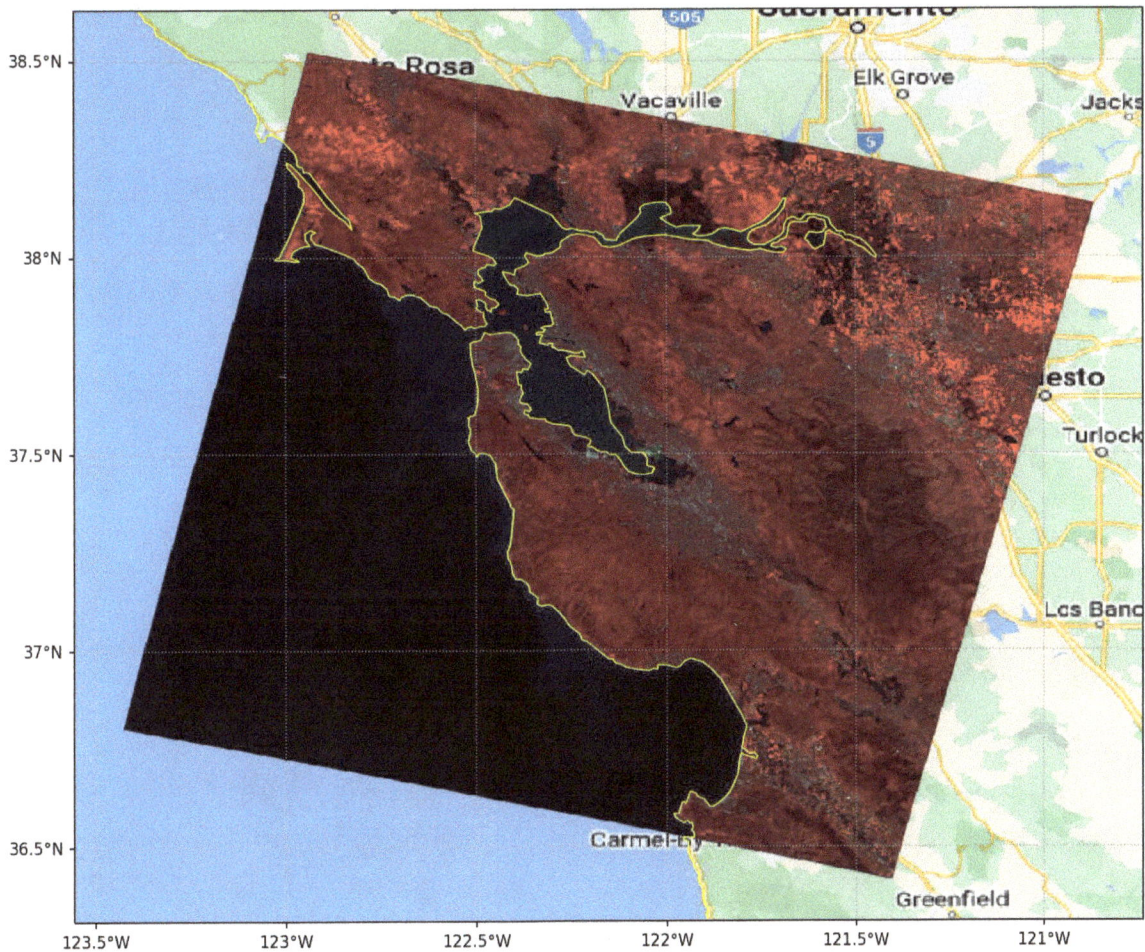

Figure 8.8: False-color composite of a Landsat scene with a Google basemap in the background.

```
fig = plt.figure(figsize=(15, 10))
ax = cartoee.get_map(image, vis_params=vis, basemap=basemap, zoom_level=8)
cartoee.pad_view(ax)
cartoee.add_gridlines(ax, interval=0.5, xtick_rotation=0, linestyle=":")
ax.coastlines(color="yellow")

plt.show()
```

8.8 Using custom projections

The PlateCarree projection

By default, cartoee uses the PlateCarree map projection, which is an equidistant cylindrical projection with the standard parallel located at the equator. A grid of parallels and meridians forms perfect squares from east to west and from pole to pole. It is one of the most common projections used in cartography.

First, let's try out the PlateCarree[168] projection. In this example, we are going to use the MODIS Ocean Color SMI[169] data to plot global sea surface temperature (SST) from January to March 2018. Note that

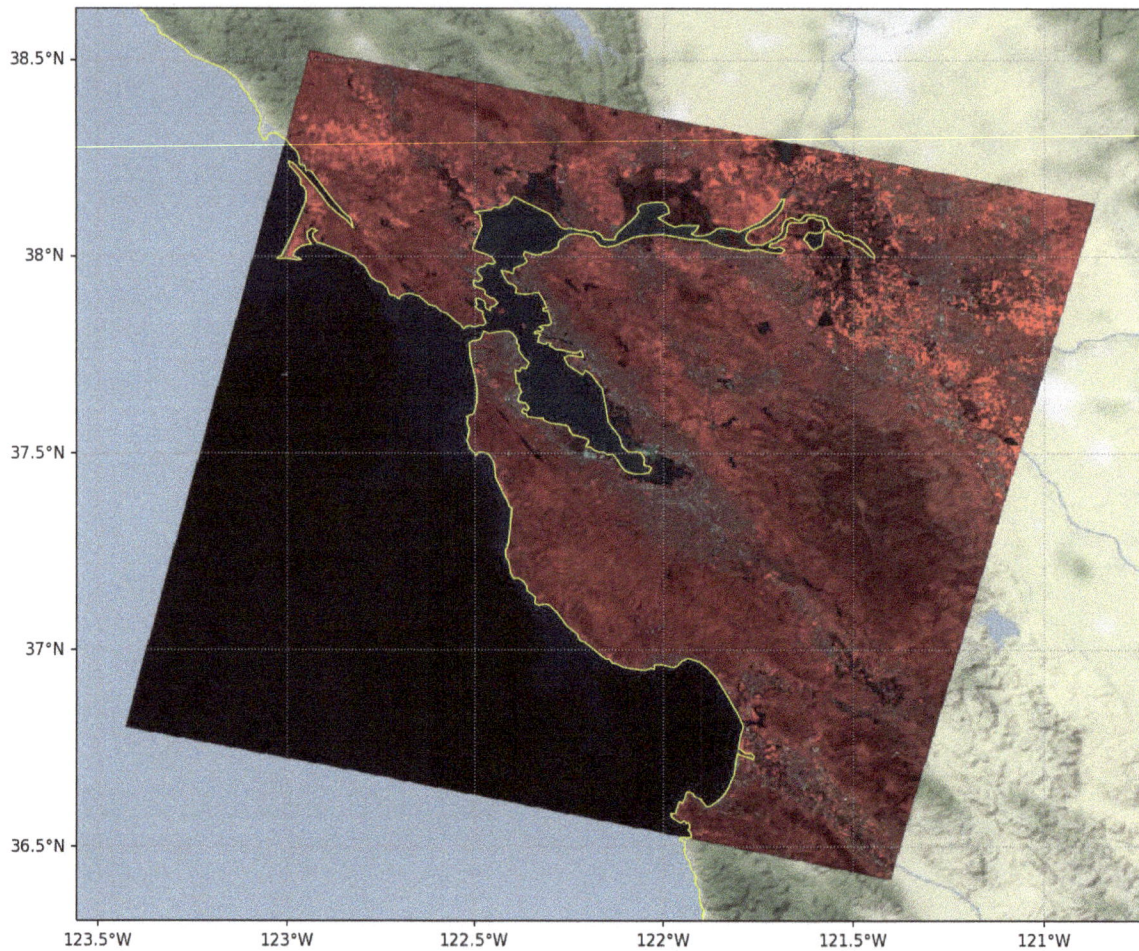

Figure 8.9: False-color composite of a Landsat scene with a Stamen basemap in the background.

the dataset contains multiple bands, but we are only interested in the `sst` band, so we will select it using the `select()` method and rename it to `SST`:

```
ocean = (
    ee.ImageCollection('NASA/OCEANDATA/MODIS-Terra/L3SMI')
    .filter(ee.Filter.date('2018-01-01', '2018-03-01'))
    .median()
    .select(["sst"], ["SST"])
)
```

Next, set the visualization parameters and the bounding box for plotting:

```
visualization = {'bands': "SST", 'min': -2, 'max': 30}
bbox = [180, -88, -180, 88]
```

Lastly, plot the global sea surface temperature using the default `PlateCarree` projection (see Fig. 8.10).

```
fig = plt.figure(figsize=(15, 10))
```

---

[168]PlateCarree: tiny.geemap.org/ssft

[169]MODIS Ocean Color SMI: tiny.geemap.org/hcli

```
ax = cartoee.get_map(ocean, cmap='plasma', vis_params=visualization, region=bbox)
cb = cartoee.add_colorbar(ax, vis_params=visualization, loc='right', cmap='plasma')

ax.set_title(label='Sea Surface Temperature', fontsize=15)

ax.coastlines()
plt.show()
```

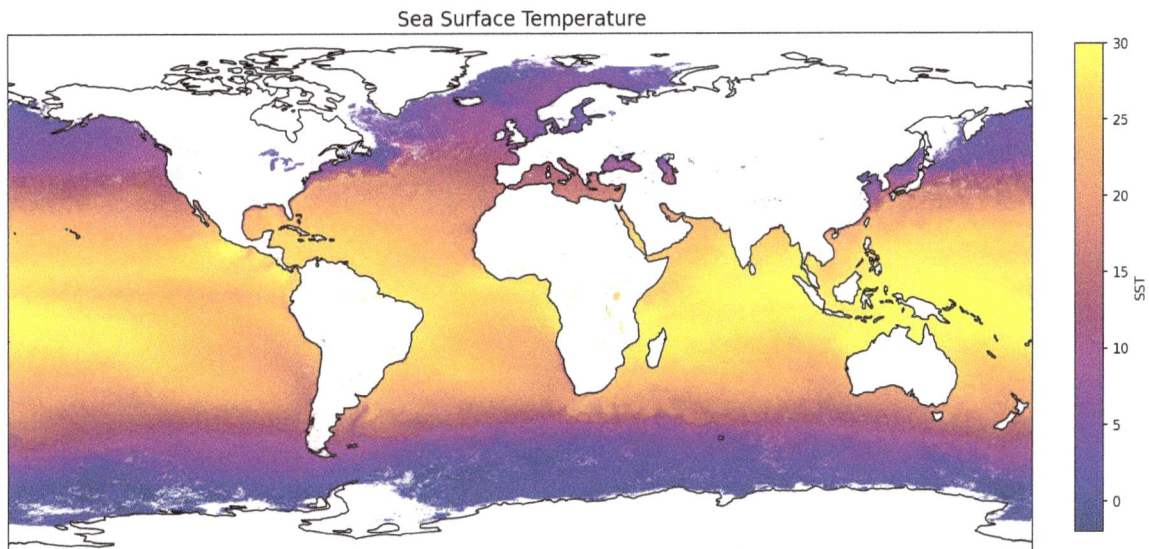

Figure 8.10: Plotting global sea surface temperature with the default PlateCarre projection.

To save the figure, use the `cartoee.savefig()` function:

```
cartoee.savefig(fig, 'SST.jpg', dpi=300)
```

Custom projections

Besides the `PlateCarree` projection, Cartopy also provides many other projections such as the `Mercator`, `Robinson`, `Mollweide`, and `Orthographic` projections. Check out the Cartopy documentation[170] for more information.

You can specify whatever projection is available within `cartopy` to plot Earth Engine data. In this section, we will try out a few of them, including the Mollweide, Robinson, InterruptedGoodeHomolosine, EqualEarth, and Orthographic projections.

First, import the `cartopy.crs` module:

```
import cartopy.crs as ccrs
```

Most projections require a central longitude and latitude. For example, the `Mollweide` projection requires a central longitude with a default value of 0. For example, ccrs.Mollweide(central_longitude=-180) creates a Mollweide projection centered on the -180 longitude. Custom projections can be specified using the `proj` parameter in the `cartoee.get_map()` function. To plot the global sea surface temperature using the Mollweide[171] projection, use the following code:

---

[170]Cartopy documentation: tiny.geemap.org/ekxl

[171]Mollweide: tiny.geemap.org/lsnd

```
fig = plt.figure(figsize=(15, 10))

projection = ccrs.Mollweide(central_longitude=-180)
ax = cartoee.get_map(
    ocean, vis_params=visualization, region=bbox, cmap='plasma', proj=projection
)
cb = cartoee.add_colorbar(
    ax, vis_params=visualization, loc='bottom', cmap='plasma', orientation='horizontal'
)
ax.set_title("Mollweide projection")
ax.coastlines()

plt.show()
```

The result is shown in Fig. 8.11.

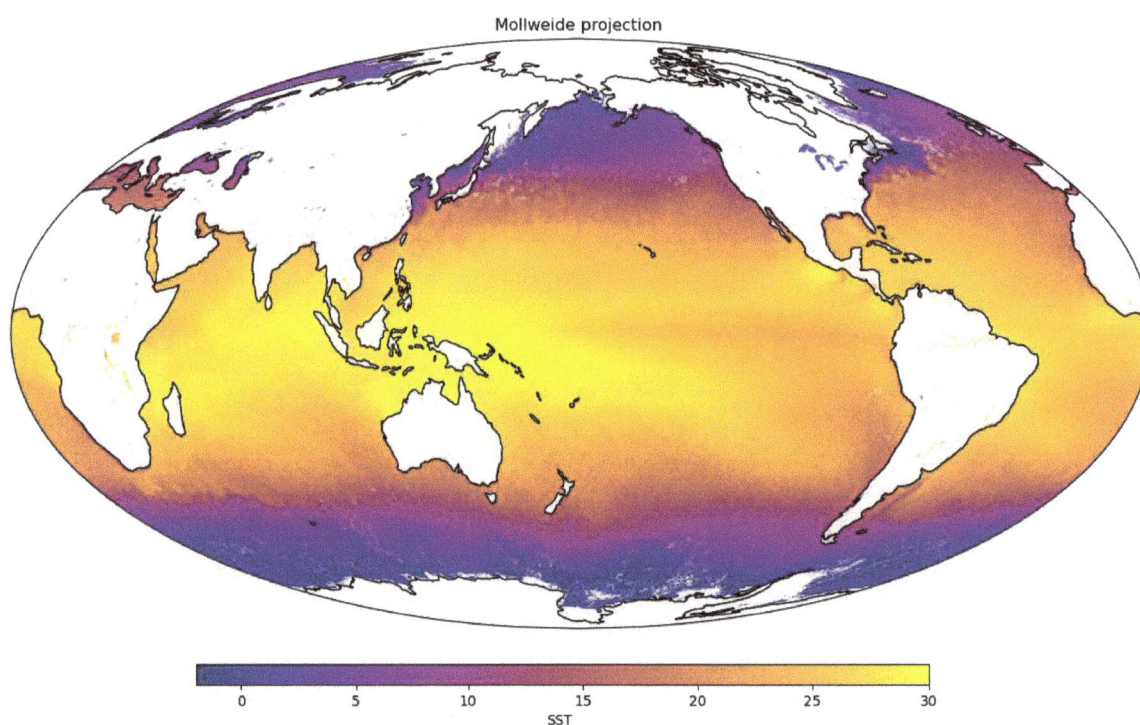

Figure 8.11: The Mollweide projection.

To plot the global sea surface temperature using the Robinson[172] projection:

```
fig = plt.figure(figsize=(15, 10))

projection = ccrs.Robinson(central_longitude=-180)
ax = cartoee.get_map(
    ocean, vis_params=visualization, region=bbox, cmap='plasma', proj=projection
)
cb = cartoee.add_colorbar(
    ax, vis_params=visualization, loc='bottom', cmap='plasma', orientation='horizontal'
)
ax.set_title("Robinson projection")
ax.coastlines()
```

---

[172]Robinson: tiny.geemap.org/xluk

```
plt.show()
```

The result is shown in Fig. 8.12.

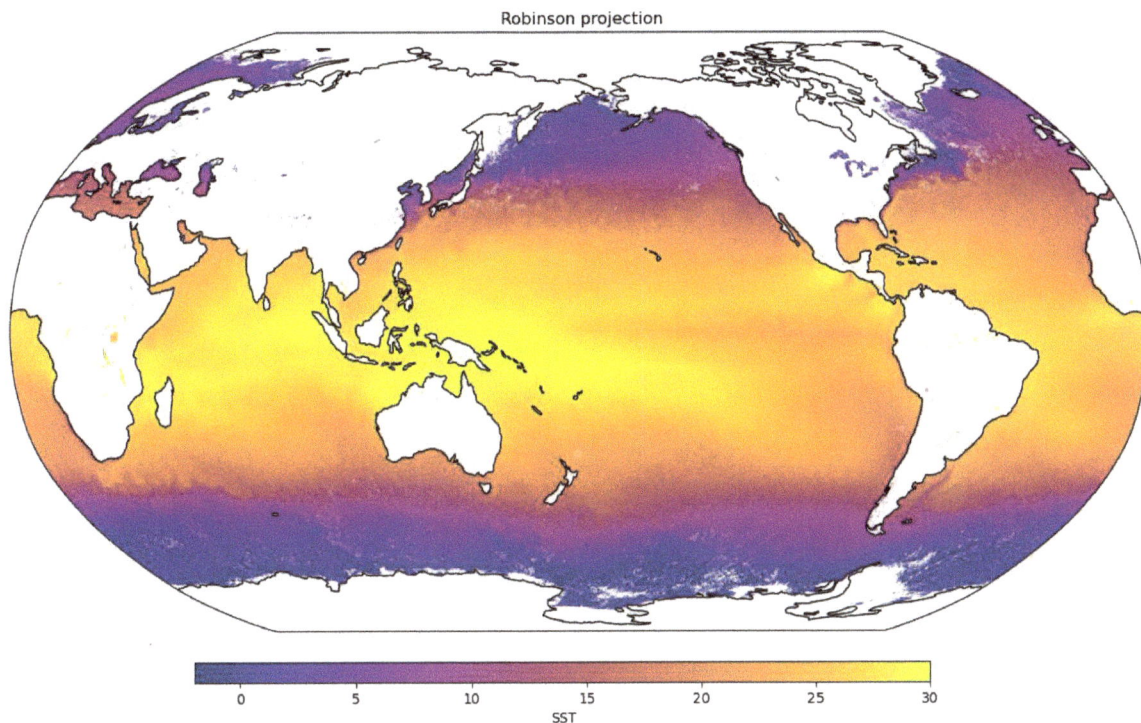

Figure 8.12: The Robinson projection.

To plot the global sea surface temperature using the InterruptedGoodeHomolosine[173] projection:

```
fig = plt.figure(figsize=(15, 10))

projection = ccrs.InterruptedGoodeHomolosine(central_longitude=-180)
ax = cartoee.get_map(
    ocean, vis_params=visualization, region=bbox, cmap='plasma', proj=projection
)
cb = cartoee.add_colorbar(
    ax, vis_params=visualization, loc='bottom', cmap='plasma', orientation='horizontal'
)
ax.set_title("Goode homolosine projection")
ax.coastlines()

plt.show()
```

To plot the global sea surface temperature using the EqualEarth[174] projection:

```
fig = plt.figure(figsize=(15, 10))

projection = ccrs.EqualEarth(central_longitude=-180)
ax = cartoee.get_map(
    ocean, vis_params=visualization, region=bbox, cmap='plasma', proj=projection
)
```

---

[173]InterruptedGoodeHomolosine: tiny.geemap.org/ukff
[174]EqualEarth: tiny.geemap.org/ypbk

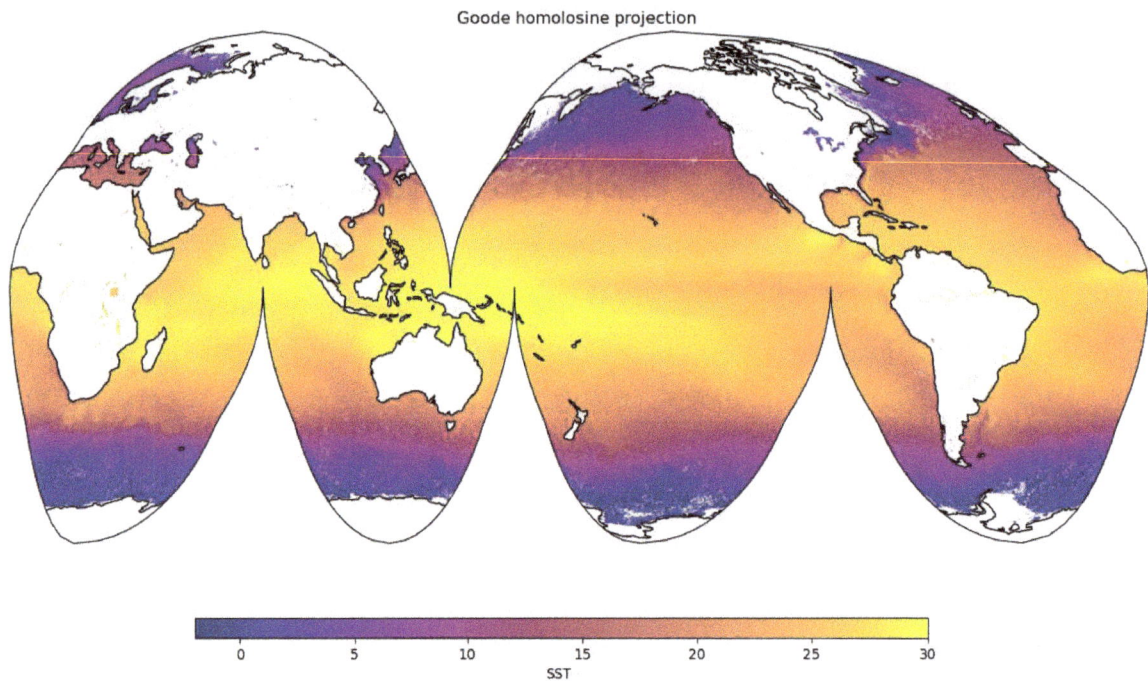

Figure 8.13: The Goode homolosine projection.

```
cb = cartoee.add_colorbar(
    ax, vis_params=visualization, loc='right', cmap='plasma', orientation='vertical'
)
ax.set_title("Equal Earth projection")
ax.coastlines()

plt.show()
```

The result is shown in Fig. 8.14.

To plot the global sea surface temperature using the Orthographic[175] projection:

```
fig = plt.figure(figsize=(11, 10))

projection = ccrs.Orthographic(-130, -10)
ax = cartoee.get_map(
    ocean, vis_params=visualization, region=bbox, cmap='plasma', proj=projection
)
cb = cartoee.add_colorbar(
    ax, vis_params=visualization, loc='right', cmap='plasma', orientation='vertical'
)
ax.set_title("Orographic projection")
ax.coastlines()

plt.show()
```

The result is shown in Fig. 8.15.

---

[175]Orthographic: tiny.geemap.org/gceb

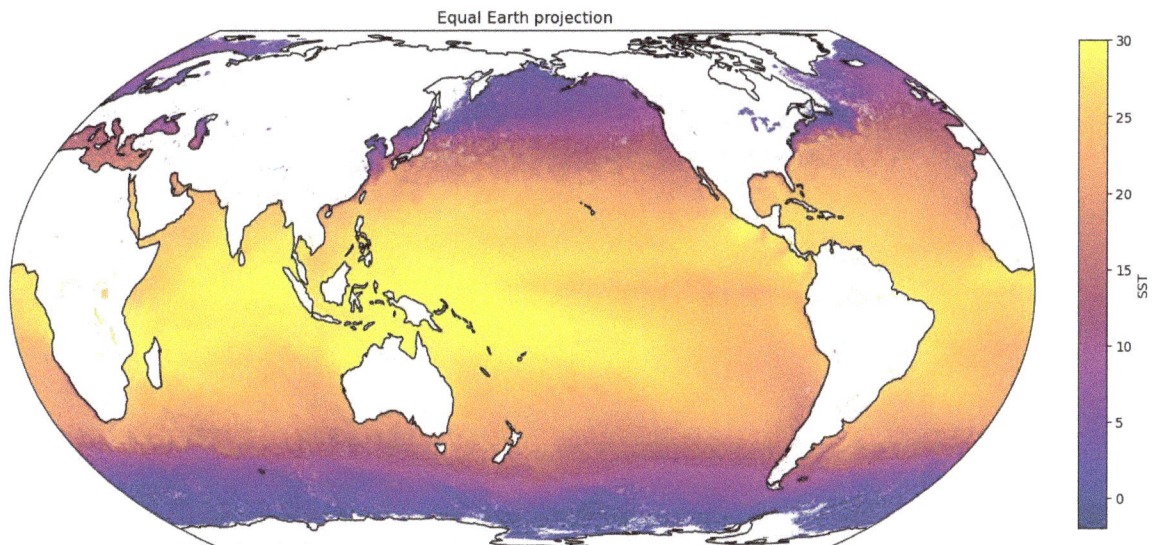

Figure 8.14: The Equal Earth projection.

The warping artifacts

Oftentimes global projections are not needed, so we use a specific projection for the map that provides the best view for the geographic region of interest. When we use these, sometimes image warping effects occur. This is because `cartoee` only requests data for the region of interest and when mapping with `cartopy` the pixels get warped to fit the view extent as best as possible. Consider the following example where we want to map SST over the South Pole using the SouthPolarStereo[176] projection:

```
fig = plt.figure(figsize=(11, 10))

spole = [180, -88, -180, 0]
projection = ccrs.SouthPolarStereo()

ax = cartoee.get_map(
    ocean, cmap='plasma', vis_params=visualization, region=spole, proj=projection
)
cb = cartoee.add_colorbar(ax, vis_params=visualization, loc='right', cmap='plasma')
ax.coastlines()
ax.set_title('The South Pole')

plt.show()
```

The result is shown in Fig. 8.16.

As you can see from the result, there are warping effects on the plotted image. There is really no way of getting around this other than requesting a larger extent of data which may not always be the case. So, what we can do is set the extent of the map to a more realistic view after plotting the image as in the following example:

```
fig = plt.figure(figsize=(11, 10))

ax = cartoee.get_map(
    ocean, cmap='plasma', vis_params=visualization, region=spole, proj=projection
)
```

---

[176]SouthPolarStereo: tiny.geemap.org/qepp

Orographic projection

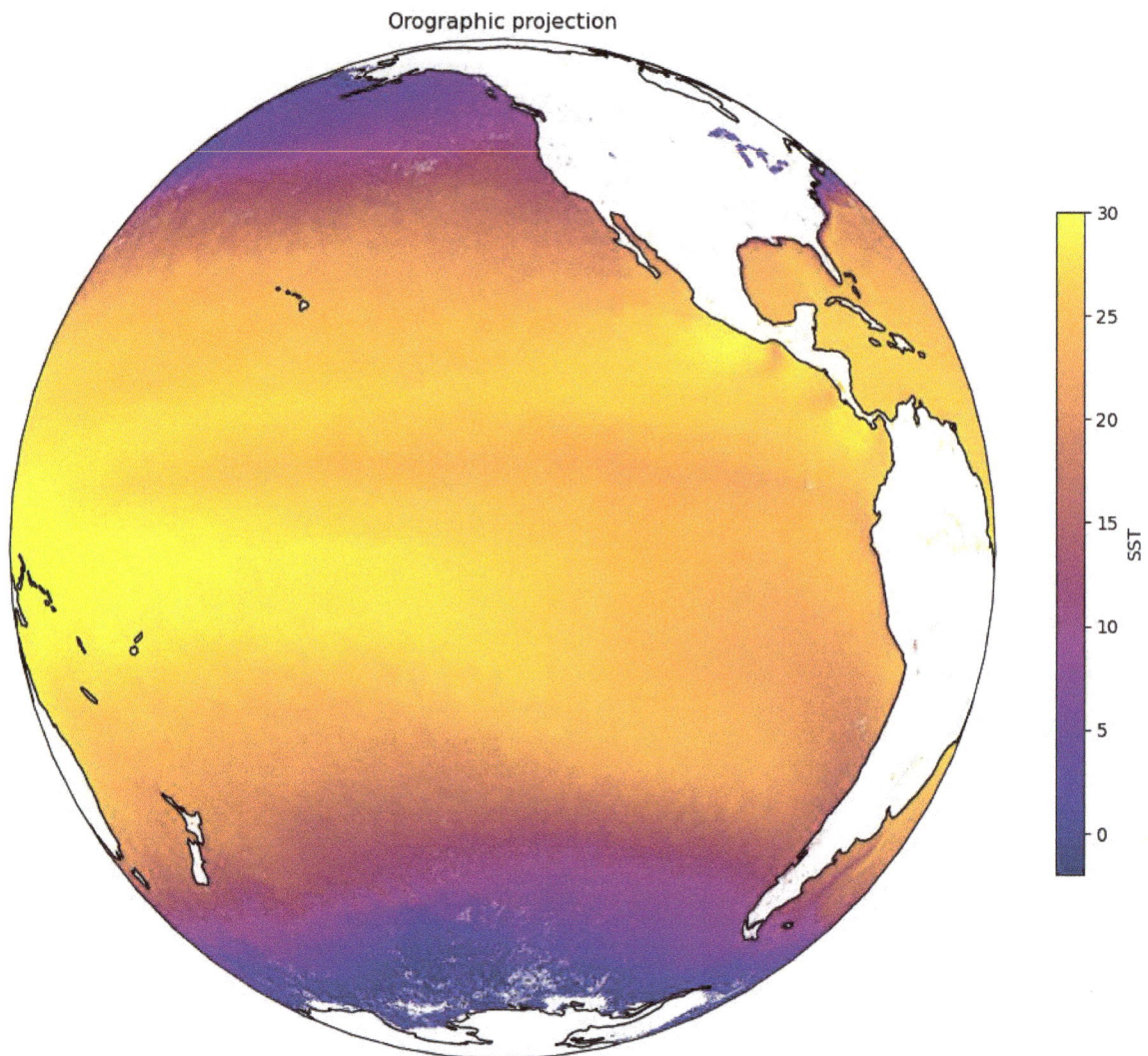

Figure 8.15: The Orthographic projection.

```
cb = cartoee.add_colorbar(ax, vis_params=visualization, loc='right', cmap='plasma')
ax.coastlines()
ax.set_title('The South Pole')

# get bounding box coordinates of a zoom area
zoom = spole
zoom[-1] = -20

# convert bbox coordinate from [W,S,E,N] to [W,E,S,N] as matplotlib expects
zoom_extent = cartoee.bbox_to_extent(zoom)

# set the extent of the map to the zoom area
ax.set_extent(zoom_extent, ccrs.PlateCarree())

plt.show()
```

The result is shown in Fig. 8.17.

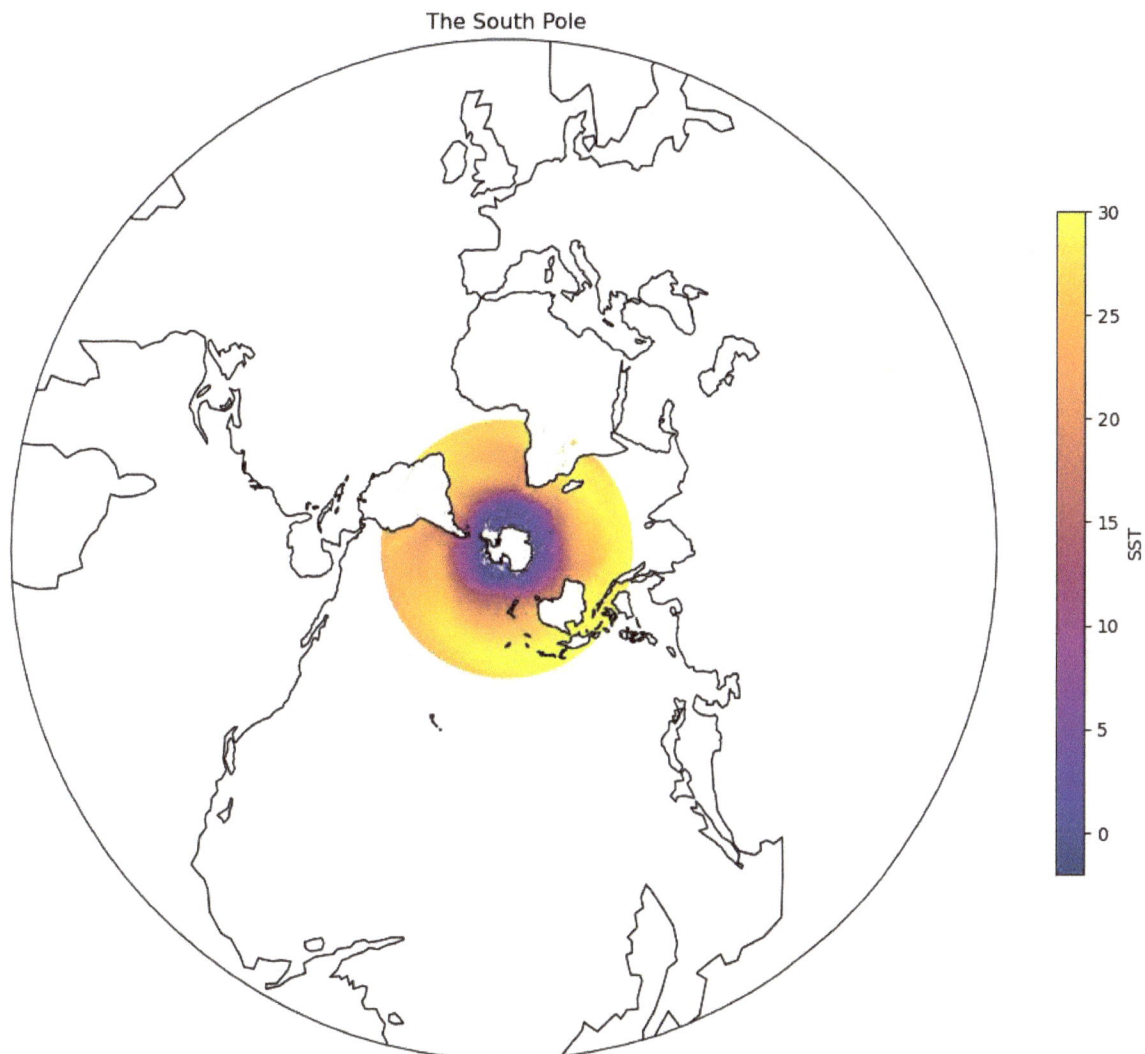

Figure 8.16: The South Polar Stereo projection.

8.9 Plotting multiple Earth Engine layers

Creating a blended image

The previous sections show how to plot a single Earth Engine layer using the `cartoee.get_map()` function. However, it is also possible to plot multiple layers on the same map by creating a blended image using the ee.Image.blend()[177] method, which overlays one image on top of another. The images are blended together using the masks as opacity. The following example shows how to blend a MODIS NDVI image with a layer of countries:

```
Map = geemap.Map()

image = (
    ee.ImageCollection('MODIS/MCD43A4_006_NDVI')
```

---

[177]ee.Image.blend(): tiny.geemap.org/snhw

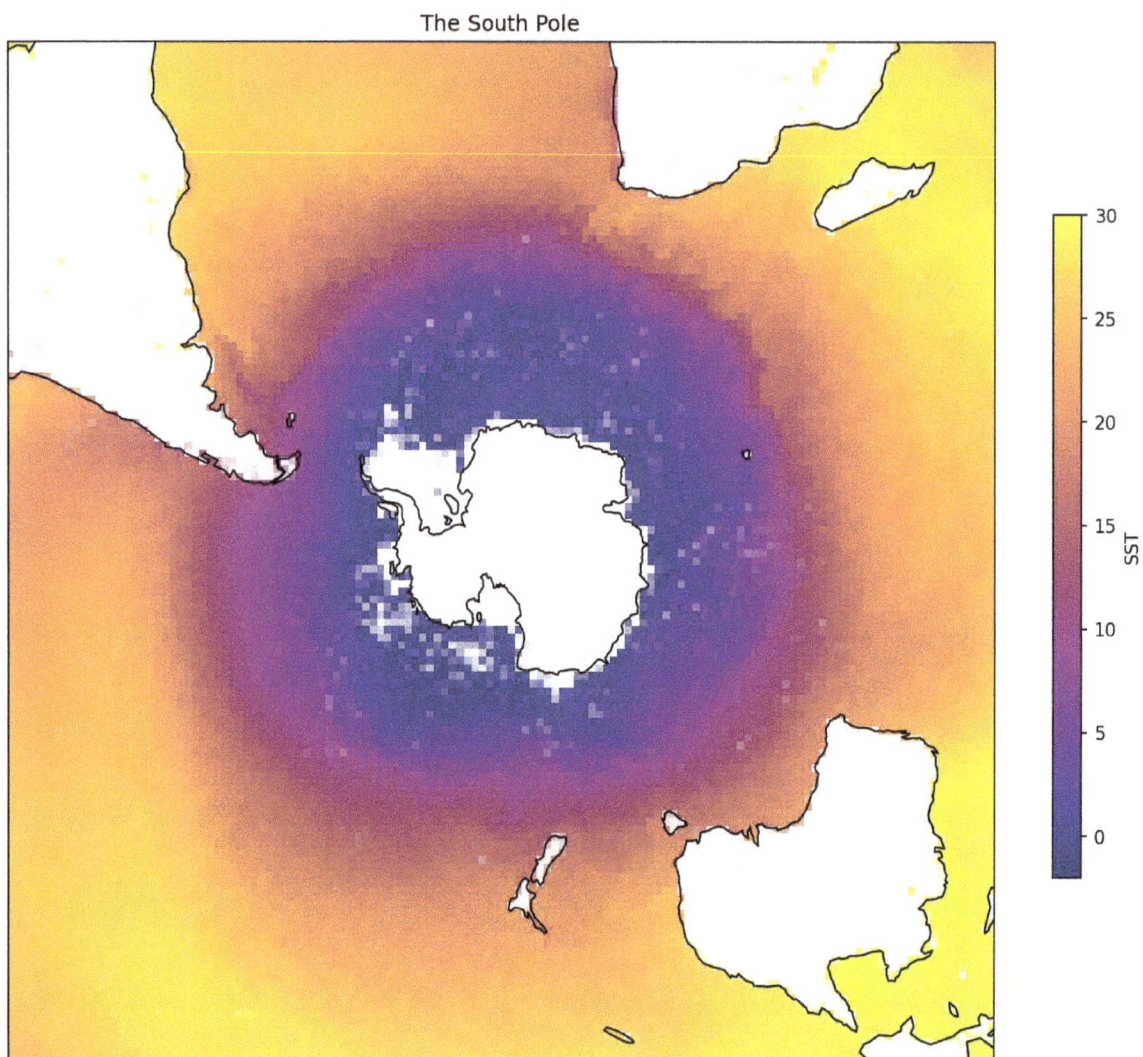

Figure 8.17: The South Polar Stereo projection with a zoomed extent.

```python
    .filter(ee.Filter.date('2022-05-01', '2022-06-01'))
    .select("NDVI")
    .first()
)

vis_params = {'min': 0.0, 'max': 1.0, 'palette': 'ndvi'}
Map.setCenter(-7.03125, 31.0529339857, 2)
Map.addLayer(image, vis_params, 'MODIS NDVI')

countries = ee.FeatureCollection(geemap.examples.get_ee_path('countries'))
style = {"color": "00000088", "width": 1, "fillColor": "00000000"}
Map.addLayer(countries.style(**style), {}, "Countries")

ndvi = image.visualize(**vis_params)
blend = ndvi.blend(countries.style(**style))

Map.addLayer(blend, {}, "Blend")
Map
```

Note that we first use the ee.Image.visualize() method to create a visualization image from the NDVI image. Then we use the ee.Image.blend() method to blend the visualization image with the countries layer. The blended image can then be displayed on the interactive map using the Map.addLayer() method.

Plotting a blended image with the default projection

With the blended image created, we can plot it using the cartoee.get_map() function just like any other Earth Engine layer. The following example shows how to plot the blended image with the default PlateCarree projection:

```
fig = plt.figure(figsize=(15, 10))

bbox = [180, -88, -180, 88]
ax = cartoee.get_map(blend, region=bbox)
cb = cartoee.add_colorbar(ax, vis_params=vis_params, loc='right')
ax.set_title(label='MODIS NDVI (May 2022)', fontsize=15)

plt.show()
```

The result is shown in Fig. 8.18.

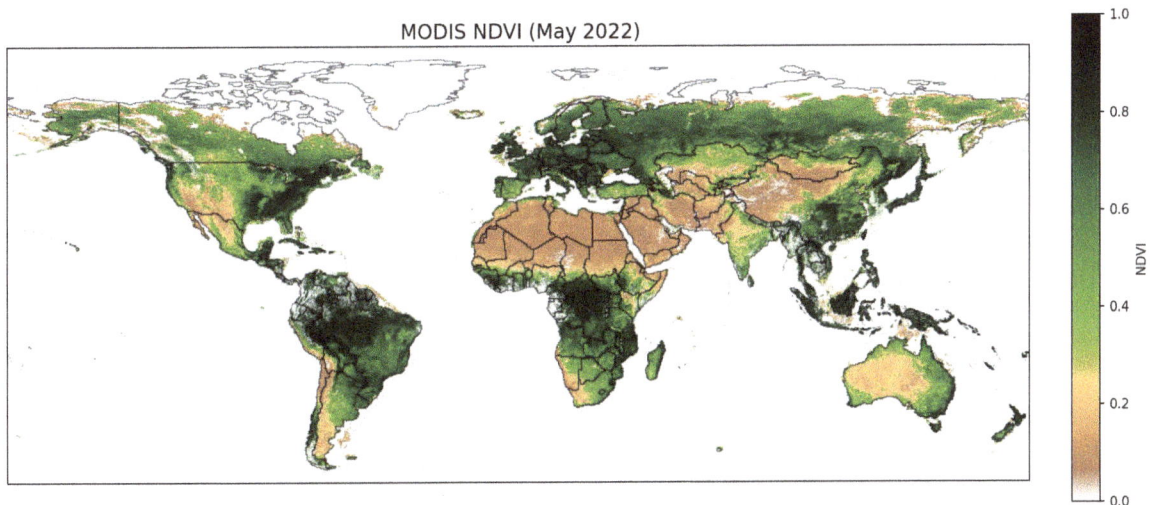

Figure 8.18: Plotting a blended image with the default projection.

Plotting a blended image with a custom projection

We can also plot the blended image with any custom projection available on the cartopy projection list[178]. First, we need to import the cartopy.crs module:

```
import cartopy.crs as ccrs
```

Then we can plot the blended image with the EqualEarth projection or any other projection of our choice:

```
fig = plt.figure(figsize=(15, 10))
```

---

[178]the cartopy projection list: tiny.geemap.org/jlqn

```
projection = ccrs.EqualEarth(central_longitude=0)
ax = cartoee.get_map(blend, region=bbox, proj=projection)
cb = cartoee.add_colorbar(ax, vis_params=vis_params, loc='right')
ax.set_title(label='MODIS NDVI (May 2022)', fontsize=15)

plt.show()
```

The result is shown in Fig. 8.19.

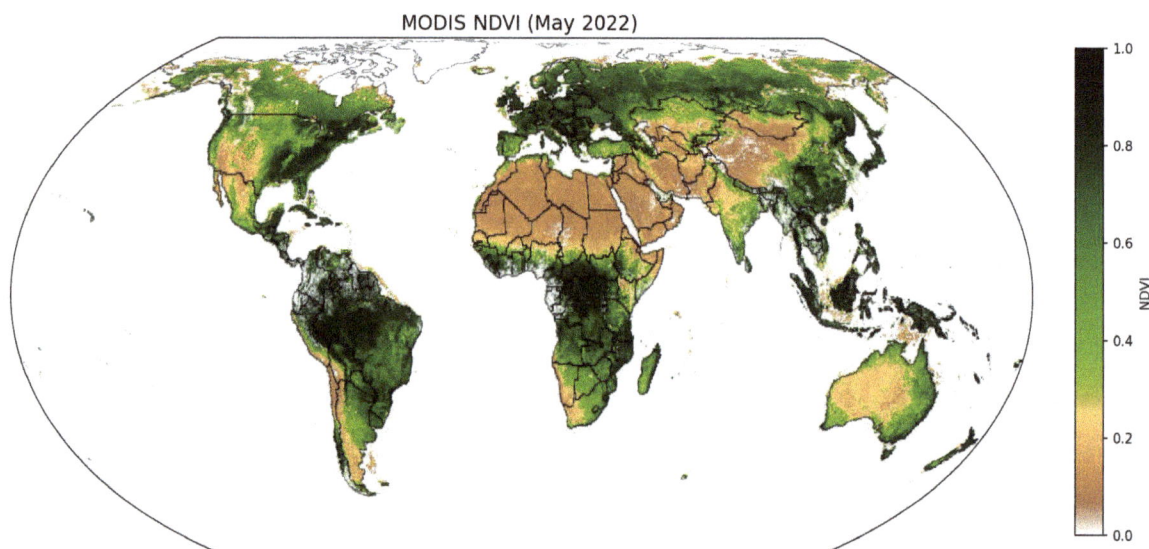

Figure 8.19: Plotting a blended image with a custom projection.

8.10 Creating timelapse animations

The `cartoee` module also provides a function to create timelapse animations from Earth Engine layers. This can be particularly useful for visualizing time series of satellite images. Different from the timelapse functionality that will be introduced in the next chapter, the `cartoee` timelapse functionality allows adding map elements to the animation, such as grid lines, north arrows, scale bars, legends, and colorbars.

Creating an ImageCollection

To create a timelapse animation, we first need to create an `ee.ImageCollection` object. The following example shows how to create a time series of Landsat 5 surface reflectance images covering the area of Las Vegas, Nevada, USA:

```
lon = -115.1585
lat = 36.1500
start_year = 1984
end_year = 2011

point = ee.Geometry.Point(lon, lat)
years = ee.List.sequence(start_year, end_year)

def get_best_image(year):

    start_date = ee.Date.fromYMD(year, 1, 1)
    end_date = ee.Date.fromYMD(year, 12, 31)
```

```
    image = (
        ee.ImageCollection("LANDSAT/LT05/C02/T1_L2")
        .filterBounds(point)
        .filterDate(start_date, end_date)
        .sort("CLOUD_COVER")
        .first()
    )
    image = (
        image.select('SR_B.')
        .multiply(0.0000275)
        .add(-0.2)
        .set({'system:time_start': image.get('system:time_start')}))
    return ee.Image(image)

collection = ee.ImageCollection(years.map(get_best_image))
```

The get_best_image() function selects the least cloudy image for each year and returns an ee.Image object. The ee.ImageCollection.map() method is used to apply the get_best_image() function to each year in the years list, resulting in a time series of high-quality images covering the area of Las Vegas.

Displaying a sample image

We can display a sample image (e.g., the first image) from the ee.ImageCollection object on the interactive map using the Map.addLayer() method:

```
Map = geemap.Map()

vis_params = {"bands": ['SR_B5', 'SR_B4', 'SR_B3'], "min": 0, "max": 0.5}
image = ee.Image(collection.first())
Map.addLayer(image, vis_params, 'First image')
Map.setCenter(lon, lat, 8)

Map
```

The image is indeed a high-quality cloud-free image covering the area of Las Vegas.

Getting a sample output image

Before creating the timelapse animation, it is recommended to get a sample output image to make sure the layout and map elements (e.g., north arrow, scale bar, grid lines) of the animation look as expected. Following the example in the previous section, we can create a sample map with a north arrow, a scale bar, and grid lines using the following code:

```
w = 0.4
h = 0.3
region = [lon + w, lat - h, lon - w, lat + h]

fig = plt.figure(figsize=(10, 8))

ax = cartoee.get_map(image, region=region, vis_params=vis_params)
cartoee.add_gridlines(ax, interval=[0.2, 0.2], linestyle=":")

# add north arrow
north_arrow_dict = {
    "text": "N",
    "xy": (0.1, 0.3),
    "arrow_length": 0.15,
    "text_color": "white",
    "arrow_color": "white",
    "fontsize": 20,
```

```
        "width": 5,
        "headwidth": 15,
        "ha": "center",
        "va": "center",
    }
    cartoee.add_north_arrow(ax, **north_arrow_dict)

    # add scale bar
    scale_bar_dict = {
        "length": 10,
        "xy": (0.1, 0.05),
        "linewidth": 3,
        "fontsize": 20,
        "color": "white",
        "unit": "km",
        "ha": "center",
        "va": "bottom",
    }
    cartoee.add_scale_bar_lite(ax, **scale_bar_dict)
    ax.set_title(label='Las Vegas, NV', fontsize=15)

    plt.show()
```

The result is shown in Fig. 8.20.

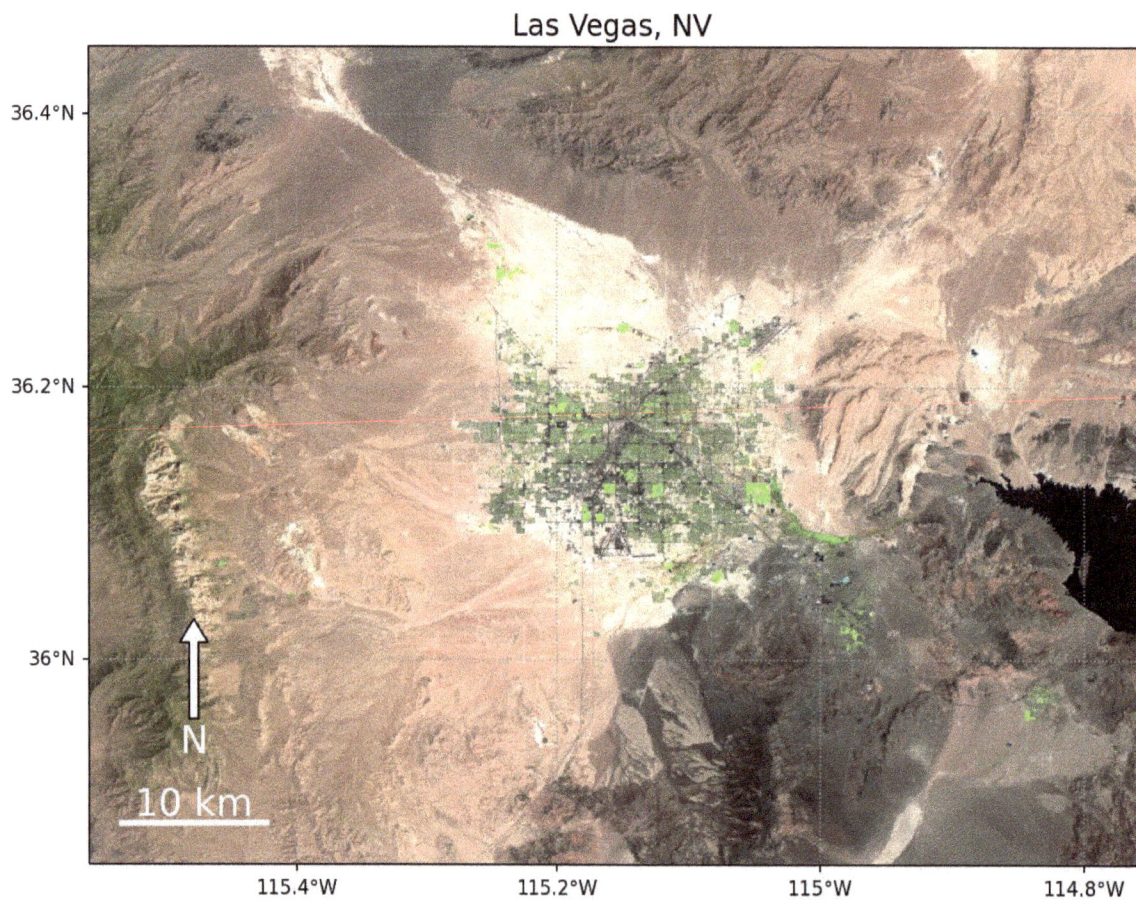

Figure 8.20: A map of Las Vegas, NV.

Creating timelapse

The sample image above looks good. Feel free to adjust the map elements (e.g., north arrow, scale bar, grid lines) as needed. Now we can create the timelapse animation using the cartoee.get_image_collection_gif() function. Under the hood, the function exports one image per year to the specified output directory, and then uses the jpg_to_gif() function to generate the timelapse animation from the exported images. The speed of the animation can be adjusted by setting the fps parameter, which stands for "frames per second". The larger the fps value, the faster the animation, and vice versa. The date_format parameter can be used to customize the date format of the plot title. It can be set to YYYY (e.g., 1984), YYYY-MM (e.g., 1984-01), or YYYY-MM-dd (e.g., 1984-01-01). Lastly, the north_arrow_dict and scale_bar_dict parameters can be used to customize the north arrow and scale bar of the animation. It might take a few minutes to generate the animation. Please be patient.

```python
cartoee.get_image_collection_gif(
    ee_ic=collection,
    out_dir='timelapse',
    out_gif="animation.gif",
    vis_params=vis_params,
    region=region,
    fps=5,
    mp4=True,
    grid_interval=(0.2, 0.2),
    plot_title="Las Vegas, NV - ",
    date_format='YYYY',
    fig_size=(10, 8),
    dpi_plot=100,
    file_format="jpg",
    north_arrow_dict=north_arrow_dict,
    scale_bar_dict=scale_bar_dict,
    verbose=True,
)
geemap.show_image('timelapse/animation.gif')
```

The result is shown in Fig. 8.21.

8.11 Summary

After analyzing Earth Engine data, we usually need to visualize the results and export them as maps to be used in project reports, presentations, or publications. In this chapter, we learned how to use the cartoee module to create publication-quality maps with key map elements, such as grid lines, colorbars, north arrows, scale bars, and legends. We also explored how to create maps with different map projections, such as the PlateCarree, Equal Earth, Robinson, and Orthographic projections. Lastly, we learned how to create a timelapse animation from a time series of satellite images.

In the next chapter, we will learn how to create cool timelapse animations from any time series of images with a few lines of code.

Las Vegas, NV - 1993

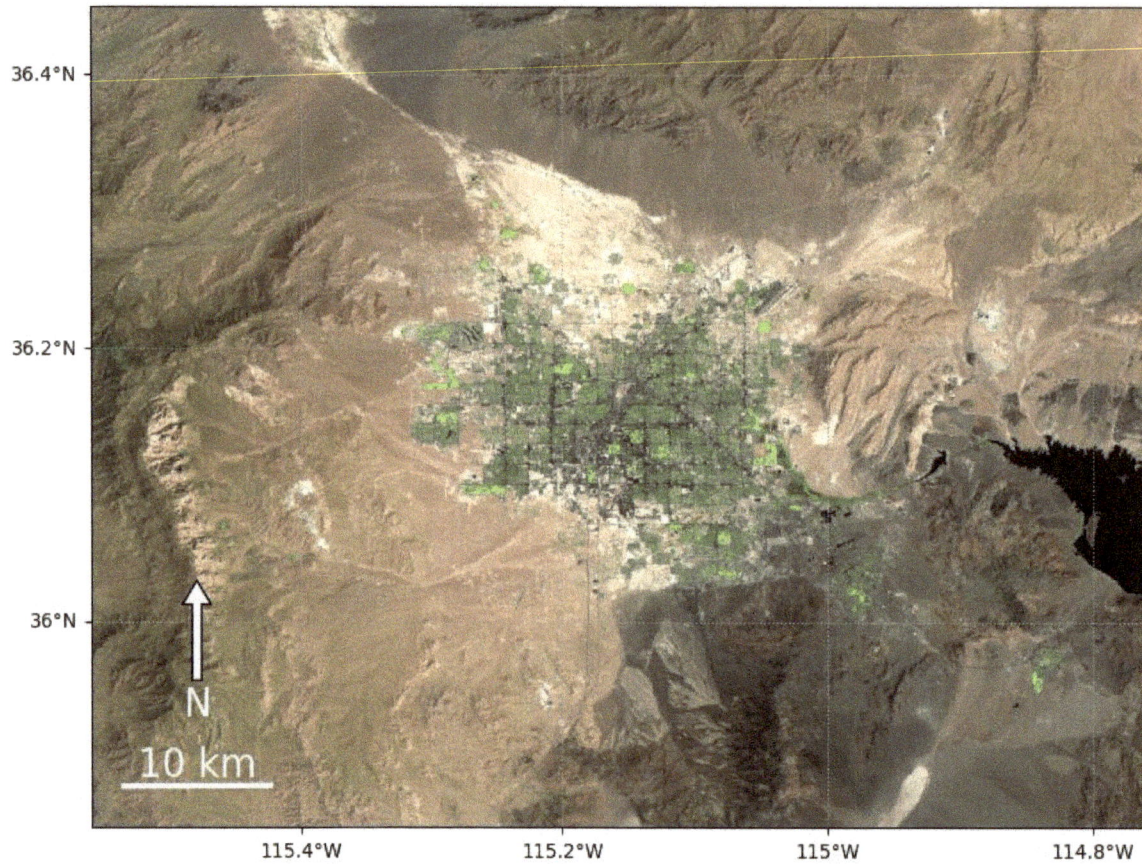

Figure 8.21: A satellite timelapse animation showing the urban expansion in Las Vegas, NV (1984-2011)

9. Creating Timelapse Animations

9.1 Introduction

Before Google Earth Engine, it was very difficult to create timelapse animations from satellite imagery. The process was very time-consuming and required a lot of manual work, such as selecting and downloading satellite imagery, processing the images, and creating a timelapse. The process could take hours to days. With Google Earth Engine, it is now possible to create timelapse animations from remote sensing imagery in a few lines of code.

In this chapter, we will learn how to create timelapse animations from different data sources, such as NAIP, Landsat, Sentinel, MODIS, and GOES. We will also walk through the process of downloading an image collection as a GIF image and adding content (e.g., text, logo, color) to the image. At the end of this chapter, you should feel comfortable creating timelapse animations from any image collection available in Google Earth Engine.

9.2 Technical requirements

To follow along with this chapter, you will need to have geemap and several optional dependencies installed. If you have already followed Section 1.5 - *Installing geemap*, then you should already have a conda environment with all the necessary packages installed. Otherwise, you can create a new conda environment and install pygis[30] with the following commands, which will automatically install geemap and all the required dependencies:

```
conda create -n gee python
conda activate gee
conda install -c conda-forge mamba
mamba install -c conda-forge pygis
```

Next, launch JupyterLab by typing the following commands in your terminal or Anaconda prompt:

```
jupyter lab
```

Alternatively, you can use geemap with a Google Colab cloud environment without installing anything on your local computer. Click 09_timelapse.ipynb[179] to launch the notebook in Google Colab.

Once in Colab, you can uncomment the following line and run the cell to install pygis, which includes geemap and all the necessary dependencies:

```
# %pip install pygis
```

The installation process may take 2-3 minutes. Once pygis has been installed successfully, click the **RESTART RUNTIME** button that appears at the end of the installation log or go to the **Runtime** menu and select **Restart runtime**. After that, you can start coding.

To begin, import the necessary libraries that will be used in this chapter:

```
import ee
```

---

[179]09_timelapse.ipynb: tiny.geemap.org/ch09

```
import geemap
```

Initialize the Earth Engine Python API:

```
geemap.ee_initialize()
```

If this is your first time running the code above, you will need to authenticate Earth Engine first. Follow the instructions in Section 1.7 - *Earth Engine authentication* to authenticate Earth Engine.

9.3   The map function

The map() function is probably one of the most important and powerful functions in Earth Engine. It can be used to apply a function to every element in ImageCollection, FeatureCollection, or List objects. The only argument to map() is a function which takes one parameter: the element in a collection/list to be processed. It then loops through the collection/list and applies the function to each element, making it possible to process large collections in parallel. Although the functionality of the map() function in Earth Engine is very similar to the for loops in Python, the use of for-loops is discouraged in Earth Engine unless you need to request data from the Earth Engine API to the client, e.g., displaying Earth Engine layers on the map or exporting Earth Engine images. Otherwise, the same results can be achieved using a map() operation where you specify a function that can be independently applied to each element. If the operation only involves a simple operation, then you might want to use a lambda function instead of a function definition.

The following example generates a list of numbers and applies a square function to each element:

```
myList = ee.List.sequence(1, 10)
myList
```

The output should look like this:

```
[1, 2, 3, 4, 5, 6, 7, 8, 9, 10]
```

Next, define a function that can be applied to the input. The map() function then takes the function that works on each element in the list independently and returns a value.

```
def computeSquares(number):
    return ee.Number(number).pow(2)


squares = myList.map(computeSquares)
squares
```

The output should look like this:

```
[1, 4, 9, 16, 25, 36, 49, 64, 81, 100]
```

Since the above function only involves a simple operation, it is possible to use a lambda function instead:

```
squares = myList.map(lambda number: ee.Number(number).pow(2))
squares
```

The output should be the same as the output in the previous step.

9.4   Creating cloud-free composites

To create a cloud-free Landsat composite, use the `ee.Algorithms.Landsat.simpleComposite()` method. The following example creates annual cloud-free Landsat composites for the Netherlands. It can be easily modified to create composites for other countries or regions.

First, use the Large Scale International Boundary (LSIB[158]) dataset and select the Netherlands out of the list of countries (see Fig. 9.1).

```python
Map = geemap.Map()
fc = ee.FeatureCollection('USDOS/LSIB_SIMPLE/2017').filter(
    ee.Filter.eq('country_na', 'Netherlands')
)

Map.addLayer(fc, {'color': 'ff000000'}, "Netherlands")
Map.centerObject(fc)
Map
```

Figure 9.1: Selecting the Netherlands from the Large Scale International Boundary (LSIB) dataset.

Next, create a list of years to be used by the `map()` function:

```python
years = ee.List.sequence(2013, 2022)
```

Define a function for creating cloud-free Landsat composites for each year:

```python
def yearly_image(year):

    start_date = ee.Date.fromYMD(year, 1, 1)
    end_date = start_date.advance(1, "year")

    collection = (
        ee.ImageCollection('LANDSAT/LC08/C02/T1')
        .filterDate(start_date, end_date)
        .filterBounds(fc)
```

```
        )

        image = ee.Algorithms.Landsat.simpleComposite(collection).clipToCollection(fc)

        return image
```

Apply the `map()` function to the list of years to create annual cloud-free Landsat composites:

```
    images = years.map(yearly_image)
```

Add each annual cloud-free Landsat composite to the map:

```
    vis_params = {'bands': ['B5', 'B4', 'B3'], 'max': 128}
    for index in range(0, 10):
        image = ee.Image(images.get(index))
        layer_name = "Year " + str(index + 2013)
        Map.addLayer(image, vis_params, layer_name)
    Map
```

The result should look like Fig. 9.2.

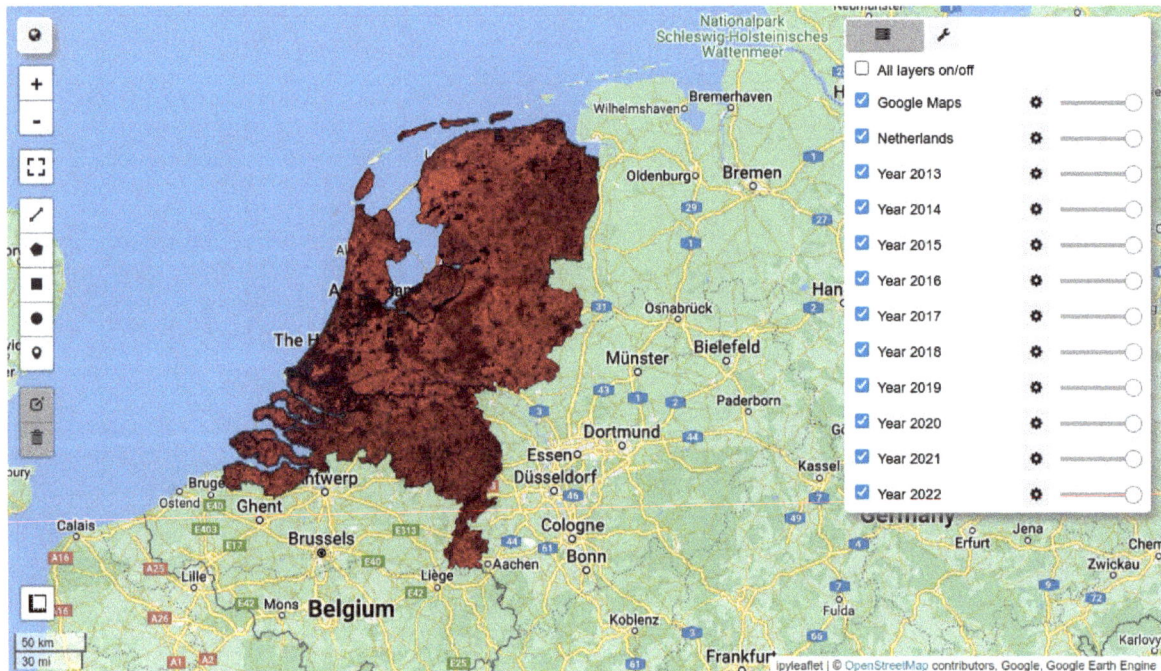

Figure 9.2: Annual cloud-free Landsat composites for Netherlands (2013-2022).

9.5 Creating time series

The previous section shows how to create annual cloud-free Landsat composites for a given region. One key step is to define a function that can be applied to each year through the `map()` function. What if we want to create monthly or quarterly cloud-free composites? We can modify the `yearly_image()` function to create monthly or quarterly composites, but we also need to create a list of monthly or quarterly time periods, which can be tedious and inefficient.

Fortunately, geemap provides a `create_timeseries()` function that can be used to create time series composites for a given region with a specified temporal frequency, such as hourly, daily, monthly, quarterly, or yearly. The input to the `create_timeseries()` function is an image collection, a start date, an end date, a region of interest (ROI), a temporal frequency, and a reducer to aggregate the images. The output is an image collection object with a single image for each time period. The following example creates yearly cloud-free Sentinel-2 composites for a given region.

First, let's filter the Harmonized Sentinel-2 image collection to select images with a cloud cover less than 10%:

```
collection = ee.ImageCollection("COPERNICUS/S2_HARMONIZED").filterMetadata(
    'CLOUDY_PIXEL_PERCENTAGE', 'less_than', 10
)
```

Next, specify the start date, end date, and region of interest:

```
start_date = '2016-01-01'
end_date = '2022-12-31'
region = ee.Geometry.BBox(-122.5549, 37.6968, -122.3446, 37.8111)
```

Lastly, create yearly cloud-free Sentinel-2 composites using the `create_timeseries()` function:

```
images = geemap.create_timeseries(
    collection, start_date, end_date, region, frequency='year', reducer='median'
)
images
```

Note that we used the `median` reducer to aggregate the images and create a time series. Other commonly used reducers include `mean`, `min`, `max`, `sum`, and `variance`. The output is an image collection with a single image for each year. Let's visualize the resulting image collection object using the time slider widget:

```
Map = geemap.Map()

vis_params = {"min": 0, "max": 4000, "bands": ["B8", "B4", "B3"]}
labels = [str(y) for y in range(2016, 2023)]

Map.addLayer(images, vis_params, "Sentinel-2", False)
Map.add_time_slider(images, vis_params, time_interval=2, labels=labels)
Map.centerObject(region)
Map
```

The result should look like Fig. 9.3.

Click on the play button or drag the time slider to animate the time series. Click on the pause button to pause the animation. Once you are done, click on the close button to remove the time slider widget from the map.

9.6 NAIP timelapse

The National Agriculture Imagery Program (NAIP) acquires aerial imagery during the agricultural growing seasons in the continental United States. The NAIP program started in 2003 and has been collecting imagery covering the continental United States on a 2-3 year cycle. Older images were collected using 3 bands (Red, Green, and Blue: RGB), but newer imagery is usually collected with an additional near-infrared band (RGBN). NAIP imagery can be accessed in Earth Engine through the image collection asset id `"USDA/NAIP/DOQQ"`. The following example illustrates how to create NAIP imagery time series and timelapse for any location in the United States.

First, create an interactive map centered on the United States:

```
Map = geemap.Map(center=[40, -100], zoom=4)
Map
```

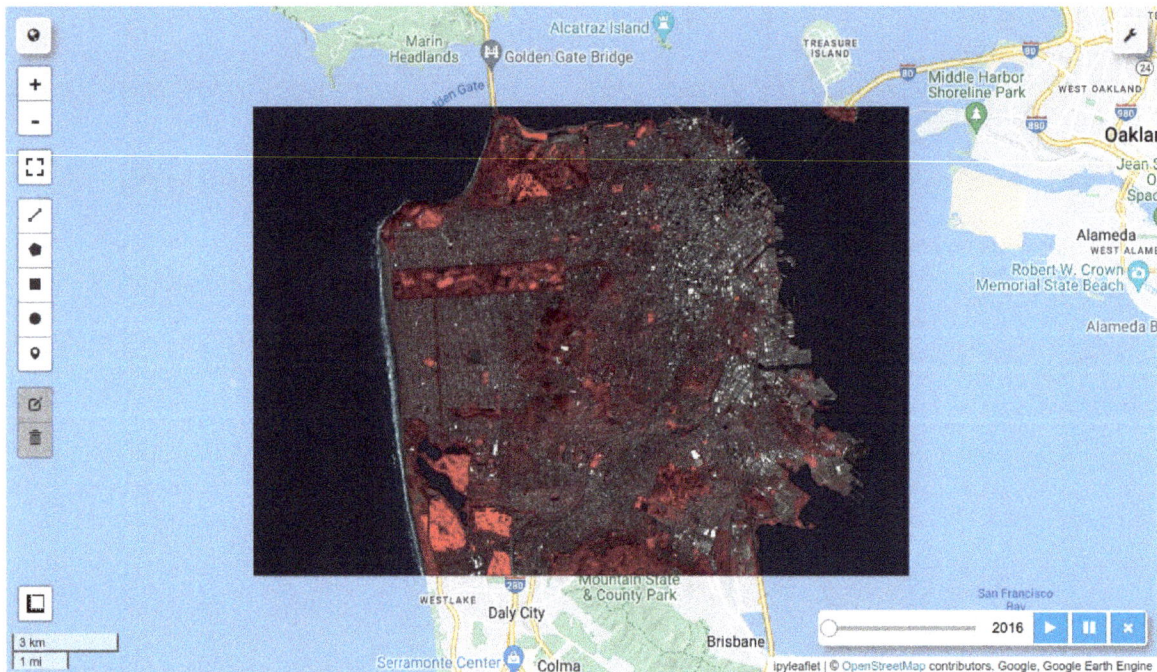

Figure 9.3: Annual Sentinel-2 time series for the San Francisco Bay Area (2016-2022).

Pan and zoom the map to the region of interest (ROI) and then use the draw tools to draw a rectangle or polygon around the ROI. If no ROI is drawn, a default ROI around central North Dakota will be used:

```
roi = Map.user_roi
if roi is None:
    roi = ee.Geometry.BBox(-99.1019, 47.1274, -99.0334, 47.1562)
    Map.addLayer(roi)
    Map.centerObject(roi)
```

Use `naip_timeseries()` to create a NAIP imagery time series. Specify `start_year` and `end_year` to create a time series for a specific range of years. Set `RGBN=True` to select 4-band imagery only, which is only available for the year 2008 and later.

```
collection = geemap.naip_timeseries(roi, start_year=2009, end_year=2022, RGBN=True)
```

Use `image_dates()` to get a list of dates for each image in the time series:

```
years = geemap.image_dates(collection, date_format='YYYY').getInfo()
print(years)
```

The output should look like the following:

```
['2009', '2010', '2012', '2014', '2015', '2016', '2017', '2018', '2019', '2020']
```

Note that NAIP imagery is available for all years except 2011 and 2013. To add the annual imagery to the map, use a for loop to iterate through the collection and add each image to the map (see Fig. 9.4).

```
size = len(years)
images = collection.toList(size)
for i in range(size):
    image = ee.Image(images.get(i))
    Map.addLayer(image, {'bands': ['N', 'R', 'G']}, years[i])
```

Map

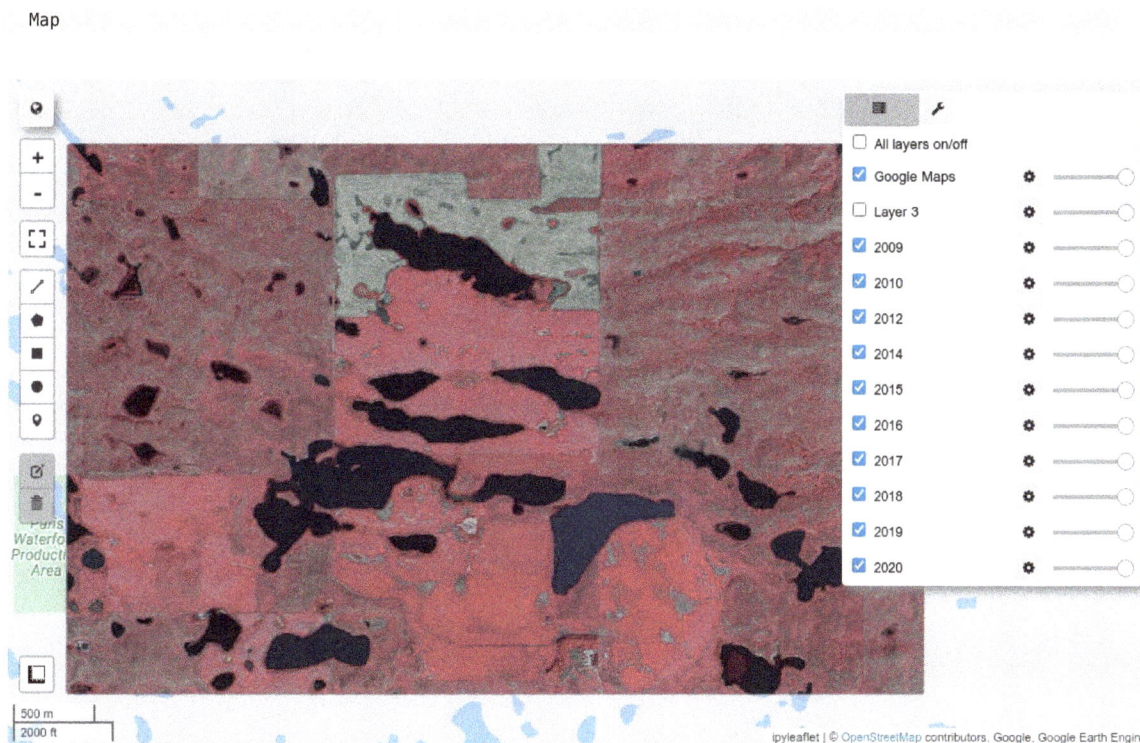

Figure 9.4: Adding NAIP imagery time series to the map.

The NAIP imagery time series on the map allows you to explore the imagery over time. Zoom in on the map to see the imagery in detail. Toggle layers on and off to compare the imagery from different years. To create a timelapse based on all available imagery for a specific location, use the `naip_timelapse()` function. Simply provide a region of interest and specify the output GIF location. The `frames_per_second` parameter controls the speed of the timelapse, i.e., how many frames are created per second. The larger the number, the faster the timelapse, and vice versa.

```
timelapse = geemap.naip_timelapse(
    roi,
    out_gif="naip.gif",
    bands=['N', 'R', 'G'],
    frames_per_second=3,
    title='NAIP Timelapse',
)
geemap.show_image(timelapse)
```

The output should look like Fig. 9.5.

9.7   Landsat timelapse

The fine-resolution NAIP imagery only covers the continental United States and is not available for other regions. Landsat imagery, on the other hand, is available for the entire globe. It has the longest time series of any satellite imagery, with imagery available since 1972. Landsat imagery is probably the most widely used satellite imagery for monitoring land cover and land use changes. In this section, we will walk through several examples of how to create a Landsat timelapse for any location in the world.

First, let's create an interactive map:

```
Map = geemap.Map()
```

Figure 9.5: A NAIP imagery timelapse.

```
Map
```

Pan and zoom the map to the region of interest (ROI) and then use the draw tools to draw a rectangle or polygon around the ROI. If no ROI is drawn, a default ROI around central Peru will be used as follows:

```
roi = Map.user_roi
if roi is None:
    roi = ee.Geometry.BBox(-74.7222, -8.5867, -74.1596, -8.2824)
    Map.addLayer(roi)
    Map.centerObject(roi)
```

With the ROI drawn, use the landsat_timeseries() function to create a Landsat timelapse. Specify start_year and end_year to create a timelapse for a specific range of years. If you are only interested in imagery acquired in a specific season, you can specify the start_date and end_date parameters. The default is 01-01 and 12-31, which means the entire year. The bands parameter allows you to select the bands to be used in the timelapse. Note that the landsat_timeseries() function incorporates imagery from Landsat 5, 7, 8, and 9. All band names have been standardized. The available bands are ['Blue', 'Green', 'Red', 'NIR', 'SWIR1', 'SWIR2'], and the default bands are ['NIR', 'Red', 'Green']. The frames_per_second parameter controls the speed of the timelapse, i.e., how many frames are created per second. The larger the number, the faster the timelapse, and vice versa. To add a progress bar to the timelapse, set progress_bar_color to a color name or hex code. The mp4 parameter controls whether to create an MP4 video file in addition to the GIF file. Check out the Landsat function documentation[180] for more details about other parameters.

Below is an example of creating a Landsat timelapse for the ROI around Pucallpa, Peru:

```
timelapse = geemap.landsat_timelapse(
    roi,
    out_gif='landsat.gif',
```

---

[180]Landsat function documentation: tiny.geemap.org/pmsf

```
    start_year=1984,
    end_year=2022,
    start_date='01-01',
    end_date='12-31',
    bands=['SWIR1', 'NIR', 'Red'],
    frames_per_second=5,
    title='Landsat Timelapse',
    progress_bar_color='blue',
    mp4=True,
)
geemap.show_image(timelapse)
```

The output should look like Fig. 9.6.

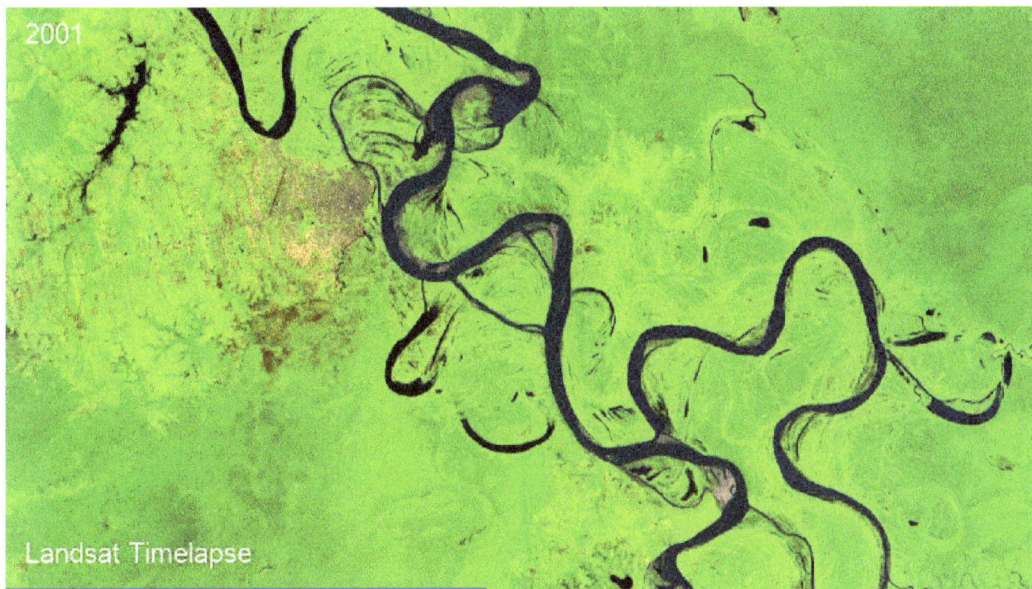

Figure 9.6: A Landsat timelapse of river dynamics in Pucallpa, Peru.

Here is another Landsat timelapse of urban expansion in Las Vegas, Nevada:

```
Map = geemap.Map()
roi = ee.Geometry.BBox(-115.5541, 35.8044, -113.9035, 36.5581)
Map.addLayer(roi)
Map.centerObject(roi)
Map

timelapse = geemap.landsat_timelapse(
    roi,
    out_gif='las_vegas.gif',
    start_year=1984,
    end_year=2022,
    bands=['NIR', 'Red', 'Green'],
    frames_per_second=5,
    title='Las Vegas, NV',
    font_color='blue',
)
geemap.show_image(timelapse)
```

The output should look like Fig. 9.7.

A Landsat timelapse of land reclamation in Hong Kong:

Figure 9.7: A Landsat timelapse of urban expansion in Las Vegas, Nevada.

```python
Map = geemap.Map()
roi = ee.Geometry.BBox(113.8252, 22.1988, 114.0851, 22.3497)
Map.addLayer(roi)
Map.centerObject(roi)
Map

timelapse = geemap.landsat_timelapse(
    roi,
    out_gif='hong_kong.gif',
    start_year=1990,
    end_year=2022,
    start_date='01-01',
    end_date='12-31',
    bands=['SWIR1', 'NIR', 'Red'],
    frames_per_second=3,
    title='Hong Kong',
)
geemap.show_image(timelapse)
```

The output should look like Fig. 9.8.

9.8   Sentinel-1 timelapse

The Sentinel-1 mission provides data from a dual-polarization C-band Synthetic Aperture Radar (SAR) instrument at 5.405GHz (C band). Sentinel-1 Ground Range Detected (GRD) scenes are available in the Earth Engine Data Catalog with image collection ID COPERNICUS/S1_GRD[181]. Sentinel-1 SAR data have been widely used for mapping the extent of floods, monitoring glacier dynamics, and detecting land subsidence.

The `sentinel1_timelapse()` function allows you to create a Sentinel-1 timelapse for any location around the world. The function is similar to the `landsat_timelapse()` function. The bands parameter allows

---

[181]COPERNICUS/S1_GRD: tiny.geemap.org/gbea

Figure 9.8: A Landsat timelapse of land reclamation in Hong Kong.

you to select the bands to be used in the timelapse. The available bands are ['HH', 'HV', 'VV', 'VH'], and the default bands are ['VV']. Since only one band is used to create timelapse, we can specify a color palette and add a color bar to the timelapse. Change the frequency parameter to day to create a daily timelapse. Since Sentinel-1 data doesn't have daily coverage, the function will automatically select all images available for the specified date range. Check out the Sentinel-1 function documentation[182] for more details about other parameters.

Below is an example of creating a Sentinel-1 timelapse for the ROI around northeastern Indonesia. First, create an interactive map:

```
Map = geemap.Map()
Map
```

Pan and zoom to the area of interest. Then, draw a rectangle to define the ROI:

```
roi = Map.user_roi
if roi is None:
    roi = ee.Geometry.BBox(117.1132, 3.5227, 117.2214, 3.5843)
    Map.addLayer(roi)
    Map.centerObject(roi)
```

Lastly, use the sentinel1_timelapse() function to create a Sentinel-1 timelapse:

```
timelapse = geemap.sentinel1_timelapse(
    roi,
    out_gif='sentinel1.gif',
    start_year=2019,
    end_year=2019,
    start_date='04-01',
    end_date='08-01',
```

---

[182]Sentinel-1 function documentation: tiny.geemap.org/rdur

```
        bands=['VV'],
        frequency='day',
        vis_params={"min": -30, "max": 0},
        palette="Greys",
        frames_per_second=3,
        title='Sentinel-1 Timelapse',
        add_colorbar=True,
        colorbar_bg_color='gray',
    )
    geemap.show_image(timelapse)
```

The output should look like Fig. 9.9.

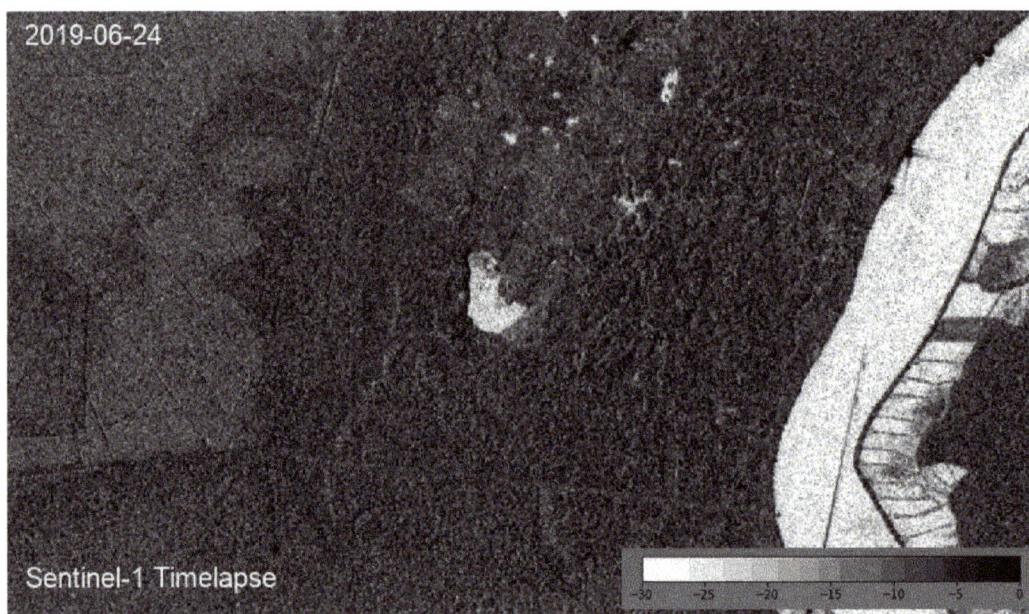

Figure 9.9: A Sentinel-1 timelapse of northeastern Indonesia.

9.9 Sentinel-2 timelapse

The Landsat mission has the longest archive of multi-spectral satellite imagery since 1984 with a 30-m resolution. In contrast, Sentinel-2 has a 10-m resolution and has been operational since 2015.

Similar to the landsat_timelapse() function, we can use the sentinel2_timelapse() function to create a Sentinel-2 timelapse for any location around the world. All the function parameters are exactly the same as the landsat_timelapse() function. The only difference is that the bands parameter allows you to select the bands to be used in the timelapse. The available bands include 'Blue', 'Green', 'Red', 'NIR', 'SWIR1', 'SWIR2', 'Red Edge 1', 'Red Edge 2', 'Red Edge 3', and 'Red Edge 4'. Note that Sentinel-2 has four red edge bands, which Landsat does not have. Check out the Sentinel-2 function documentation[183] to learn more about the other parameters.

Below is an example of creating a Sentinel-2 timelapse for the ROI around Pucallpa, Peru. First, create an interactive map:

```
Map = geemap.Map()
Map
```

---

[183]Sentinel-2 function documentation: tiny.geemap.org/zkxl

Pan and zoom to the area of interest. Then, draw a rectangle to define the ROI:

```python
roi = Map.user_roi
if roi is None:
    roi = ee.Geometry.BBox(-74.7222, -8.5867, -74.1596, -8.2824)
    Map.addLayer(roi)
    Map.centerObject(roi)
```

Lastly, use the `sentinel2_timelapse()` function to create a Sentinel-2 timelapse:

```python
timelapse = geemap.sentinel2_timelapse(
    roi,
    out_gif='sentinel2.gif',
    start_year=2016,
    end_year=2021,
    start_date='01-01',
    end_date='12-31',
    frequency='year',
    bands=['SWIR1', 'NIR', 'Red'],
    frames_per_second=3,
    title='Sentinel-2 Timelapse',
)
geemap.show_image(timelapse)
```

The output should look like Fig. 9.10.

Figure 9.10: A Sentinel-2 timelapse of river dynamics in Pucallpa, Peru.

9.10 MODIS timelapse

The Landsat and Sentinel timelapse functions introduced above are suitable for creating timelapse animations from local to regional scales. However, they are not suitable for creating timelapse animations at continental to global scales due to the large volume of data and the intensive computing resources needed to process these data. In contrast, Moderate Resolution Imaging Spectroradiometer (MODIS) data have a coarse resolution of 1,000 meters and are more suitable for creating timelapse animations at continental to global scales. MODIS is a key instrument aboard the Terra and Aqua satellites, which have been operational since 2000 and 2002, respectively.

In this section, we will learn how to create MODIS NDVI and temperature timelapse animations at continental to global scales.

MODIS vegetation indices

The MOD13A2 product[184] provides two Vegetation Indices (VI): the Normalized Difference Vegetation Index (NDVI) and the Enhanced Vegetation Index (EVI). Both indices can be used to monitor vegetation health. Geemap has a `modis_ndvi_timelapse()` function for timelapse animations of long-term average NDVI/EVI.

First, create an interactive map:

```
Map = geemap.Map()
Map
```

Pan and zoom to the area of interest, such as Africa. Then, draw a rectangle to define the ROI:

```
roi = Map.user_roi
if roi is None:
    roi = ee.Geometry.BBox(-18.6983, -36.1630, 52.2293, 38.1446)
    Map.addLayer(roi)
    Map.centerObject(roi)
```

Lastly, use the `modis_ndvi_timelapse()` function to create a MODIS NDVI timelapse. The `data` parameter allows you to select either a Terra or Aqua satellite. The band parameter allows you to select either NDVI or EVI. The `start_date` and `end_date` parameters specify the time range of MODIS data used to compute the long-term average NDVI/EVI.

For continental to global scale timelapse animations, it might be helpful to overlay the administrative boundaries on the timelapse. You can do so by setting the `overlay_data` parameter, which accepts a string, an `ee.Geometry` object, or an `ee.FeatureCollection` object. The string can be one of the following: `continents`, `countries`, `us_states`, or an HTTP link to a GeoJSON file.

Here, we create a MODIS NDVI timelapse of Africa with country boundaries overlaid on it:

```
timelapse = geemap.modis_ndvi_timelapse(
    roi,
    out_gif='ndvi.gif',
    data='Terra',
    band='NDVI',
    start_date='2000-01-01',
    end_date='2022-12-31',
    frames_per_second=3,
    title='MODIS NDVI Timelapse',
    overlay_data='countries',
)
geemap.show_image(timelapse)
```

The output should look like Fig. 9.11.

You can clearly see how the vegetation on the continent changes over time depending on the season, which is quite interesting. You can also create a timelapse of EVI by setting the band parameter to EVI.

[184]MOD13A2 product: tiny.geemap.org/icwu

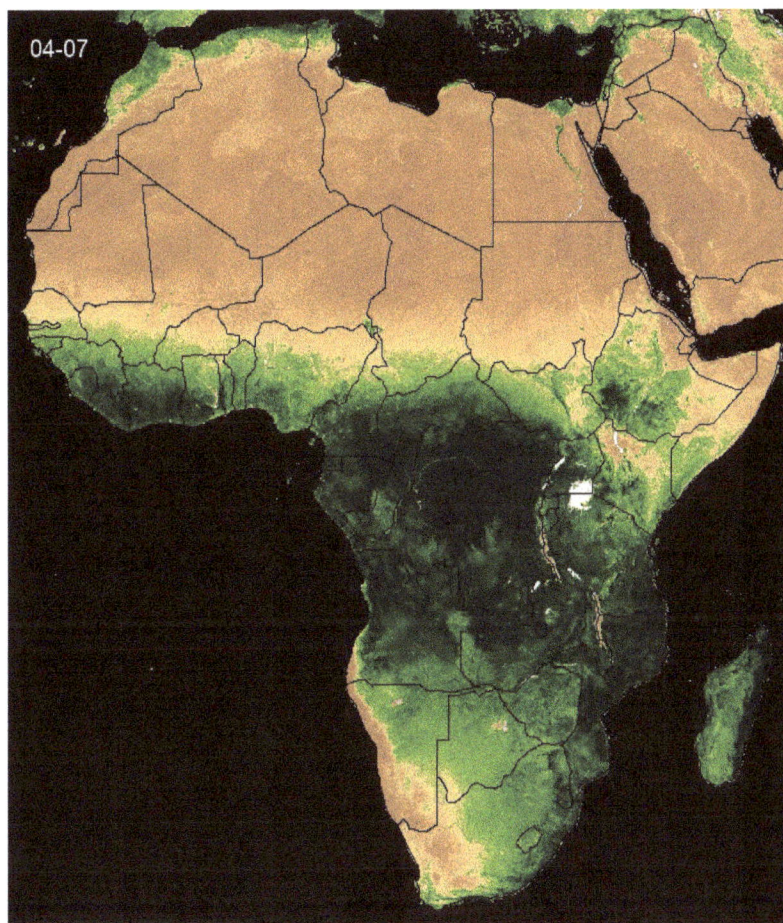

Figure 9.11: A MODIS NDVI timelapse of Africa.

MODIS temperature

The MODIS Ocean Color Standard Mapped Image (SMI)[185] product provides monthly average sea surface temperature (SST) data. The following example shows how to create a timelapse of global SST using the `modis_ocean_color_timelapse()` function.

First, create an interactive map:

```
Map = geemap.Map()
Map
```

Pan and zoom to the area of interest. Then, draw a rectangle to define the ROI. If you don't draw a rectangle, the function will use the default ROI, which is the entire globe:

```
roi = Map.user_roi
if roi is None:
    roi = ee.Geometry.BBox(-171.21, -57.13, 177.53, 79.99)
    Map.addLayer(roi)
    Map.centerObject(roi)
```

---

[185]MODIS Ocean Color Standard Mapped Image (SMI): tiny.geemap.org/ilia

Lastly, use the `modis_ocean_color_timelapse()` function to create a monthly MODIS temperature time-lapse. The `satellite` parameter allows you to select either Aqua or Terra satellite. Set the `overlay_data` parameter to `continents` to overlay the continents on the timelapse. By default, the `add_colorbar` parameter is set to `True`, which adds a colorbar to the timelapse.

```
timelapse = geemap.modis_ocean_color_timelapse(
    satellite='Aqua',
    start_date='2018-01-01',
    end_date='2020-12-31',
    roi=roi,
    frequency='month',
    out_gif='temperature.gif',
    overlay_data='continents',
    overlay_color='yellow',
    overlay_opacity=0.5,
)
geemap.show_image(timelapse)
```

The output should look like Fig. 9.12.

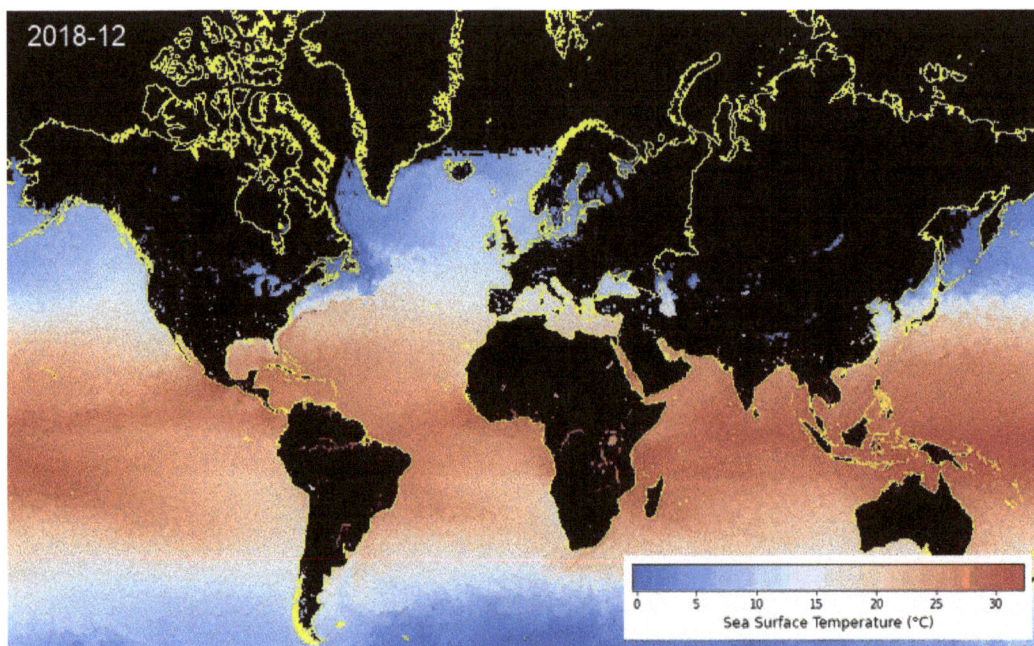

Figure 9.12: A MODIS sea surface temperature timelapse of the world.

9.11 GOES timelapse

The Geostationary Operational Environmental Satellites (GOES) provide advanced imagery and atmospheric measurements of Earth's weather, oceans, and environment. There are two GOES satellites in orbit at any given time: GOES-16 and GOES-17. The Earth Engine Data Catalog hosts the Cloud and Moisture Imagery (CMI) and Fire/Hotspot Characterization (FHS) data products from GOES satellites. The Advanced Baseline Imager (ABI) instrument on board the GOES satellites have three scan types: full disk, CONUS, and mesoscale. For more information about the GOES datasets on Earth Engine, check out the GOES in Earth Engine[186] post by Justin Braaten.

---

[186]GOES in Earth Engine: tiny.geemap.org/bjfn

The GOES full disk scan mode can acquire data from the entire Earth every 10 minutes, and the data is generally available for access in Earth Engine within an hour of observation. The high temporal frequency and global coverage make GOES data particularly useful for weather monitoring and disaster response in near real-time. In this section, we will learn how to create GOES timelapse animations of volcanic eruptions, hurricanes, and wildfires.

First, let's create a timelapse of the volcanic eruption in Hunga Tonga[187] on January 15, 2022.

```
roi = ee.Geometry.BBox(167.1898, -28.5757, 202.6258, -12.4411)
start_date = "2022-01-15T03:00:00"
end_date = "2022-01-15T07:00:00"
data = "GOES-17"
scan = "full_disk"
```

Keep in mind that GOES data have a very high temporal frequency (every 10 minutes), so you may want to limit the ROI and time period to avoid computation timeout. Avoid selecting a time period (the time difference between the start_date and end_date) longer than 24 hours.

```
timelapse = geemap.goes_timelapse(
    roi, "goes.gif", start_date, end_date, data, scan, framesPerSecond=5
)
geemap.show_image(timelapse)
```

The output should look like Fig. 9.13.

Figure 9.13: A GOES timelapse of the Hunga Tonga volcanic eruption.

A GOES timelapse of the hurricane hitting the West Coast of the United States on October 24, 2021:

```
roi = ee.Geometry.BBox(-159.5954, 24.5178, -114.2438, 60.4088)
start_date = "2021-10-24T14:00:00"
end_date = "2021-10-25T01:00:00"
data = "GOES-17"
scan = "full_disk"
```

---

[187]volcanic eruption in Hunga Tonga: tiny.geemap.org/efrt

```
timelapse = geemap.goes_timelapse(
    roi, "hurricane.gif", start_date, end_date, data, scan, framesPerSecond=5
)
geemap.show_image(timelapse)
```

The output should look like Fig. 9.14.

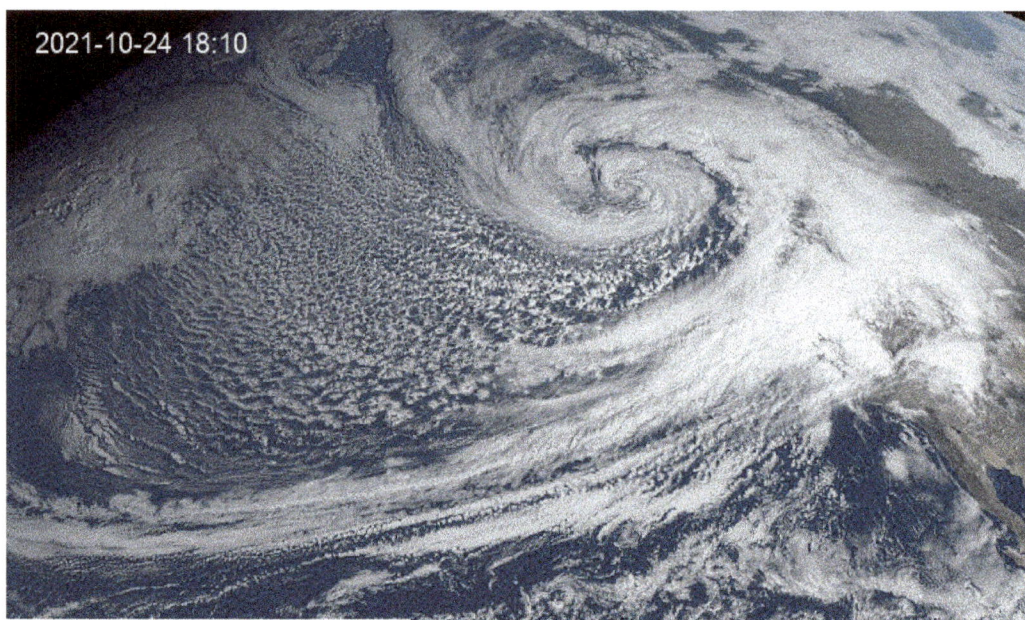

Figure 9.14: A GOES timelapse of a hurricane.

In addition to hosting the Cloud and Moisture Imagery (CMI) data products, the Earth Engine Data Catalog hosts the Fire/Hotspot Characterization (FHS)[188] data product, which includes estimates of wildland fire area, temperature, and power. Google uses Earth Engine and the GOES Fire/Hotspot Characterization data operationally to develop near real-time fire boundary layers that are displayed in alerts on Google Search and Google Maps. The following code creates a timelapse of the Creek Fire[189] in California:

```
roi = ee.Geometry.BBox(-121.0034, 36.8488, -117.9052, 39.0490)
start_date = "2020-09-05T15:00:00"
end_date = "2020-09-06T02:00:00"
data = "GOES-17"
scan = "full_disk"

timelapse = geemap.goes_fire_timelapse(
    roi, "fire.gif", start_date, end_date, data, scan, framesPerSecond=5
)
geemap.show_image(timelapse)
```

The output should look like Fig. 9.15.

---

[188]Fire/Hotspot Characterization (FHS): tiny.geemap.org/iwgi

[189]Creek Fire: tiny.geemap.org/sbyb

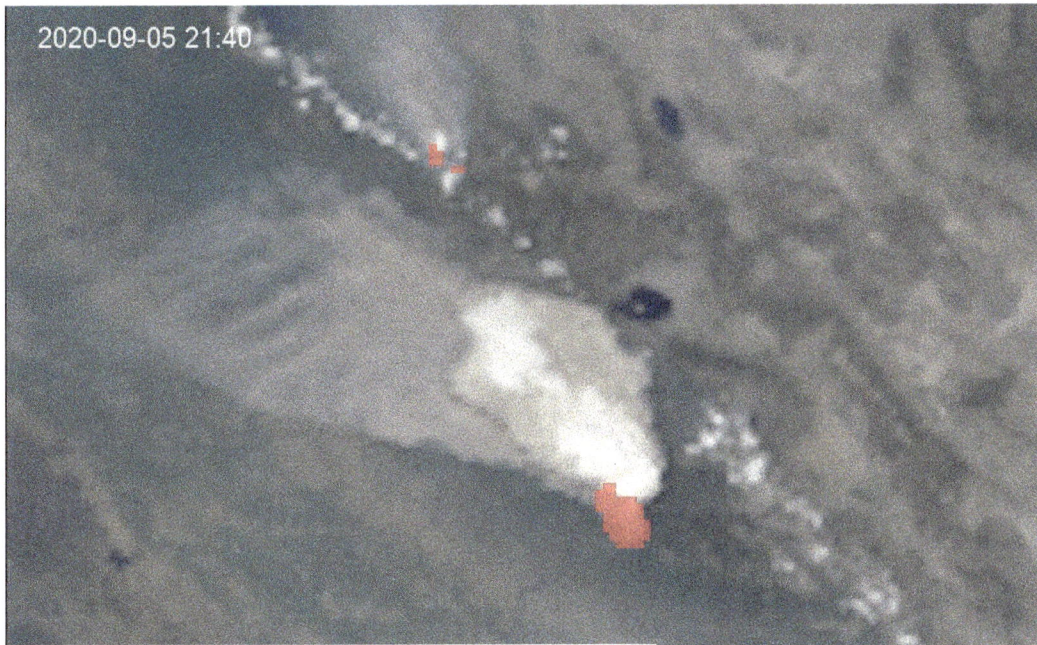

Figure 9.15: A GOES timelapse of the Creek Fire in California.

9.12 Fading effects

If a timelapse animation is too abrupt, you can add fading effects to make the transition smoother. You can add fading effects to existing GIF images or create a timelapse from scratch using the timelapse functions introduced in the previous sections.

Here is a Landsat timelapse without fading effect:

```
in_gif = "https://i.imgur.com/ZWSZC5z.gif"
geemap.show_image(in_gif)
```

The output should look like Fig. 9.16.

Use the gif_fading() function to add fading effects to an existing GIF image:

```
out_gif = "gif_fading.gif"
geemap.gif_fading(in_gif, out_gif, verbose=False)
geemap.show_image(out_gif)
```

The output should look like Fig. 9.17.

Alternatively, you can create a timelapse from scratch using the landsat_timelapse() function and set the fading parameter to True:

```
roi = ee.Geometry.BBox(-69.3154, -22.8371, -69.1900, -22.7614)
timelapse = geemap.landsat_timelapse(
    roi,
    out_gif='mines.gif',
    start_year=2004,
    end_year=2010,
    frames_per_second=1,
    title='Copper mines, Chile',
    fading=True,
)
```

Figure 9.16: A Landsat timelapse without fading effect.

Figure 9.17: A Landsat timelapse with fading effect.

```
geemap.show_image(timelapse)
```

The output should look like Fig. 9.18.

Figure 9.18: A Landsat timelapse of copper mines in Chile with fading effect.

9.13 Adding text to timelapse

The timelapse functions introduced above allow adding animated text to the GIF images. If you have some existing GIF images without animated text content, you can use the add_text_to_gif() function to add text and labels to the GIF images with only one line of code.

Let's download a sample GIF image of a Landsat timelapse (1984-2022) of urban expansion in Las Vegas, which contains 39 frames.

```
url = 'https://i.imgur.com/Rx0wjSw.gif'
in_gif = 'animation.gif'
geemap.download_file(url, in_gif)
geemap.show_image(in_gif)
```

The GIF image without text is shown in Fig. 9.19.

To add animated text to the image, use the add_text_to_gif() function. The text_sequence parameter accepts an integer, a string, or a list of strings. If an integer is provided, a list of text labels will be automatically generated by increasing the integer by 1. For example, if you set text_sequence to 1984, the function will generate a list of text labels from 1984 to the current year. If a list of strings is provided, the function will use the list of strings as the text labels. For example, if you set text_sequence to ["1984", "1985", "1986", ..., "2022"], the function will use the list of strings as the text labels. The xy parameter specifies the location of the text label. Optionally, you can add an animated progress bar to the image and customize the progress bar height and color.

Figure 9.19: A GIF image without text.

```
out_gif = 'las_vegas.gif'
geemap.add_text_to_gif(
    in_gif,
    out_gif,
    xy=('3%', '5%'),
    text_sequence=1984,
    font_size=30,
    font_color='#0000ff',
    add_progress_bar=True,
    progress_bar_color='#ffffff',
    progress_bar_height=5,
    duration=100,
    loop=0,
)
geemap.show_image(out_gif)
```

Besides adding animated text to a GIF image, you can add static text (e.g., title, footnote) to a GIF image using the add_text_to_gif() function. Instead of providing an integer or a list of strings to text_sequence, you can provide a string to text_sequence. The string will be added to the GIF image as static text. For example, if you set text_sequence to "Las Vegas", the function will add the text "Las Vegas" to the GIF image.

```
geemap.add_text_to_gif(
    out_gif, out_gif, xy=('45%', '90%'), text_sequence="Las Vegas", font_color='black'
)
geemap.show_image(out_gif)
```

The GIF image with animated text is shown in Fig. 9.20.

9.14   Adding image and colorbar to timelapse

In this section, we will learn how to download an ImageCollection object as a GIF image and add some text, a logo, and a colorbar to the GIF image.

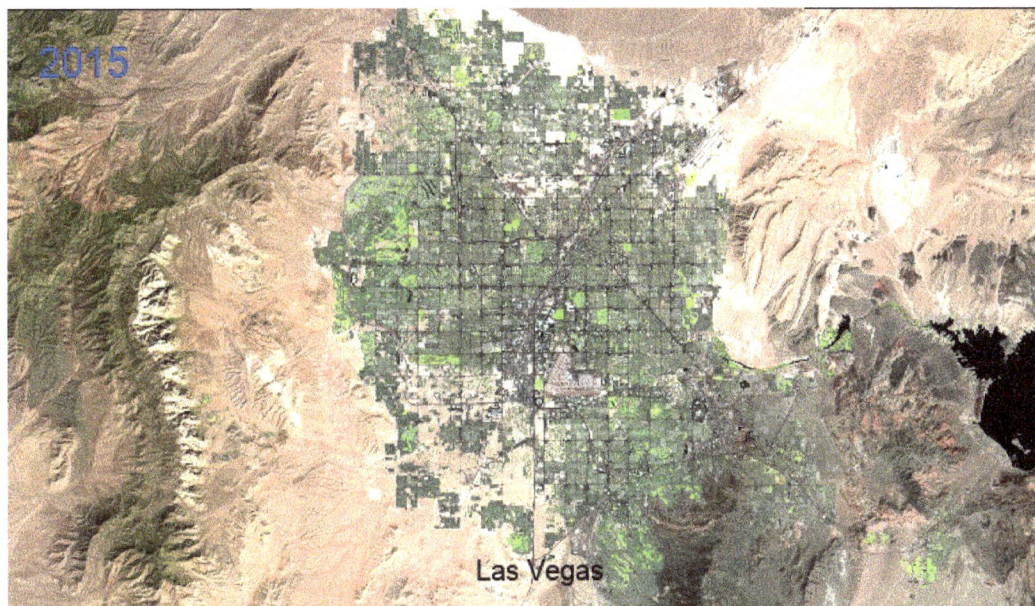

Figure 9.20: A GIF image with animated text.

Preparing data

Contrary to the `create_timelapse()` function that requires specifying a temporal frequency (e.g., yearly, monthly, daily), the `download_ee_video()` function allows downloading an image collection as a GIF image without having to create a time series first. One of the required parameters is `video_args`, which is passed to the `ee.ImageCollection.getVideoThumbURL()` function to generate the GIF image. The `video_args` parameter is a dictionary that contains keys such as `dimensions`, `region`, `framesPerSecond`, `crs`, `min`, `max`, and `palette`.

In this example, we will use the Global Forecast System (GFS)[190] Predicted Atmosphere Data which consists of selected model outputs as gridded forecast variables. The `temperature_2m_above_ground` band represents the predicted temperature at 2 meters above the ground. Let's filter the image collection to select the images of predicted temperature at 2 meters above the ground on December 22, 2018, which contain 24 images (one image per hour). The `limit()` function is used to limit the number of images to 24. Then, define the `video_args` parameter to generate the image thumbnails. Lastly, download the thumbnails as a GIF image using the `download_ee_video()` function:

```
aoi = ee.Geometry.Polygon(
    [[[-179.0, 78.0], [-179.0, -58.0], [179.0, -58.0], [179.0, 78.0]]], None, False
)

collection = (
    ee.ImageCollection('NOAA/GFS0P25')
    .filterDate('2018-12-22', '2018-12-23')
    .limit(24)
    .select('temperature_2m_above_ground')
)

video_args = {
    'dimensions': 768,
    'region': aoi,
    'framesPerSecond': 10,
```

---

[190]Global Forecast System (GFS): tiny.geemap.org/lsbv

```
    'crs': 'EPSG:3857',
    'min': -35.0,
    'max': 35.0,
    'palette': ['blue', 'purple', 'cyan', 'green', 'yellow', 'red'],
}

saved_gif = 'temperature.gif'
geemap.download_ee_video(collection, video_args, saved_gif)
geemap.show_image(saved_gif)
```

The downloaded GIF image is shown in Fig. 9.21.

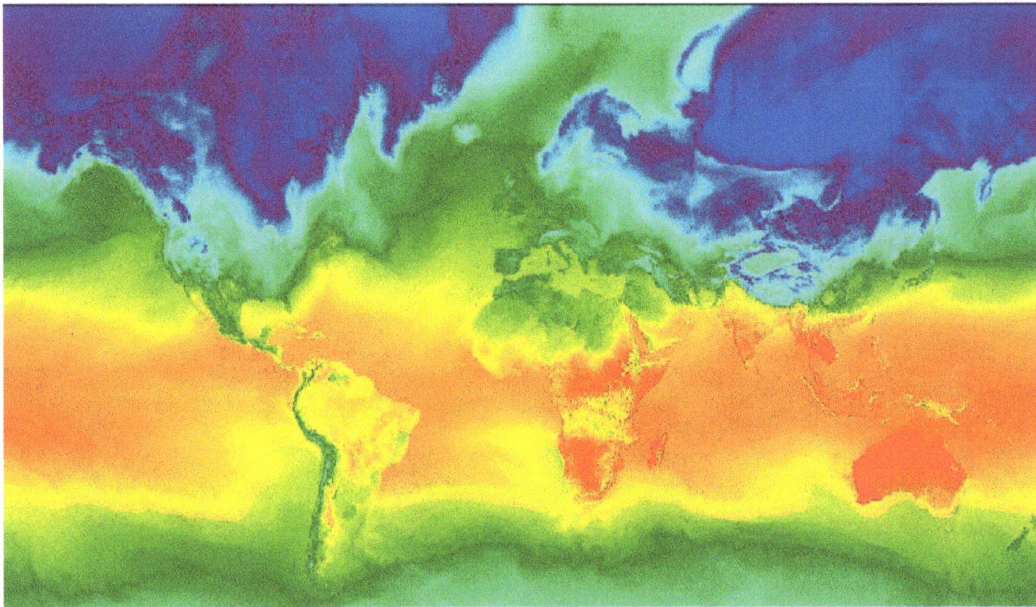

Figure 9.21: A timelapse of predicted global temperature on December 22, 1018.

Adding animated text

The downloaded GIF image does not have any text. We can use the `add_text_to_gif()` function to add text to the GIF image. Let's add the animated timestamps and title to the image:

```
text = [str(n).zfill(2) + ":00" for n in range(0, 24)]
out_gif = 'temperature_v2.gif'
geemap.add_text_to_gif(
    saved_gif,
    out_gif,
    xy=('3%', '5%'),
    text_sequence=text,
    font_size=30,
    font_color='#ffffff',
)

geemap.add_text_to_gif(
    out_gif,
    out_gif,
    xy=('32%', '92%'),
    text_sequence='NOAA GFS Hourly Temperature',
    font_color='white',
)
geemap.show_image(out_gif)
```

Adding logo

NOAA produces the GFS dataset and Google Earth Engine provides the computing power to process the data. To give credits to them, let's add the NOAA and Google Earth Engine logos to the GIF image using the `add_image_to_gif()` function:

```
noaa_logo = 'https://i.imgur.com/gZ6BYZB.png'
ee_logo = 'https://i.imgur.com/Qbvacvm.jpg'

geemap.add_image_to_gif(
    out_gif, out_gif, in_image=noaa_logo, xy=('2%', '80%'), image_size=(80, 80)
)

geemap.add_image_to_gif(
    out_gif, out_gif, in_image=ee_logo, xy=('13%', '79%'), image_size=(85, 85)
)
```

Adding colorbar

To add a colorbar to the GIF image, we first need to generate a colorbar and save it as an image using the `save_colorbar()` function:

```
palette = ['blue', 'purple', 'cyan', 'green', 'yellow', 'red']
colorbar = geemap.save_colorbar(width=2.5, height=0.3, vmin=-35, vmax=35, palette=palette, transparent=True)
geemap.show_image(colorbar)
```

If no output file path is specified, the colorbar will be saved to the system temporary directory as a PNG image, which can then be added to the GIF image just like the logos we added above:

```
geemap.add_image_to_gif(
    out_gif, out_gif, in_image=colorbar, xy=('72%', '85%'), image_size=(250, 250)
)
geemap.show_image(out_gif)
```

The final GIF image is shown in Fig. 9.22.

9.15   Summary

In this chapter, we learned the powerful map function for processing image collections very efficiently. We also walked through many practical examples of creating timelapse animations from various remote sensing datasets. You should now feel comfortable creating satellite timelapse animations from any image collection in Earth Engine. Feel free to explore the Earth Engine Data Catalog and find some interesting datasets to create your own timelapse animations.

In the next chapter, we will learn how to create interactive Earth Engine web apps with various web app frameworks, such as Voila, Streamlit, and Solara.

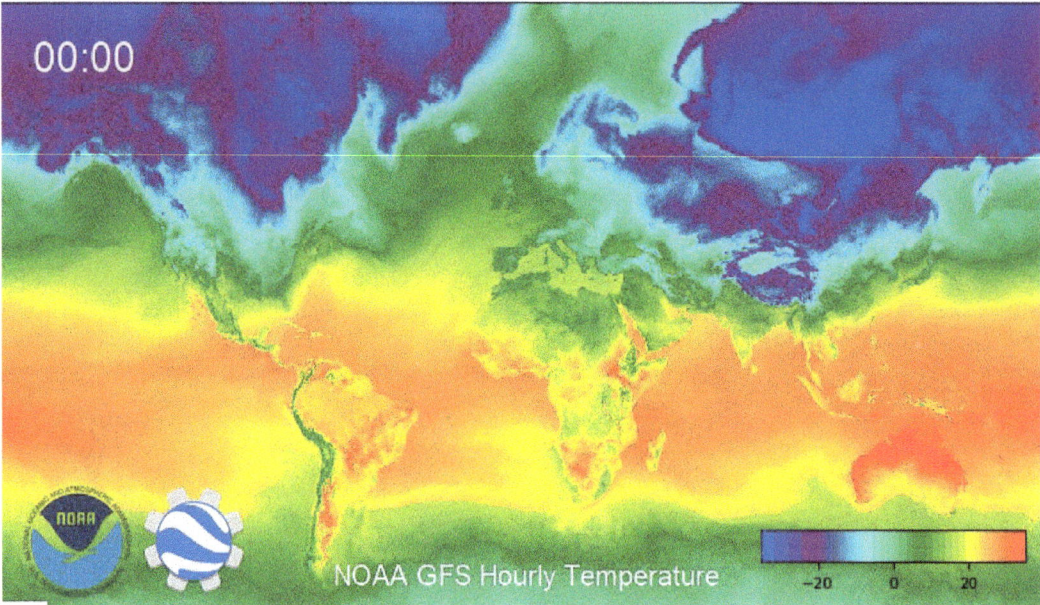

Figure 9.22: A timelapse of predicted global temperature on December 22, 1018 with logo and colorbar.

10. Building Interactive Web Apps

10.1 Introduction

Earth Engine has a User Interface API that allows users to build and publish interactive web apps directly from the Earth Engine JavaScript Code Editor. Users can utilize the ui package to construct a graphical user interface (GUI) for their Earth Engine script. The GUI may include simple widgets (e.g., Labels, Buttons, Checkboxes, Sliders, Textboxes) as well as more complex widgets (e.g., Charts, Maps, Panels) for controlling the GUI layout. See ui widgets[191] and Panels[192] for more information. Once a GUI is constructed, users can publish their app from the JavaScript Code Editor by clicking the **Apps** button above the script section in the Code Editor.

Contrary to the Google Earth Engine JavaScript API, the GEE Python API does not provide functionality for building interactive user interfaces. Fortunately, the Jupyter ecosystem has **ipywidgets**, an architecture for creating interactive user interface controls (e.g., Buttons, Sliders, Checkboxes, Textboxes, Dropdowns) in Jupyter notebooks that communicate with Python code. This integration of graphical widgets into the notebook workflow allows users to configure ad-hoc control panels to interactively sweep over parameters using graphical widget controls. One very powerful widget is the Output widget[193] , which can be used to display rich output generated by IPython, such as text, images, charts, videos. By integrating ipyleaflet (for creating interactive maps) and ipywidgets (for designing interactive user interfaces), the geemap Python package makes it much easier to explore and analyze massive Earth Engine datasets via a web browser, in a Jupyter environment suitable for interactive exploration, teaching, and sharing. Users can build interactive Earth Engine web apps using geemap with minimal coding (see Fig. 10.1).

In this chapter, you will learn how to build interactive web apps using the Earth Engine JavaScript API, geemap, Streamlit, and Solara. You will also learn how to deploy web apps with Voila, ngrok, Streamlit Cloud, and Hugging Face Spaces.

10.2 Technical requirements

To follow along with this chapter, you will need to have geemap and several optional dependencies installed. If you have already followed Section 1.5 - *Installing geemap*, then you should already have a conda environment with all the necessary packages installed. Otherwise, you can create a new conda environment and install pygis[30] with the following commands, which will automatically install geemap and all the required dependencies:

```
conda create -n gee python
conda activate gee
conda install -c conda-forge mamba
mamba install -c conda-forge pygis
```

Additionally, you will need to install Voila[194] , Streamlit[195] , and Solara[196] for building Earth Engine

---

[191]widgets: tiny.geemap.org/zldc

[192]Panels: tiny.geemap.org/ctkx

[193]Output widget: tiny.geemap.org/lclt

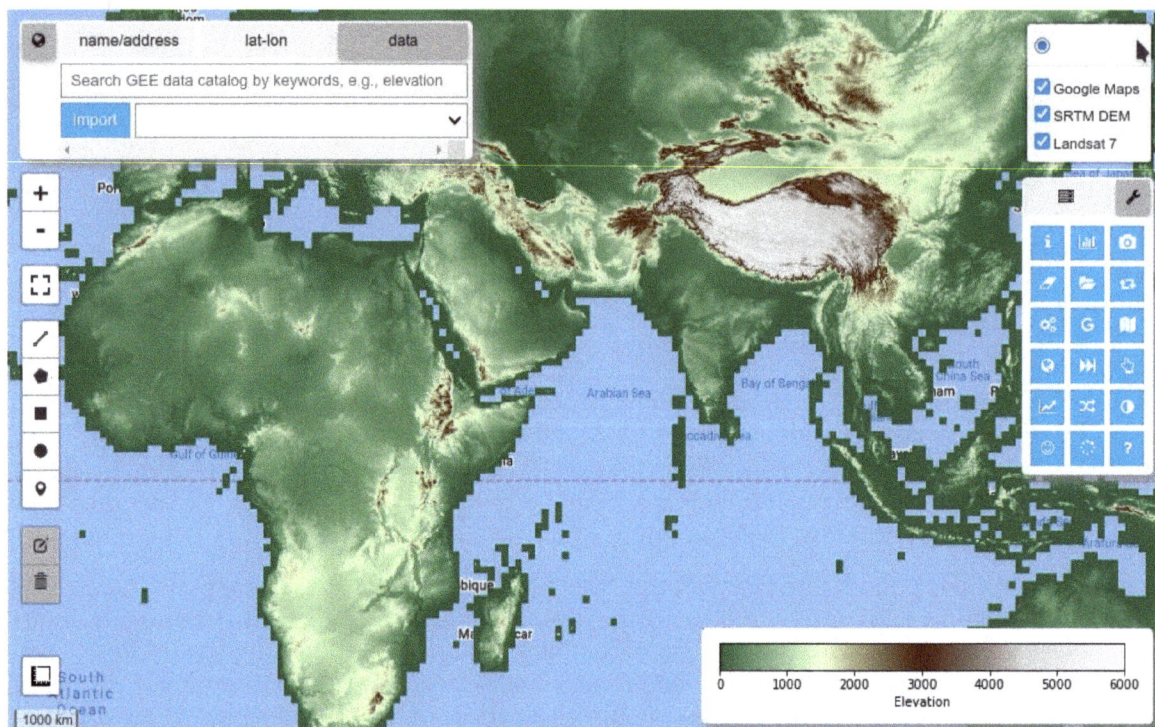

Figure 10.1: The graphical user interface of geemap in a Jupyter environment.

web apps. You can install them using the following commands:

```
pip install "geemap[apps]"
```

Next, launch JupyterLab by typing the following commands in your terminal or Anaconda prompt:

```
jupyter lab
```

Alternatively, you can use geemap with a Google Colab cloud environment without installing anything on your local computer. Click 10_webapps.ipynb[197] to launch the notebook in Google Colab.

Once in Colab, you can uncomment the following line and run the cell to install pygis, which includes geemap and all the necessary dependencies:

```
# %pip install pygis
```

The installation process may take 2-3 minutes. Once pygis has been installed successfully, click the **RESTART RUNTIME** button that appears at the end of the installation log or go to the **Runtime** menu and select **Restart runtime**. After that, you can start coding.

To begin, import the necessary libraries that will be used in this chapter:

```
import ee
import geemap
```

---

[194]Voila: tiny.geemap.org/vbem

[195]Streamlit: streamlit.io

[196]Solara: github.com/widgetti/solara

[197]10_webapps.ipynb: tiny.geemap.org/ch10

Initialize the Earth Engine Python API:

```
geemap.ee_initialize()
```

If this is your first time running the code above, you will need to authenticate Earth Engine first. Follow the instructions in Section 1.7 - *Earth Engine authentication* to authenticate Earth Engine.

10.3   Building JavaScript web apps

In this section, you will learn how to design user interfaces for an **Earth Engine App** using JavaScript and the **Earth Engine User Interface API**. Upon completion of this section, you will have an Earth Engine App with a split-panel map for visualizing land cover change using the Landsat-based U.S. National Land Cover Database (NLCD). Go to the GEE Code Editor at code.earthengine.google.com[198] and follow the steps below to create an Earth Engine App using JavaScript.

First, let's define a function called `getNLCD` for filtering the NLCD image collection by year and select the `landcover` band. The `getNLCD` function returns an Earth Engine `ui.Map.Layer` of the `landcover` band of the selected NLCD image. Note that NLCD spans nine epochs at the time of writing, including 1992, 2001, 2004, 2006, 2008, 2011, 2013, 2016, and 2019. The 1992 data are primarily based on unsupervised classification of Landsat data, while the rest of the images rely on the imperviousness data layer for the urban classes and on a decision-tree classification for the rest. The 1992 image is not directly comparable to any later editions of NLCD (see Earth Engine Data Catalog - NLCD[199]). Therefore, we will only use the eight epochs after 2000 in this chapter.

```
var getNLCD = function (year) {
  var dataset = ee.ImageCollection("USGS/NLCD_RELEASES/2019_REL/NLCD");
  var nlcd = dataset.filter(ee.Filter.eq("system:index", year)).first();
  var landcover = nlcd.select("landcover");
  return ui.Map.Layer(landcover, {}, year);
};
```

Next, define a dictionary with each NLCD epoch as the key and its corresponding NLCD image layer as the value. The keys of the dictionary (i.e., the eight NLCD epochs) will be used as the input to the dropdown list (`ui.Select`) on the map. When an epoch is selected from the dropdown list, the corresponding NLCD image layer will be displayed on the map.

```
var images = {
  2001: getNLCD("2001"),
  2004: getNLCD("2004"),
  2006: getNLCD("2006"),
  2008: getNLCD("2008"),
  2011: getNLCD("2011"),
  2013: getNLCD("2013"),
  2016: getNLCD("2016"),
  2019: getNLCD("2019"),
};
```

The split-panel map is composed of two individual maps (i.e., `leftMap` and `rightMap`). The map controls (e.g., `zoomControl`, `scaleControl`, `mapTypeControl`) will only be shown on rightMap. A control panel (`ui.Panel`) composed of a label (`ui.Label`) and a dropdown list (`ui.Select`) is added to each map. When an epoch is selected from the dropdown list, the function `updateMap` will be called to show the corresponding image layer of the selected NLCD epoch.

```
var leftMap = ui.Map();
```

---

[198]code.earthengine.google.com: code.earthengine.google.com

[199]Earth Engine Data Catalog - NLCD: tiny.geemap.org/uctg

```
leftMap.setControlVisibility(false);
var leftSelector = addLayerSelector(leftMap, 0, "top-left");

var rightMap = ui.Map();
rightMap.setControlVisibility(true);
var rightSelector = addLayerSelector(rightMap, 7, "top-right");

function addLayerSelector(mapToChange, defaultValue, position) {
  var label = ui.Label("Select a year:");

  function updateMap(selection) {
    mapToChange.layers().set(0, images[selection]);
  }

  var select = ui.Select({ items: Object.keys(images), onChange: updateMap });
  select.setValue(Object.keys(images)[defaultValue], true);

  var controlPanel = ui.Panel({
    widgets: [label, select],
    style: { position: position },
  });

  mapToChange.add(controlPanel);
}
```

When displaying a land cover classification image on the map, it would be useful to add a legend to
the map, making it easier for users to interpret the land cover type associated with each color. Let's
define a dictionary that will be used to construct the legend. The dictionary contains two keys: names
(a list of land cover types) and colors (a list of colors associated with each land cover type). The legend
will be placed in the bottom right of the map.

```
var title = "NLCD Land Cover Classification";
var position = "bottom-right";
var dict = {
  names: [
    "11 Open Water",
    "12 Perennial Ice/Snow",
    "21 Developed, Open Space",
    "22 Developed, Low Intensity",
    "23 Developed, Medium Intensity",
    "24 Developed, High Intensity",
    "31 Barren Land (Rock/Sand/Clay)",
    "41 Deciduous Forest",
    "42 Evergreen Forest",
    "43 Mixed Forest",
    "51 Dwarf Scrub",
    "52 Shrub/Scrub",
    "71 Grassland/Herbaceous",
    "72 Sedge/Herbaceous",
    "73 Lichens",
    "74 Moss",
    "81 Pasture/Hay",
    "82 Cultivated Crops",
    "90 Woody Wetlands",
    "95 Emergent Herbaceous Wetlands",
  ],

  colors: [
    "#466b9f",
    "#d1def8",
    "#dec5c5",
    "#d99282",
    "#eb0000",
    "#ab0000",
```

```
      "#b3ac9f",
      "#68ab5f",
      "#1c5f2c",
      "#b5c58f",
      "#af963c",
      "#ccb879",
      "#dfdfc2",
      "#d1d182",
      "#a3cc51",
      "#82ba9e",
      "#dcd939",
      "#ab6c28",
      "#b8d9eb",
      "#6c9fb8",
    ],
};
```

With the legend dictionary defined above, we can now create a panel to hold the legend widget and add it to the map. Each row on the legend widget is composed of a color box followed by its corresponding land cover type.

```
var legend = ui.Panel({
  style: {
    position: position,
    padding: "8px 15px",
  },
});

function addCategoricalLegend(panel, dict, title) {
  var legendTitle = ui.Label({
    value: title,
    style: {
      fontWeight: "bold",
      fontSize: "18px",
      margin: "0 0 4px 0",
      padding: "0",
    },
  });
  panel.add(legendTitle);

  var loading = ui.Label("Loading legend...", { margin: "2px 0 4px 0" });
  panel.add(loading);

  var makeRow = function (color, name) {
    var colorBox = ui.Label({
      style: {
        backgroundColor: color,
        padding: "8px",
        margin: "0 0 4px 0",
      },
    });
    var description = ui.Label({
      value: name,
      style: { margin: "0 0 4px 6px" },
    });

    return ui.Panel({
      widgets: [colorBox, description],
      layout: ui.Panel.Layout.Flow("horizontal"),
    });
  };

  var palette = dict.colors;
  var names = dict.names;
```

```
    loading.style().set("shown", false);

    for (var i = 0; i < names.length; i++) {
      panel.add(makeRow(palette[i], names[i]));
    }

    rightMap.add(panel);
  }

  addCategoricalLegend(legend, dict, title);
```

The last step is to create a split-panel map to hold the linked maps (i.e., leftMap and rightMap) and tie everything together. When users pan and zoom one map, the other map will also be panned and zoomed to the same extent automatically. Select a year from the dropdown lists and the image layer will be updated accordingly. Users can use the slider to swipe through and visualize land cover change easily (see Fig. 10.2).

```
var splitPanel = ui.SplitPanel({
  firstPanel: leftMap,
  secondPanel: rightMap,
  wipe: true,
  style: { stretch: "both" },
});

ui.root.widgets().reset([splitPanel]);
var linker = ui.Map.Linker([leftMap, rightMap]);
leftMap.setCenter(-100, 40, 4);
```

The complete script can be found at https://bit.ly/3XQ8JHD. You can run the script in the Earth Engine Code Editor. Move the slider up at the top of the map to increase the size of the map so that you can see the dropdown lists and the legend. The output should look like Fig. 10.2.

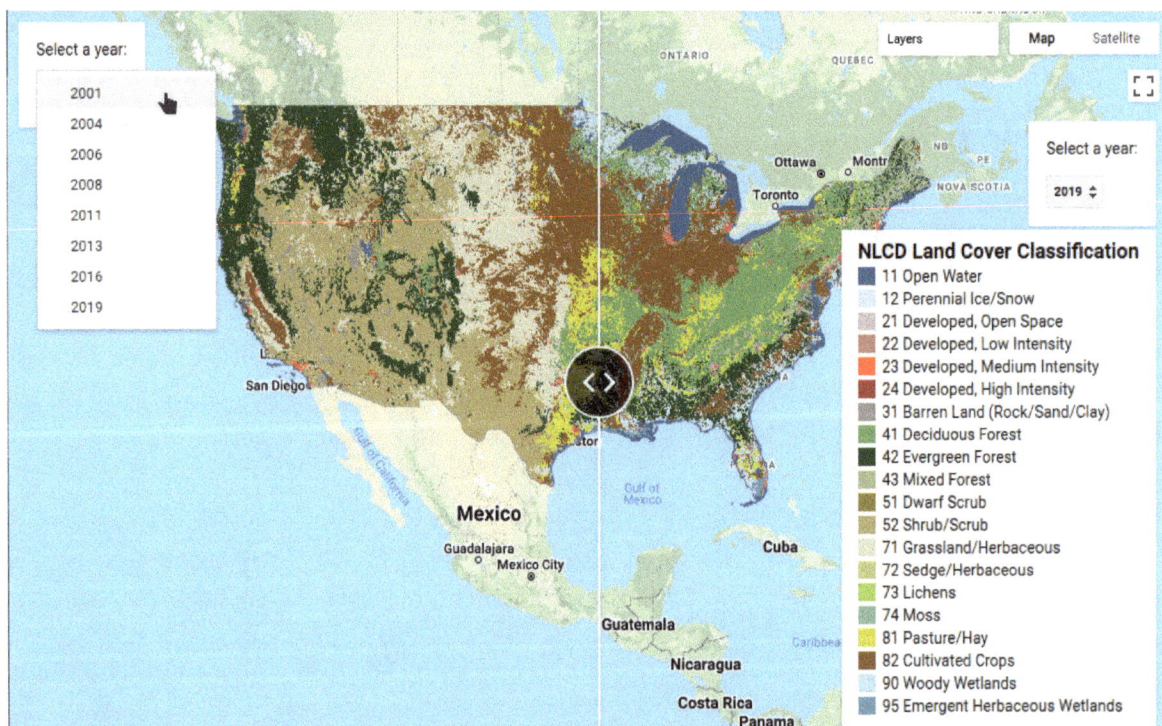

Figure 10.2: A split-panel map for visualizing land cover change using NLCD.

10.4   Publishing JavaScript web apps

The goal of this section is to publish the Earth Engine App that we have created in the previous section. First, let's load the script (see the app source code[200]) into the Code Editor. Then, open the App Management panel by clicking the **Apps** button above the script section in the Code Editor (Fig. 10.3).

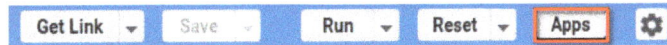

Figure 10.3: The Manage Apps Button in the JavaScript Code Editor.

Then click on the **NEW APP** button (Fig. 10.4).

Figure 10.4: The New App Button.

In the *Publish New App* dialog (Fig. 10.5), choose a name for the App (e.g., NLCD Land Cover Change), specify the app owner, and select a Google Cloud Project. Then click on the **NEXT** button.

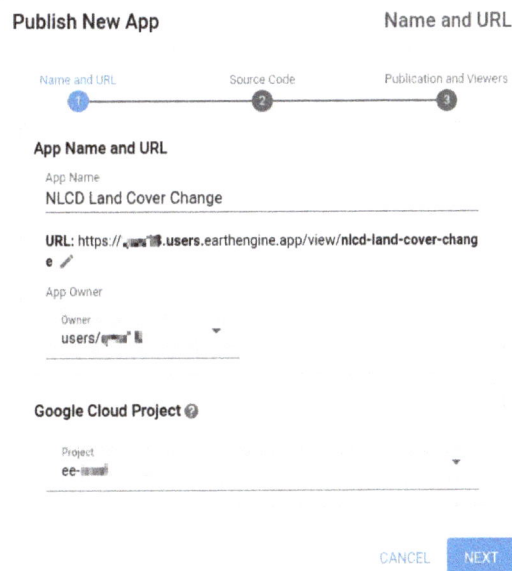

Figure 10.5: Specify the name of the App and select a Google Cloud Project.

Next, specify the App's source code. You can either select the current script or specify a different script. Click **NEXT** when you are done (Fig. 10.6).

---

[200]app source code: bit.ly/3XQ8JHD

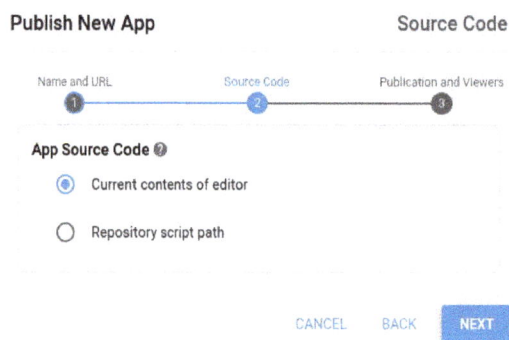

Figure 10.6: Specify the App's source code.

Lastly, you can provide a thumbnail to be shown in the Public Apps Gallery if needed (Fig. 10.7). You may restrict access to the App to a particular Google Group or make it publicly accessible. Toggle on **Feature this app in your Apps Gallery** if you would like this App to appear on your public gallery of apps available at `https://USERNAME.users.earthengine.app`. Click the **PUBLISH** button to complete publishing the App.

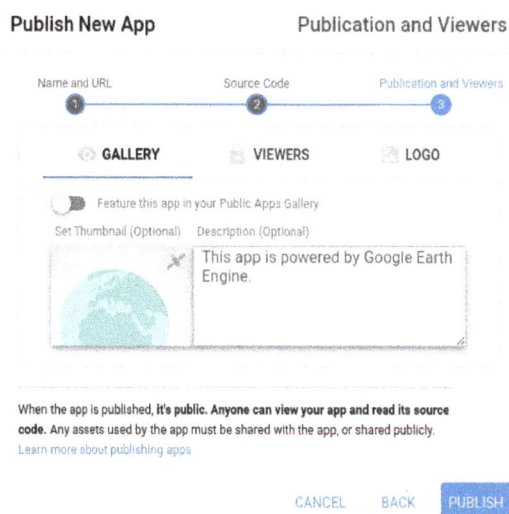

Figure 10.7: Specify the App's thumbnail, access, and visibility.

Within a few minutes, the App will be published. Click on the app name under the list of Apps to launch the App. The App URL should look like this:

```
https://USERNAME.users.earthengine.app/view/nlcd-land-cover-change
```

To manage an App from the Code Editor, open the App Management panel by clicking on the **Apps** button above the script section in the Code Editor. From there you can update your App's configuration or delete the App (Fig. 10.8).

10.5   Building Python Web Apps

In this section, you will learn how to develop an Earth Engine App using geemap and Voila within a Jupyter environment. First, launch JupyterLab and create a new Jupyter notebook named `nlcd_`

Manage Apps

App Name (click to launch)	ID (click to update app)	Delete
⊙ NLCD Land Cover Change	users/▓▓▓▓/nlcd-land-cover-change ✏	🗑

CLOSE

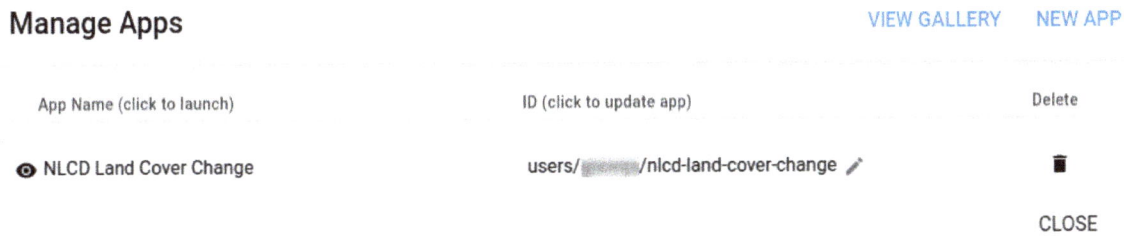

Figure 10.8: Manage Earth Engine Apps.

`app.ipynb`. Then, import the necessary libraries that will be used in this chapter:

```python
import ee
import geemap
```

Create an interactive map by specifying the map center and zoom level.

```python
Map = geemap.Map(center=[40, -100], zoom=4)
```

Retrieve the NLCD 2019 image by filtering the NLCD image collection and selecting the landcover band. Display the NLCD 2019 image on the interactive map by using the `Map.addLayer()` method.

```python
dataset = ee.ImageCollection('USGS/NLCD_RELEASES/2019_REL/NLCD')
nlcd2019 = dataset.filter(ee.Filter.eq('system:index', '2019')).first()
landcover = nlcd2019.select('landcover')
Map.addLayer(landcover, {}, 'NLCD 2019')
Map
```

Next, add the NLCD legend to the map. Geemap has several built-in legends, including the NLCD legend. Therefore, you can add the NLCD legend to the map by using just one line of code (i.e., the `Map.add_legend()` method).

```python
title = 'NLCD Land Cover Classification'
Map.add_legend(title=title, builtin_legend='NLCD')
```

The map with the NLCD 2019 image and legend should look like Fig. 10.9.

The map above only shows the NLCD 2019 image. To create an Earth Engine App for visualizing land cover change, we need a stack of NLCD images. Let's print the list of system IDs of all available NLCD images:

```python
dataset.aggregate_array("system:id")
```

The output should look like this:

```
['USGS/NLCD_RELEASES/2019_REL/NLCD/2001',
 'USGS/NLCD_RELEASES/2019_REL/NLCD/2004',
 'USGS/NLCD_RELEASES/2019_REL/NLCD/2006',
 'USGS/NLCD_RELEASES/2019_REL/NLCD/2008',
 'USGS/NLCD_RELEASES/2019_REL/NLCD/2011',
 'USGS/NLCD_RELEASES/2019_REL/NLCD/2013',
 'USGS/NLCD_RELEASES/2019_REL/NLCD/2016',
 'USGS/NLCD_RELEASES/2019_REL/NLCD/2019']
```

Select the eight NLCD epochs after 2000:

```python
years = ['2001', '2004', '2006', '2008', '2011', '2013', '2016', '2019']
```

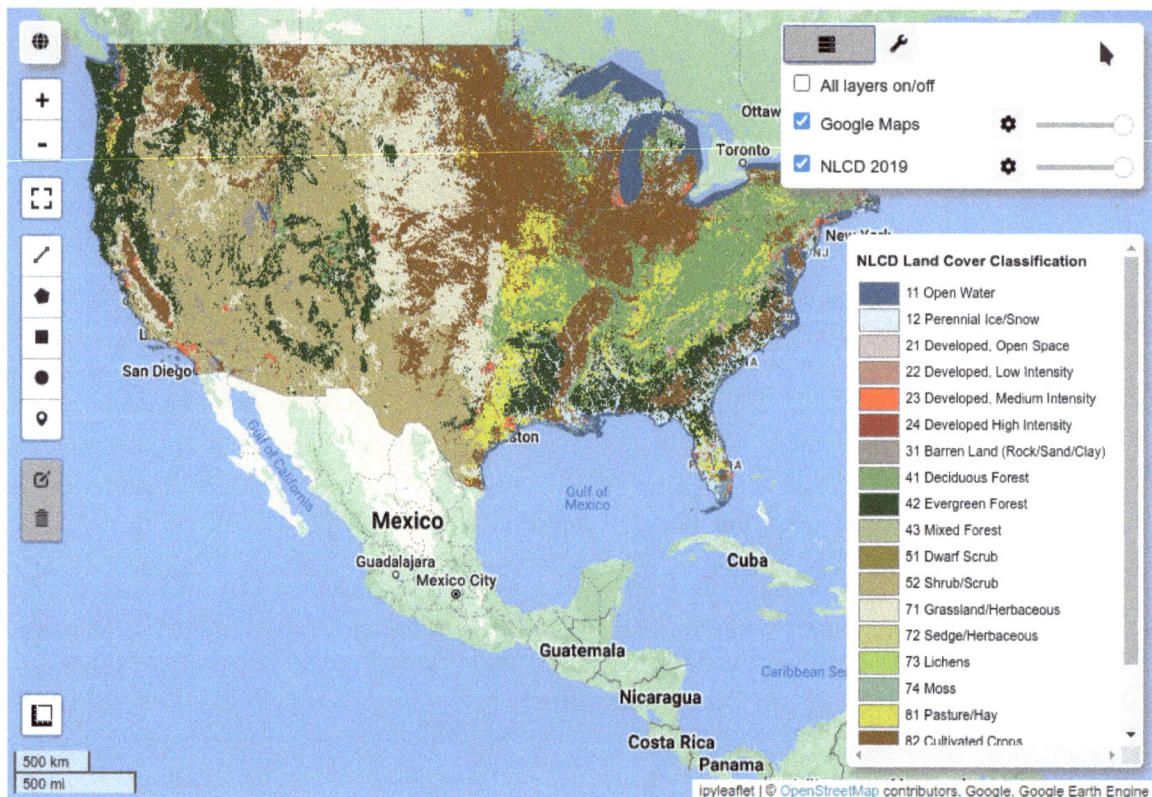

Figure 10.9: The NLCD 2019 image layer displayed in geemap.

Define a function for filtering the NLCD image collection by year and select the `landcover` band:

```python
def getNLCD(year):
    dataset = ee.ImageCollection('USGS/NLCD_RELEASES/2019_REL/NLCD')
    nlcd = dataset.filter(ee.Filter.eq('system:index', year)).first()
    landcover = nlcd.select('landcover')
    return landcover
```

Create an NLCD image collection to be used in the split-panel map:

```python
collection = ee.ImageCollection(ee.List(years).map(lambda year: getNLCD(year)))
```

Next, create a list of labels to populate the dropdown list:

```python
labels = [f'NLCD {year}' for year in years]
labels
```

The output should look like this:

```python
['NLCD 2001',
 'NLCD 2004',
 'NLCD 2006',
 'NLCD 2008',
 'NLCD 2011',
 'NLCD 2013',
 'NLCD 2016',
 'NLCD 2019']
```

The last step is to create a split-panel map by passing the NLCD image collection and the list of labels to the `Map.ts_inspector()` method.

```
Map.ts_inspector(
    left_ts=collection,
    right_ts=collection,
    left_names=labels,
    right_names=labels
)
Map
```

The split-panel map should look like Fig. 10.10.

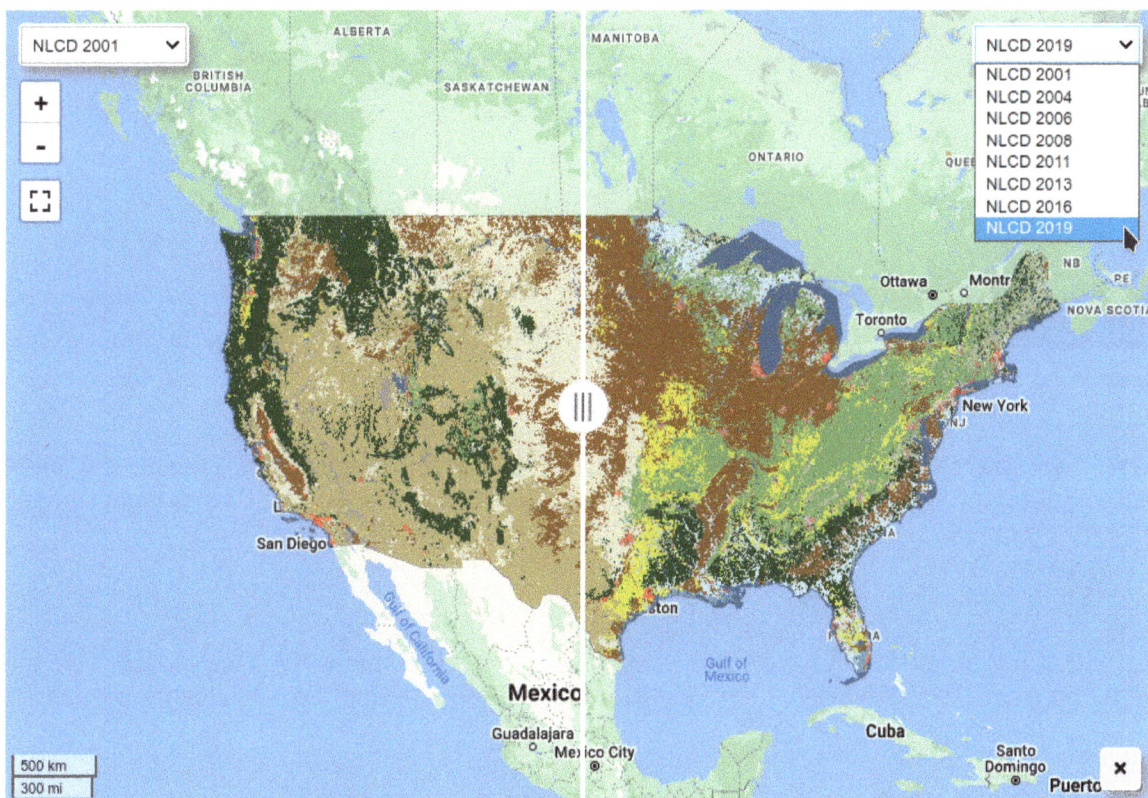

Figure 10.10: A split-panel map for visualizing land cover change with geemap.

To visualize land cover change, choose one NLCD image from the left dropdown list and another image from the right dropdown list. Then use the slider to swipe through to visualize land cover change interactively. With the web app ready, we can now move on to the next section to learn how to publish the app.

10.6   Using Voila to deploy web apps

In this section, you will learn how to deploy an Earth Engine App using your local computer as a web server with ngrok[201]. Assuming that you have completed the previous section and created a Jupyter notebook named `nlcd_app.ipynb`. First, sign up for a free account on ngrok and log in to your account to get your authentication token Fig. 10.11.

---

[201]ngrok: ngrok.com

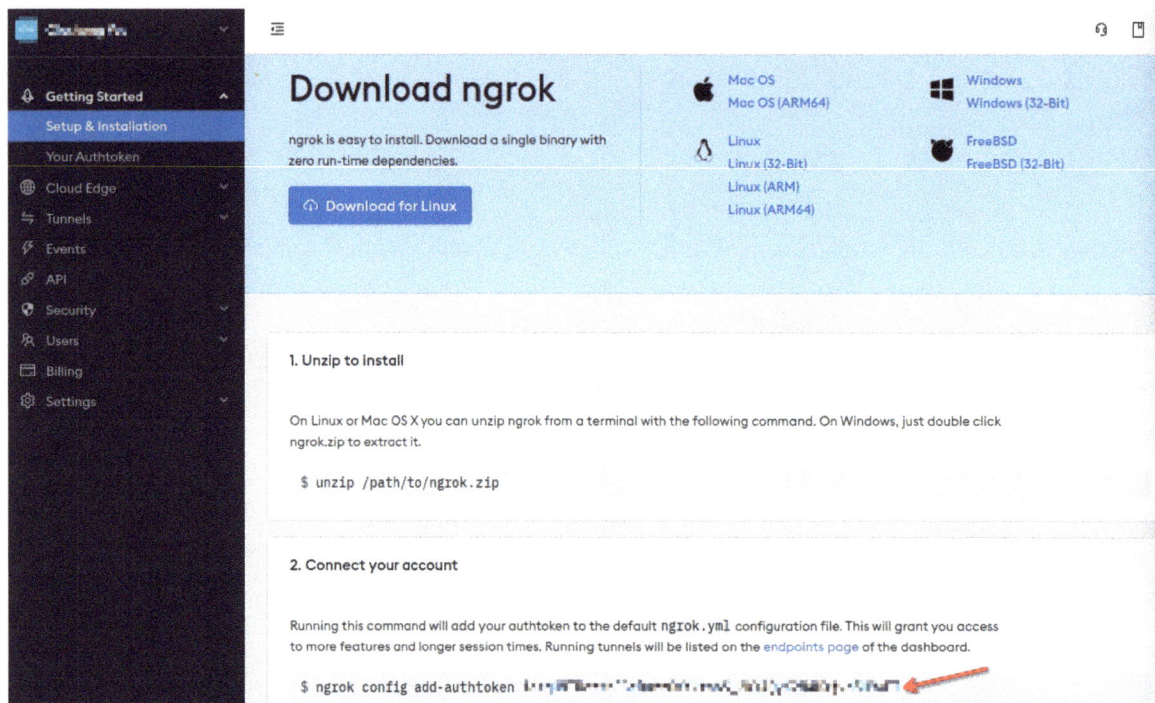

Figure 10.11: The ngrok authentication token.

Next, download the ngrok program for your operating system. Unzip the downloaded file to a directory on your computer. Open the Anaconda Prompt or Terminal and enter the following commands to authenticate ngrok. Replace <token> with the authentication token you obtained earlier.

```
cd /path/to/ngrok/dir
ngrok config add-authtoken <token>
```

If you get an error message saying command not found: ngrok, you may need to prepend ./ to the ngrok command. For example, if you are using macOS or Linux, you can run the following command:

```
./ngrok config add-authtoken <token>
```

Once ngrok has been authenticated, copy nlcd_app.ipynb to the same directory where ngrok is located. You can then launch the Voila dashboard by running the following commands:

```
conda activate gee
voila --no-browser nlcd_app.ipynb
```

The output of the terminal should look like this:

Voila can be used to run, convert, and serve a Jupyter notebook as a standalone app. Click the link (e.g., http://localhost:8866) shown in the terminal window to launch the interactive dashboard. Note that the port number is 8866, which is required for the next step to launch ngrok. Open another terminal and enter the following commands:

```
cd /path/to/ngrok/dir
ngrok http 8866
```

The output of the terminal should look like Fig. 10.13. Click the link shown in the terminal window to launch the interactive dashboard. The link should look like https://random-string.ngrok-free.app,

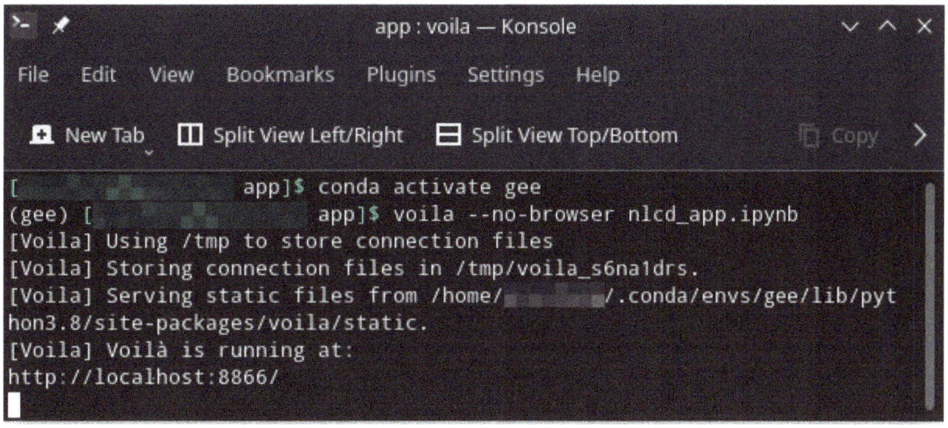

Figure 10.12: The output of the terminal running Voila.

which is publicly accessible. Anyone with the link will be able to launch the interactive dashboard and use the split-panel map to visualize NLCD land cover change.

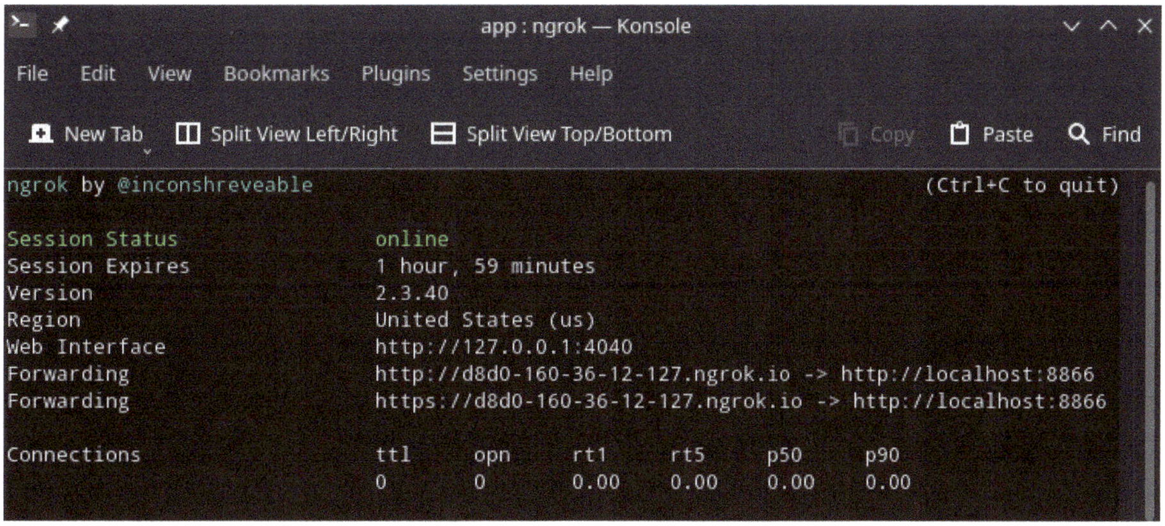

Figure 10.13: The output of the terminal running ngrok.

To stop the web server, press **Ctrl+C** on both terminal windows. See below for some optional settings for running Voila and ngrok. To show code cells from your app, run the following command from the terminal:

```
voila --no-browser --strip_sources=False nlcd_app.ipynb
```

To protect your app with a password, run this:

```
ngrok http -auth="username:password" 8866
```

10.7 Building Streamlit web apps

In this section, you will learn how to design and deploy an interactive Earth Engine web app using Streamlit[195] and geemap. Streamlit is an open-source Python package that can turn Python scripts into shareable web apps in minutes. To test if Streamlit has been installed on your computer, enter the following command in the terminal:

```
streamlit hello
```

If you get an error message saying command not found: streamlit, you may need to install Streamlit first. Enter the following command to install Streamlit:

```
pip install streamlit
```

Then run the streamlit hello command again. Streamlit's Hello web app should appear in a new tab in your web browser as shown below. You can click on the menu on the left sidebar to explore this simple web app Fig. 10.14.

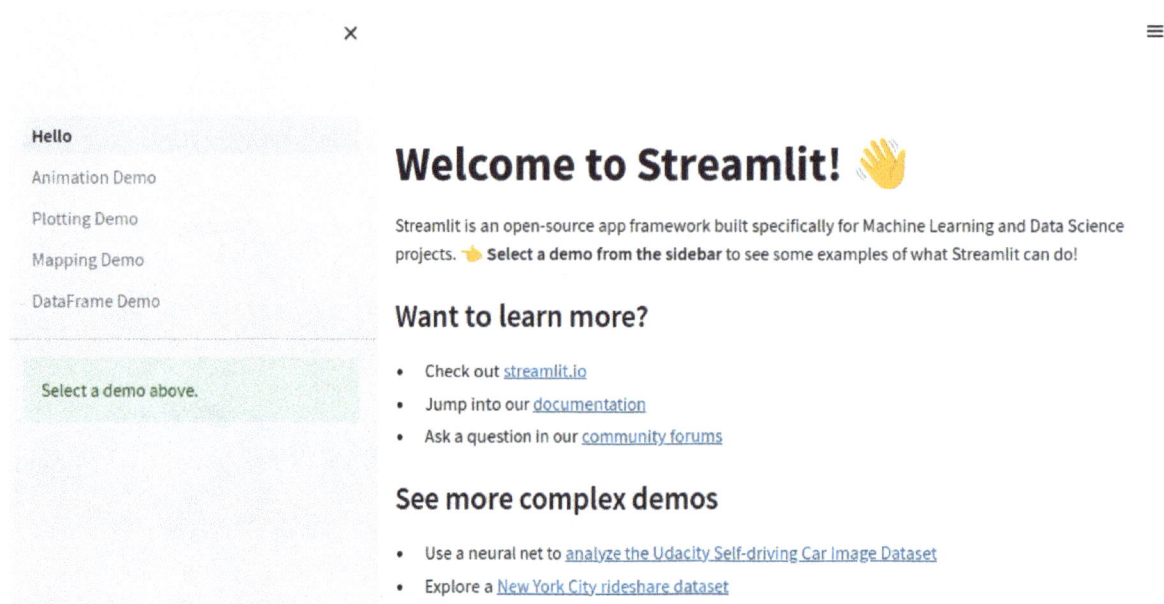

Figure 10.14: Streamlit's Hello web app.

One advantage of Streamlit is that it allows you to write web apps the same way you write plain Python scripts. Note that in previous sections we used Jupyter notebook to write code and develop web apps. Compared with other Python IDEs (e.g., Visual Studio Code[202] , PyCharm[203]), Jupyter notebook has relatively limited interactive programming and debugging capabilities. Also, Streamlit builds web apps based on Python scripts, not Jupyter notebooks. Therefore, we can use Visual Studio Code to develop an interactive web app in this section. In addition, we need to use Git to push code to GitHub. Download and install Git (git-scm.com/downloads[204]) and Visual Studio Code (code.visualstudio.com[205]) on your computer if you don't have them installed yet.

After installing Git, you need to set your user name and email address to be associated with each Git commit. Enter the following commands into the terminal or Git Bash. Make sure you replace **Firstname Lastname** and **user@example.com** with your real name and the email address for your GitHub account.

---

[202]Visual Studio Code: code.visualstudio.com

[203]PyCharm: www.jetbrains.com/pycharm

[204]git-scm.com/downloads: git-scm.com/downloads

[205]code.visualstudio.com: code.visualstudio.com

```
git config --global user.name "Firstname Lastname"
git config --global user.email user@example.com
```

Next, navigate to github.com/giswqs/geemap-apps[206] and fork the repo to your GitHub account. Click the green **Code** button on the forked repo page and copy the Git URL to the repo under your account.

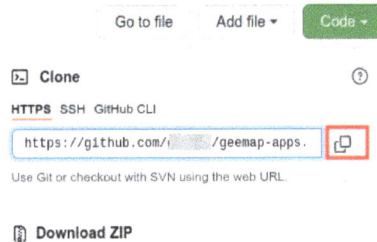

Figure 10.15: Clone a Git repo on GitHub.

Open the terminal or Git Bash and navigate to a directory on your computer where you want to save the repo. Enter the following command to clone the repo to your computer:

```
git clone https://github.com/USERNAME/geemap-apps.git
```

Use Visual Studio Code to open the repo directory, i.e., `geemap-apps`. Note that Streamlit and geemap are listed in `requirements.txt`. If you plan to develop a web app that requires additional Python packages, you can add them to `requirements.txt`. All package listed in this file will be installed automatically when the app is deployed. In this section, we are going to deploy the app on Streamlit Cloud[207] , a workspace that allows users to deploy and manage Streamlit apps. You can connect your Streamlit Cloud account directly to your GitHub repo and then Streamlit Cloud can launch the app directly from the Python scripts stored on GitHub. Whenever you update the code on GitHub, the web app will automatically refresh and update for you. In that way, you can focus on writing and pushing code to GitHub without worrying about the app deployment.

Next, navigate to share.streamlit.io[208] to log in with your GitHub or Google account. It is recommended that you start with GitHub the first time you log in, and then you can connect to your Google account afterwards. Connecting your GitHub account allows your Streamlit Cloud workspace to launch apps directly from the app files you store in your repositories. It can also check for updates to those app files so that your apps can automatically update.

Since we are deploying an Earth Engine app, we need to authenticate each Earth Engine session. Locate the Earth Engine `credentials` file on your computer in the following file path, depending on the operating system you are using. Note that you might need to show the hidden directories on your computer in order to see the `.config` folder under the home directory.

```
Windows: C:\Users\USERNAME\.config\earthengine\credentials
Linux: /home/USERNAME/.config/earthengine/credentials
MacOS: /Users/USERNAME/.config/earthengine/credentials
```

Open the `credentials` file to copy all the content to the clipboard. To deploy a web app, click the "**New app**" button in the upper-right corner of your workspace in Streamlit Cloud, then fill in your repo path

[206]github.com/giswqs/geemap-apps: github.com/giswqs/geemap-apps
[207]Streamlit Cloud: streamlit.io/cloud
[208]share.streamlit.io: share.streamlit.io

(e.g., USERNAME/geemap-apps), branch (e.g., master/main), and main file path (e.g., app.py).

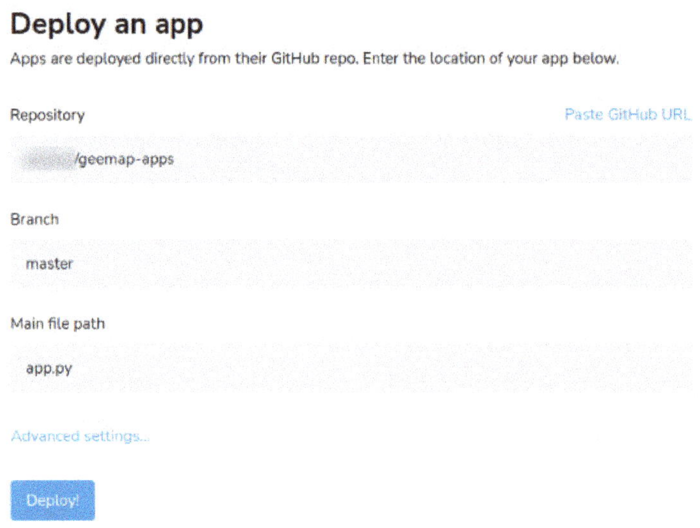

Deploy an app

Apps are deployed directly from their GitHub repo. Enter the location of your app below.

Repository   Paste GitHub URL

    /geemap-apps

Branch

    master

Main file path

    app.py

Advanced settings...

Deploy!

Figure 10.16: Deploy a web app in Streamlit Cloud.

Next, click on **Advanced settings** to set EARTHENGINE_TOKEN as an environment variable. The Earth Engine token can be found in the credentials file as shown above. Copy all the content in the credentials file and paste it into the triple quotes under the **Secrets** section as shown in Fig. 10.17. Click on the **Save** button to save the secret, which can be accessed by the web app to authenticate Earth Engine. Lastly, click on the "**Deploy!**" button to deploy the web app.

Depending on the number of dependencies specified in requirements.txt, the app might take a couple of minutes to install all the dependencies and deploy the app. Once the app is deployed, you can find app URL in the Streamlit Cloud workspace, which will follow a standard structure based on your GitHub repo, such as:

```
https://share.streamlit.io/[user name]/[repo name]/[branch name]/[app path]
```

For example, you can reach the sample app at the URL below:

share.streamlit.io/giswqs/geemap-apps/app.py[209]

Once the sample multi-page web app has been deployed successfully, we can add a new app for displaying NLCD data, similar to what we did in previous sections. First, open the cloned GitHub repo using Visual Studio Code. Locate the pages directory under the repo. You should see several Python scripts there. Each script contains code for one app shown on the left sidebar menu. To add a new app, right-click the pages directory within Visual Studio Code and select New File. Enter a filename for the new file (e.g., nlcd.py). Double click on the newly created file to open it in the Editor area to the right.

First, let's import the libraries needed for this web app.

```
import ee
import streamlit as st
import geemap.foliumap as geemap
```

Next, create a function for retrieving an NLCD image by year.

---

[209]share.streamlit.io/giswqs/geemap-apps/app.py: tiny.geemap.org/osft

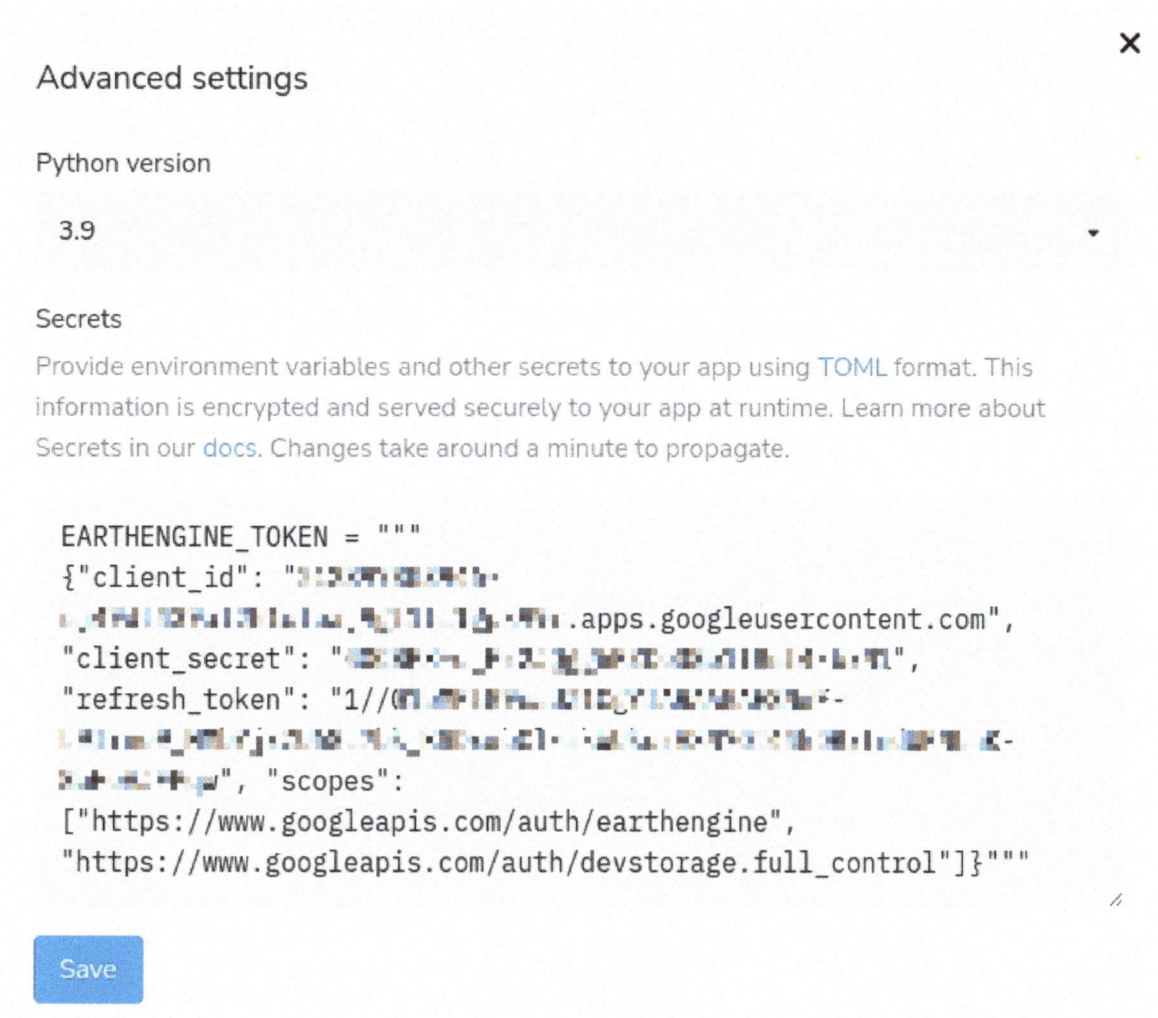

Advanced settings

Python version

3.9 ▾

Secrets

Provide environment variables and other secrets to your app using TOML format. This
information is encrypted and served securely to your app at runtime. Learn more about
Secrets in our docs. Changes take around a minute to propagate.

```
EARTHENGINE_TOKEN = """
{"client_id": "▓▓▓▓▓▓▓▓▓▓▓▓▓▓▓▓▓▓.apps.googleusercontent.com",
"client_secret": "▓▓▓▓▓▓▓▓▓▓▓▓▓▓▓",
"refresh_token": "1//▓▓▓▓▓▓▓▓▓▓▓▓▓▓▓▓▓▓▓▓▓▓▓▓▓▓▓▓▓▓▓▓▓▓▓▓▓▓▓▓", "scopes":
["https://www.googleapis.com/auth/earthengine",
"https://www.googleapis.com/auth/devstorage.full_control"]}"""
```

Save

Figure 10.17: Set environment variables on Streamlit Cloud.

```python
def getNLCD(year):
    dataset = ee.ImageCollection("USGS/NLCD_RELEASES/2019_REL/NLCD")
    nlcd = dataset.filter(ee.Filter.eq("system:index", year)).first()
    landcover = nlcd.select("landcover")
    return landcover
```

Create an app layout with two columns, one for displaying the map and the other for hosting the
dropdown list of years and the checkbox to show the NLCD legend. `st.columns([3, 1])` creates two
columns where the first column is 3 times the width of the second. When a specific year is selected
from the dropdown list, the corresponding NLCD image will be displayed on the map. Note that the
dropdown list is a multi-select widget, allowing users to select multiple years of images to display on
the map. If the `Show the legend` checkbox is checked, the NLCD legend will be added to the map.

```python
st.header("National Land Cover Database (NLCD)")
row1_col1, row1_col2 = st.columns([3, 1])
Map = geemap.Map()
years = ["2001", "2004", "2006", "2008", "2011", "2013", "2016", "2019"]
with row1_col2:
    selected_year = st.multiselect("Select a year", years)
```

```
        add_legend = st.checkbox("Show legend")
if selected_year:
    for year in selected_year:
        Map.addLayer(getNLCD(year), {}, "NLCD " + year)

    if add_legend:
        Map.add_legend(title="NLCD Land Cover", builtin_legend="NLCD")
    with row1_col1:
        Map.to_streamlit(height=600)
else:
    with row1_col1:
        Map.to_streamlit(height=600)
```

Finally, we can now preview the app. Under the **TERMINAL** tab with Visual Studio Code, enter the following commands:

```
conda activate gee
streamlit run app.py
```

The app should appear in a new tab in your web browser as shown in Fig. 10.18. To stop the app, press Ctrl+C within the terminal. The last step is to commit changes and push code to GitHub, which will trigger the app deployed on Streamlit Cloud to update automatically. That's it. You now have an interactive web app up and running on Streamlit Cloud. You can share the permanent app URL with others.

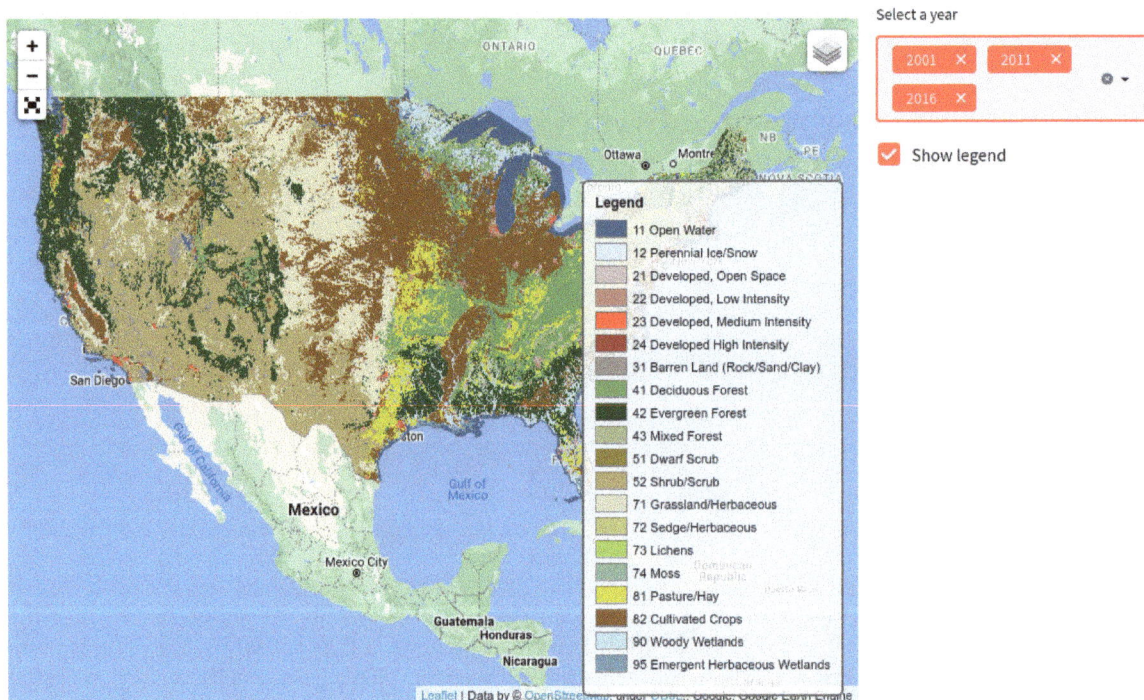

Figure 10.18: The user interface of the Streamlit web app for visualizing NLCD layers.

This section demonstrates how to build a simple interactive web app using Streamlit and geemap. A more advanced example can be found on GitHub at github.com/opengeos/streamlit-geospatial[210]. The permanent URL to this web app is streamlit.gishub.org[211]. This multi-page web app provides many examples for using Streamlit and geemap. For example, the timelapse web app (Fig. 10.19) allows users to create timelapse animations from any Earth Engine ImageCollection (e.g., Landsat, Sentinel, MODIS) at a specified temporal frequency (e.g., year, quarter, month). The timelapse animation can be downloaded in both GIF and MP4 formats, which can be useful for sharing on social media.

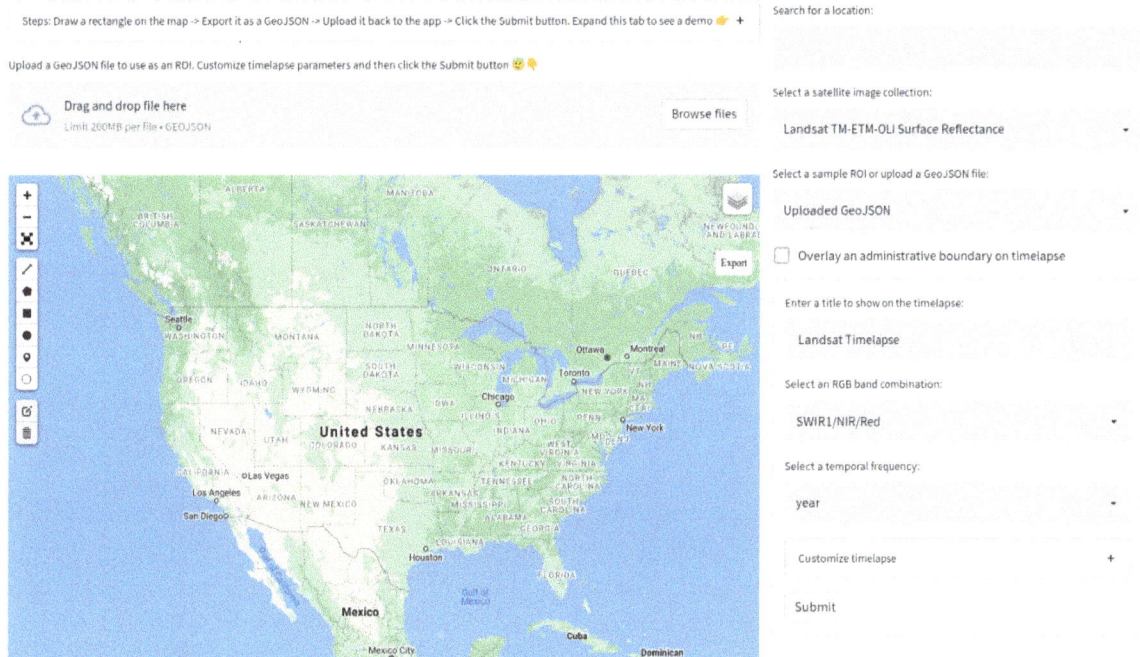

Figure 10.19: An interactive web app for creating satellite timelapse animations.

10.8   Building Solara web apps

In the previous section, we learned how to build an interactive web app using Streamlit. Under the hood, the geemap `Map.to_streamlit()` method converts the ipyleaflet map to a static HTML file, which is then displayed in the Streamlit web app. The disadvantage of this approach is that the map is static and lacks bi-directional communication functionality. In this section, you will learn how to build an interactive web app with bi-directional communication functionality using Solara. Solara lets you build web apps from pure Python using ipywidgets or a React-like API on top of ipywidgets. These apps work both inside the Jupyter notebook and as standalone web apps with frameworks like FastAPI. For more information about Solara, please visit the Solara website[212]. Since Solara has native support for ipywidgets, all the geemap interactive widgets can be used in Solara web apps. Solara should have been installed if have followed the instructions in the previous section. If not, Solara can be installed using the following command:

```
pip install solara
```

To make it easier to build Solara web apps, a template repo named solara-geemap[213] has been created. You can fork this repo to your GitHub account and then clone it to your computer. Open the downloaded repo using Visual Studio Code. Under the **pages** directory, you should see several Python scripts. Open the **04_split_map.py** script, which contains the code for the web app for visualizing land

---

[210]github.com/opengeos/streamlit-geospatial: tiny.geemap.org/vtjp

[211]streamlit.gishub.org: streamlit.gishub.org

[212]Solara website: solara.dev

cover change, similar to the Voila web app we created in the previous section. The source code for this web app is shown below:

```python
import ee
import geemap
import solara


class Map(geemap.Map):
    def __init__(self, **kwargs):
        super().__init__(**kwargs)
        self.add_ee_data()

    def add_ee_data(self):
        years = ['2001', '2004', '2006', '2008', '2011', '2013', '2016', '2019']
        def getNLCD(year):
            dataset = ee.ImageCollection('USGS/NLCD_RELEASES/2019_REL/NLCD')
            nlcd = dataset.filter(ee.Filter.eq('system:index', year)).first()
            landcover = nlcd.select('landcover')
            return landcover

        collection = ee.ImageCollection(ee.List(years).map(lambda year: getNLCD(year)))
        labels = [f'NLCD {year}' for year in years]
        self.ts_inspector(
            left_ts=collection,
            right_ts=collection,
            left_names=labels,
            right_names=labels,
        )
        self.add_legend(
            title='NLCD Land Cover Type',
            builtin_legend='NLCD',
            height="460px",
            add_header=False
        )


@solara.component
def Page():
    with solara.Column(style={"min-width": "500px"}):
        Map.element(
            center=[40, -100],
            zoom=4,
            height="800px",
        )
```

In the code above, we first create a custom Map class that inherits the geemap.Map class. The Map class is used to create a split-panel map for visualizing land cover change. The add_ee_data() method is used to retrieve the NLCD images and add them to the map. The ts_inspector() method is used to create a split-panel map. The add_legend() method is used to add the NLCD legend to the map. Next, we create a Page component that contains the Map component. The Page component is used to create a web app page. The Column component is used to create a column layout. The Map component is used to create a split-panel map. The center and zoom parameters are used to specify the map center and zoom level. The height parameter is used to specify the map height. The Page component is then rendered as a web app page. To preview the web app, enter the following commands in the terminal:

```
conda activate gee
solara run ./pages
```

The web app should appear in a new tab in your web browser with the URL http://localhost:8765. You should see a menu at the top of the web app page as shown in Fig. 10.20. Click on the **SPLIT-MAP**

---

[213]solara-geemap: tiny.geemap.org/fqrm

menu item to launch the split-panel map for visualizing land cover change with NLCD layers.

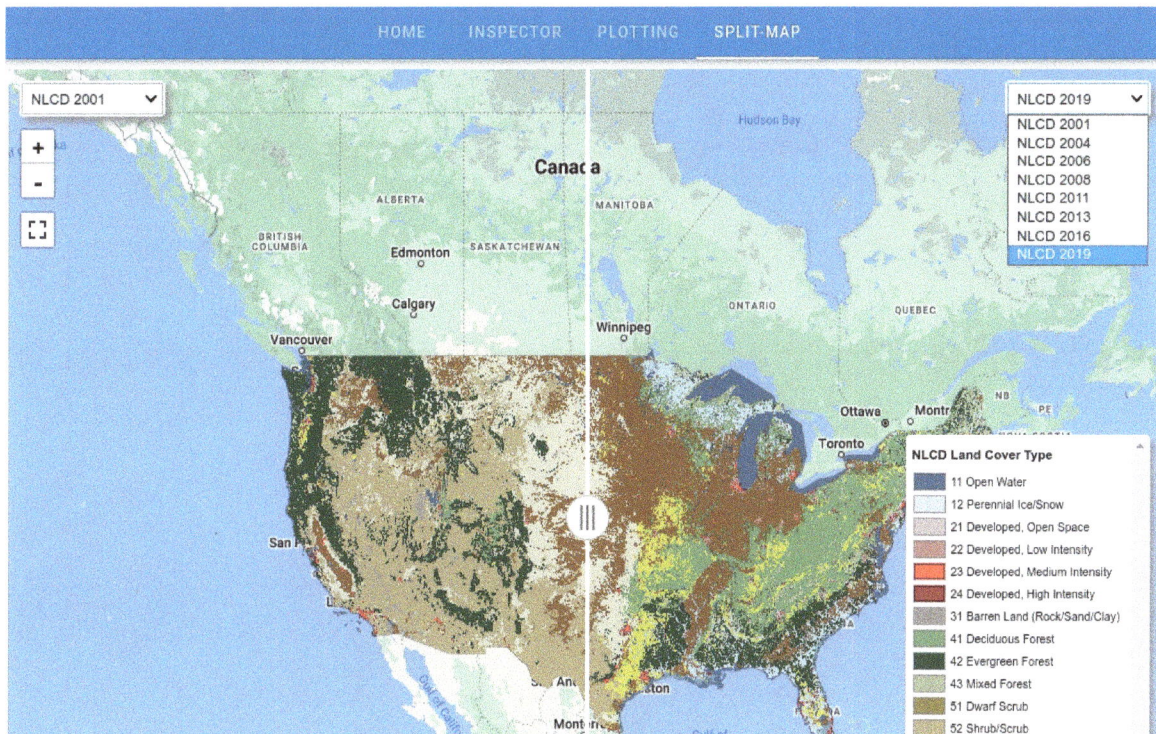

Figure 10.20: An interactive Solara web app for visualizing NLCD layers.

Click on the **INSPECTOR** menu item to play with the geemap inspector tool in the Solara web app. Hover the mouse over the map and click on any location to inspect Earth Engine objects at the clicked location (Fig. 10.21). Unlike Streamlit web apps, Solara web apps are interactive and bi-directional.

You are welcome to review the source code of the other web apps under the **pages** directory. Note that each script contains a Map class and a Page component. The Map class is used to create a custom map, while the Page component is used to render the map. Feel free to explore the scripts and add additional Earth Engine datasets to the scripts. To stop the app, press Ctrl+C within the terminal.

10.9 Deploying web apps on Hugging Face

Hugging Face[214] provides free cloud computing resources for hosting web apps. In this section, you will learn how to deploy an Earth Engine App on Hugging Face. First, sign up for a free account on Hugging Face and log into your account. To make it easier to deploy Earth Engine Apps on Hugging Face, a template repo named solara-geemap on Hugging Face[215] has been created. Navigate to the template repo and duplicate it to your Hugging Face account by clicking on the **Duplicate this Space** menu item (Fig. 10.22).

Select **Private** for the visibility of the duplicated repo on the pop-up window. Change the Space name if needed. Lastly, click on the **Duplicate Space** button to duplicate the repo. The duplicated repo should appear under the **Spaces** tab in your Hugging Face account. Click on the duplicated repo name to open it.

---

[214]Hugging Face: huggingface.co

[215]solara-geemap on Hugging Face: bit.ly/solara-geemap

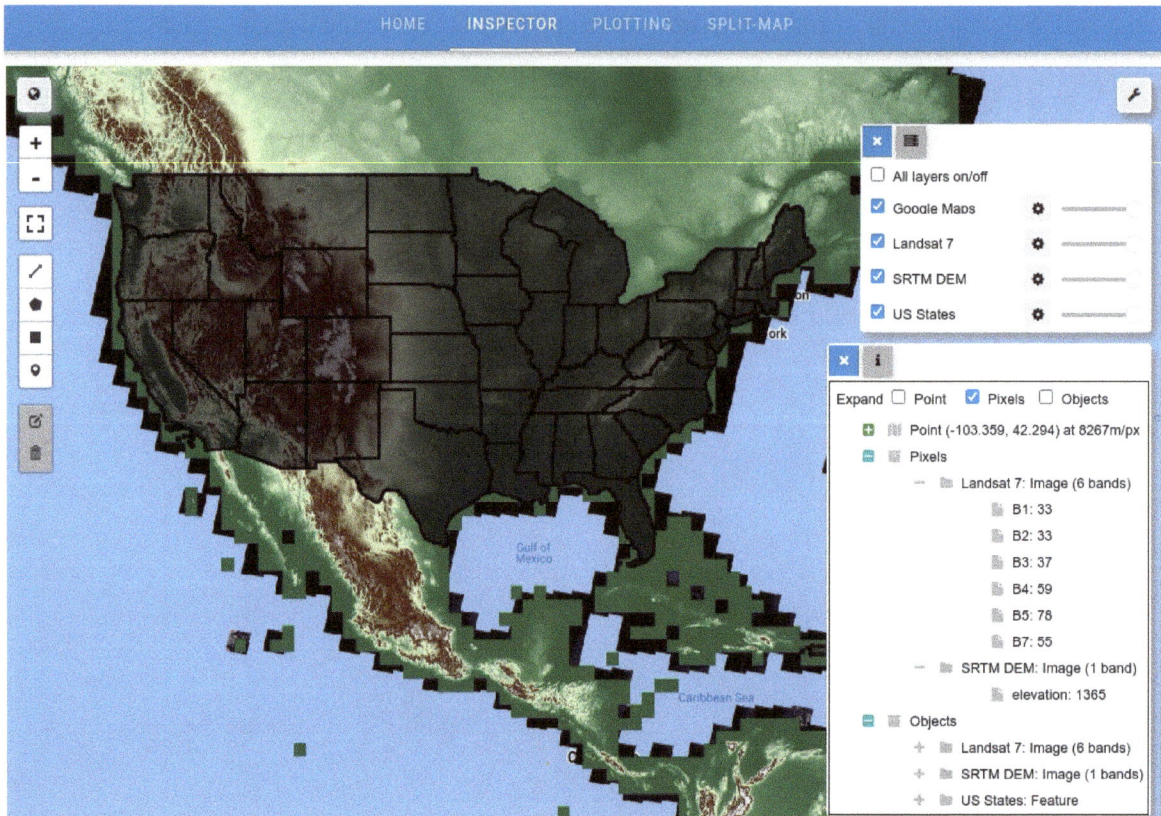

Figure 10.21: An interactive Solara web app for inspecting Earth Engine objects.

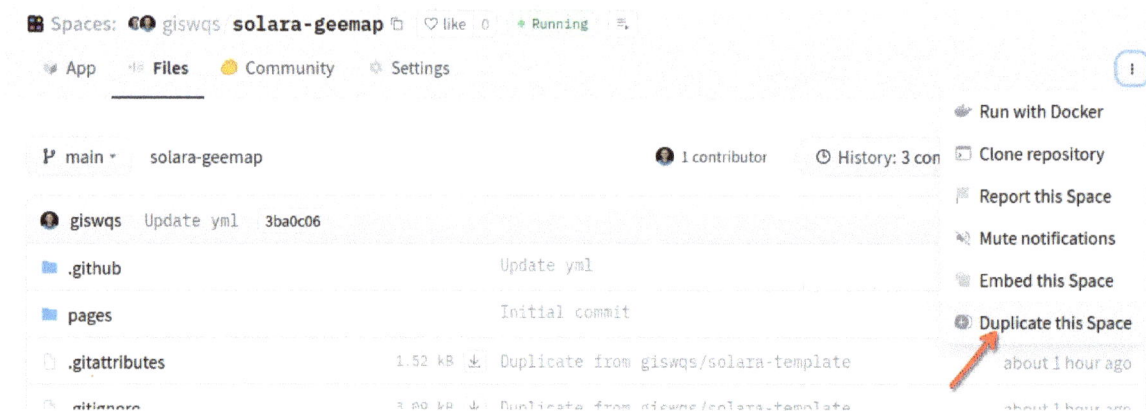

Figure 10.22: Duplicate the solara-geemap repo on Hugging Face.

Next, retrieve the Earth Engine token stored on your computer by running the following code in a Jupyter notebook or Python script:

```
import geemap
geemap.get_ee_token()
```

Copy the printed Earth Engine token to the clipboard. Then, open the duplicated repo on Hugging Face and click on the **Settings** tab. Scroll down to the **Secrets** section and click on the **New secret** button. Enter EARTHENGINE_TOKEN as the secret name and paste the Earth Engine token into the **Value** field. Click on the **Save** button to save the secret. The Earth Engine token will be used to authenticate Earth Engine in the web app.

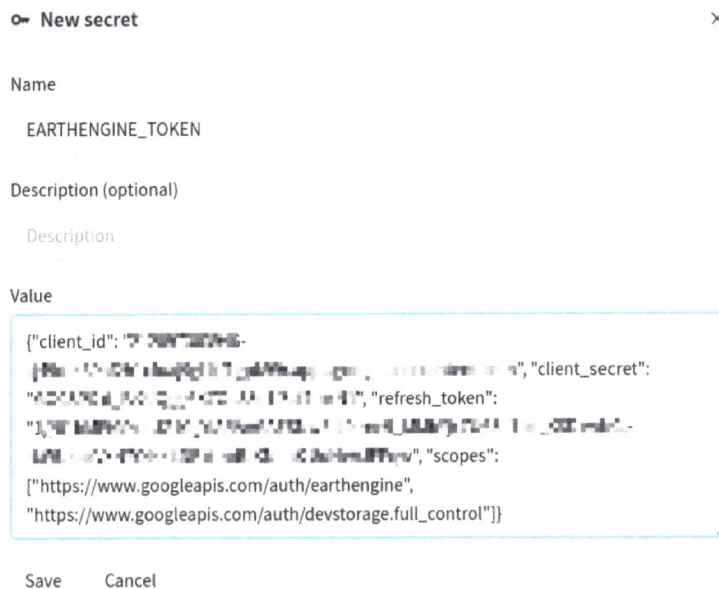

Figure 10.23: Set the Earth Engine token as a secret on Hugging Face.

After the space is built successfully, click the **Embed this Space** menu item to find the Direct URL for the app (Fig. 10.24).

The Direct URL to the web app should look like this:

```
https://USERNAME-solara-geemap.hf.space
```

The web app should look like Fig. 10.20. Feel free to explore the web app and make changes to the source code. The web app will be updated automatically whenever you push changes to the repo on Hugging Face.

10.10 Summary

You learned how to design Earth Engine Apps using both the Earth Engine User Interface API (JavaScript) and geemap (Python). You also learned how to deploy Earth Engine Apps on multiple platforms, such as the JavaScript Code Editor, a local web server, Streamlit Cloud, and Hugging Face. The skills of designing and deploying interactive Earth Engine Apps are essential for making your research and data products more accessible to the scientific community and the general public. Anyone with the link to your web app can analyze and visualize Earth Engine datasets without needing an Earth Engine account.

Embed this Space   ×

Iframe

```
<iframe
    src="https://giswqs-solara-geemap.hf.space"
    frameborder="0"
    width="850"
    height="450"
></iframe>
```
Copy

Direct URL

https://giswqs-solara-geemap.hf.space   Copy

Figure 10.24: Embed the Hugging Face space.

11. Earth Engine Applications

11.1 Introduction

In this chapter, we will explore some real-world applications of Earth Engine. We will learn how to use Earth Engine to analyze surface water dynamics, map land cover change, monitor forest cover change, and create global land cover maps. This is an opportunity to apply the visualization and analytical skills you have learned in the previous chapters to solve real-world problems. Keep in mind that the examples in this chapter are just a small fraction of the applications that can be built with Earth Engine. Check out the review papers below for more examples of Earth Engine applications:

- Amani, M., Ghorbanian, A., Ahmadi, A., Kakooei, M., ..., Wu, Q., & Brisco, B. (2020). Google Earth Engine Cloud Computing Platform for Remote Sensing Big Data Applications: A Comprehensive Review. *IEEE Journal of Selected Topics in Applied Earth Observations and Remote Sensing.* `https://doi.org/10.1109/JSTARS.2020.3021052`
- Boothroyd, R., Williams, R., Hoey, T., Barrett, B., & Prasojo, O. (2020). Applications of Google Earth Engine in fluvial geomorphology for detecting river channel change. *WIREs Water.* `https://doi.org/10.1002/wat2.1496`
- Tamiminia, H., Salehi, B., Mahdianpari, M., Quackenbush, L., Adeli, S., Brisco, B., 2020. Google Earth Engine for geo-big data applications: A meta-analysis and systematic review. *ISPRS J. Photogramm. Remote Sens.* 164, 152–170. `https://doi.org/10.1016/j.isprsjprs.2020.04.001`
- Yang, L., Driscol, J., Sarigai, S., Wu, Q., Chen, H., & Lippitt, C. D. (2022). Google Earth Engine and Artificial Intelligence (AI): A Comprehensive Review. *Remote Sensing*, 14(14), 3253. `https://doi.org/10.3390/rs14143253`

11.2 Technical requirements

To follow along with this chapter, you will need to have geemap and several optional dependencies installed. If you have already followed Section 1.5 - *Installing geemap*, then you should already have a conda environment with all the necessary packages installed. Otherwise, you can create a new conda environment and install pygis[30] with the following commands, which will automatically install geemap and all the required dependencies:

```
conda create -n gee python
conda activate gee
conda install -c conda-forge mamba
mamba install -c conda-forge pygis
```

Next, launch JupyterLab by typing the following commands in your terminal or Anaconda prompt:

```
jupyter lab
```

Alternatively, you can use geemap with a Google Colab cloud environment without installing anything on your local computer. Click 11_applications.ipynb[216] to launch the notebook in Google Colab.

Once in Colab, you can uncomment the following line and run the cell to install pygis, which includes

---

[216]11_applications.ipynb: tiny.geemap.org/ch11

geemap and all the necessary dependencies:

```
# %pip install pygis
```

The installation process may take 2-3 minutes. Once pygis has been installed successfully, click the **RESTART RUNTIME** button that appears at the end of the installation log or go to the **Runtime** menu and select **Restart runtime**. After that, you can start coding.

To begin, import the necessary libraries that will be used in this chapter:

```
import ee
import geemap
```

Initialize the Earth Engine Python API:

```
geemap.ee_initialize()
```

If this is your first time running the code above, you will need to authenticate Earth Engine first. Follow the instructions in Section 1.7 - *Earth Engine authentication* to authenticate Earth Engine.

11.3   Analyzing surface water dynamics

Surface water is a critical component of the Earth's hydrological cycle and is a key indicator of environmental change. Mapping surface water dynamics at the global scale used to be very challenging because of the dynamic nature of water bodies and the need for intensive computing resources. Empowered by Earth Engine, the JRC Global Surface Water[217] dataset maps the location and temporal distribution of surface water from 1984 to 2021 and provides statistics on the extent and change of those water surfaces. In this section, we will explore the JRC Global Surface Water dataset and learn how to use it to map surface water dynamics.

Surface water occurrence

First, let's load the JRC Global Surface Water dataset and display the band names:

```
dataset = ee.Image('JRC/GSW1_4/GlobalSurfaceWater')
dataset.bandNames()
```

```
0:occurrence
1:change_abs
2:change_norm
3:seasonality
4:recurrence
5:transition
6:max_extent
```

The occurrence band represents the frequency of water occurrence between 1984 and 2021 in a given pixel. It is an 8-bit integer image with a value range of 0-100. The value of 0 represents no water, and the value of 100 represents 100% water. The occurrence band is useful for analyzing temporal frequency of surface water, ranging from permanent water bodies to seasonal water bodies. Let's create an interactive map to display the surface water occurrence:

```
Map = geemap.Map()
Map.add_basemap('HYBRID')
Map
```

---

[217]JRC Global Surface Water: tiny.geemap.org/szil

Select the `occurrence` band from the dataset and specify a region of interest (ROI) to display the surface water occurrence of the ROI. The ROI can be specified by an `ee.Geometry` object or by drawing a polygon on the map. A color bar corresponding to the surface water occurrence values can also be added to the map using the `Map.add_colorbar()` method.

```python
image = dataset.select(['occurrence'])
region = Map.user_roi # Draw a polygon on the map
if region is None:
    region = ee.Geometry.BBox(-99.957, 46.8947, -99.278, 47.1531)
vis_params = {'min': 0.0, 'max': 100.0, 'palette': ['ffffff', 'ffbbbb', '0000ff']}
Map.addLayer(image, vis_params, 'Occurrence')
Map.addLayer(region, {}, 'ROI', True, 0.5)
Map.centerObject(region)
Map.add_colorbar(vis_params, label='Water occurrence (%)', layer_name='Occurrence')
```

The surface water occurrence map is shown in Fig. 11.1.

Figure 11.1: Map of surface water occurrence.

The water occurrence values range from 0 to 100. To calculate the number of pixels for each water occurrence value, use the `geemap.image_histogram()` function, which can return a histogram or a Pandas DataFrame containing the water occurrence values and their corresponding number of pixels. Set the `return_df` parameter `True` to return a Pandas DataFrame:

```python
df = geemap.image_histogram(
    image,
    region,
    scale=30,
    return_df=True,
)
df

key, value
0, 7.007843
1, 67.749020
```

```
2, 181.384314
3, 223.658824
4, 262.384314
..., ...
```

Set the `return_df` parameter `False` to return a histogram. Additionally, you can specify the x and y-axis labels and the title of the histogram using the `x_label`, `y_label`, and `title` parameters. You can also specify the layout arguments using the `layout_args` parameter. For example, the following code will create a histogram with a title centered at the top of the figure:

```python
hist = geemap.image_histogram(
    image,
    region,
    scale=30,
    x_label='Water Occurrence (%)',
    y_label='Pixel Count',
    title='Surface Water Occurrence',
    layout_args={'title': dict(x=0.5)},
    return_df=False,
)
hist
```

The histogram is shown in Fig. 11.2.

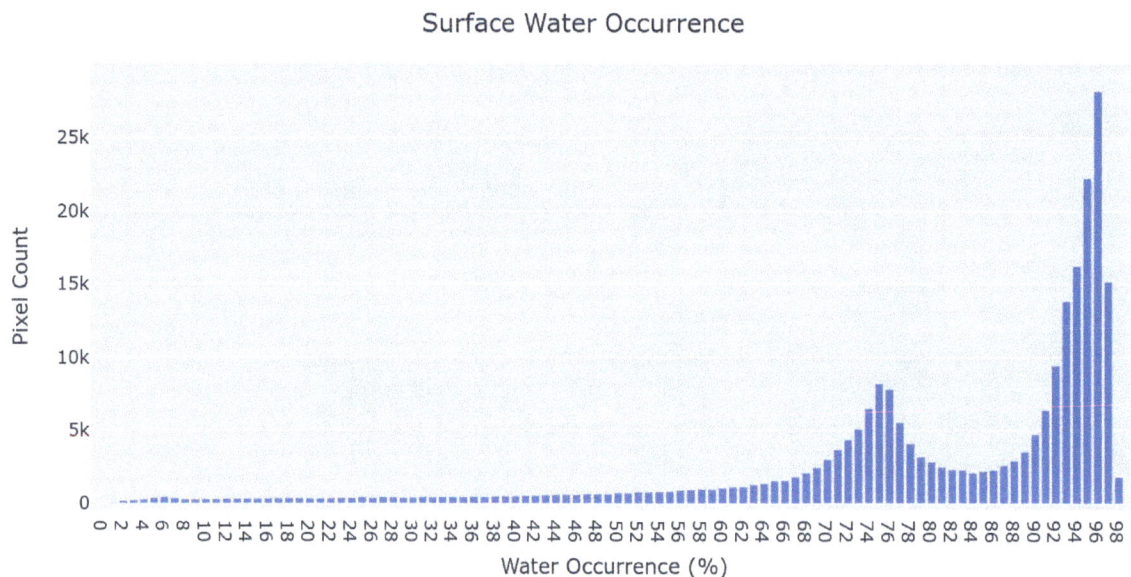

Figure 11.2: Histogram of surface water occurrence.

To export the histogram as a high-quality image, use the `write_image()` method of the histogram object. First, update the layout of the histogram object and then export it as a high-quality image:

```python
hist.update_layout(
    autosize=False,
    width=800,
    height=400,
    margin=dict(l=30, r=20, b=10, t=50, pad=4)
)
hist.write_image('water_occurrence.jpg', scale=2)
```

The `scale` parameter specifies the resolution of the exported image. The default value is 1, which means the exported image will have the same resolution as the original histogram. If you want to export a high-quality image, you can set the scale parameter to a larger value, such as 2 or 3.

Surface water monthly history

The JRC Water Occurrence dataset used above is aggregated from the JRC Monthly Water History[218] dataset, and contains 454 images at the time of writing. Each image represents the water extent of a month between 1984 and 2022.

```
dataset = ee.ImageCollection('JRC/GSW1_4/MonthlyHistory')
dataset.size()
```

```
454
```

Each image represents the water extent of a month between 1984 and 2021. The system:index property of each image contains the year and month of the image. You can use the aggregate_array() method to get a list of all the system:index values:

```
dataset.aggregate_array("system:index")
```

```
0:1984_03
1:1984_04
2:1984_05
3:1984_06
4:1984_07
...
```

Create an interactive map to display the water extent of a specific month. You can specify a region of interest as an ee.Geometry object or by drawing a polygon on the map:

```
Map = geemap.Map()
Map
```

The following code will display the water extent of August 2020:

```
image = dataset.filterDate('2020-08-01', '2020-09-01').first()
region = Map.user_roi # Draw a polygon on the map
if region is None:
    region = ee.Geometry.BBox(-99.957, 46.8947, -99.278, 47.1531)
vis_params = {'min': 0.0, 'max': 2.0, 'palette': ['ffffff', 'fffcb8', '0905ff']}

Map.addLayer(image, vis_params, 'Water')
Map.addLayer(region, {}, 'ROI', True, 0.5)
Map.centerObject(region)
```

The geemap.jrc_hist_monthly_history() function can be used to create an interactive plot of monthly water extents. The default area unit is square meters. You can change the area unit by specifying the denominator parameter, which will be used to divide the area values. For example, the following code creates a plot of monthly water extent in hectares (see Fig. 11.3).

```
geemap.jrc_hist_monthly_history(
    region=region, scale=30, frequency='month', denominator=1e4, y_label='Area (ha)'
)
```

Note that monthly water history data are generally not available for winter months due to the frozen water bodies, which are not detected by the algorithm. Specify the start_month and end_month param-

---

[218]JRC Monthly Water History: tiny.geemap.org/srgr

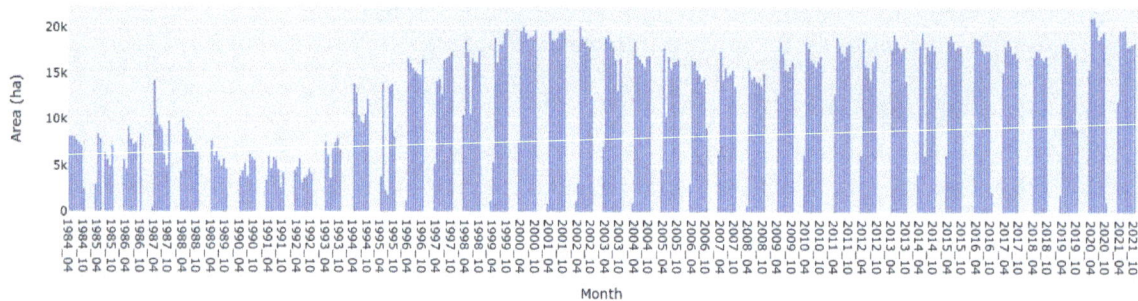

Figure 11.3: Histogram of monthly water history (January - December).

eters to create a plot of water extent for a specific season. For example, to create a plot of water extent for summer (from June to September):

```
geemap.jrc_hist_monthly_history(
    region=region,
    start_month=6,
    end_month=9,
    scale=30,
    frequency='month',
    denominator=1e4,
    y_label='Area (ha)',
)
```

The result is shown in Fig. 11.4.

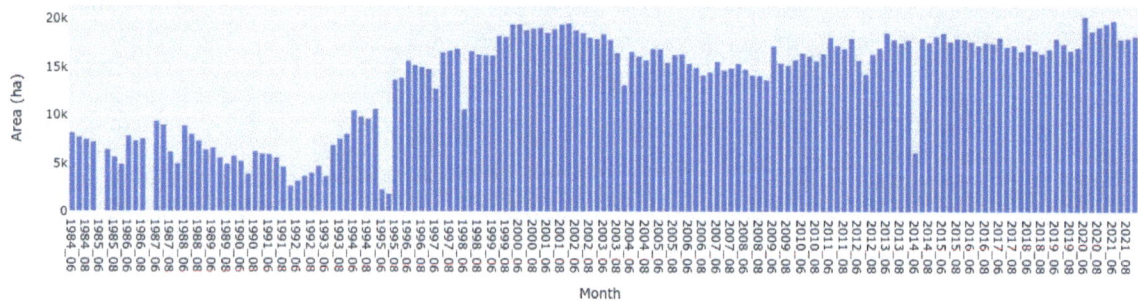

Figure 11.4: Histogram of monthly water history (June - September).

The `frequency` parameter can be set to `year` to create a plot of annual average water extent for specific months. For example, to create a plot of annual average water extent for summer (from June to September):

```
geemap.jrc_hist_monthly_history(
    region=region,
    start_month=6,
    end_month=9,
    scale=30,
    frequency='year',
    reducer='mean',
    denominator=1e4,
    y_label='Area (ha)',
)
```

The result is shown in Fig. 11.5.

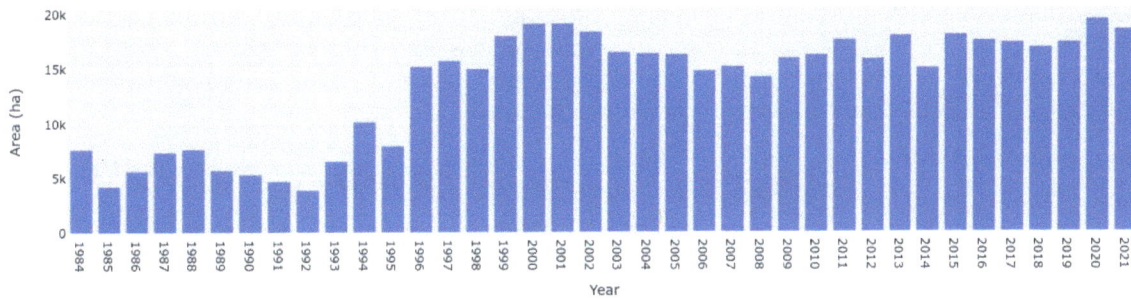

Figure 11.5: A histogram of annual average water extent (June - September).

Similarly, you can create a plot of annual maximum water extent for specific months by setting the reducer parameter to max:

```
geemap.jrc_hist_monthly_history(
    region=region,
    start_month=6,
    end_month=9,
    scale=30,
    frequency='year',
    reducer='max',
    denominator=1e4,
    y_label='Area (ha)',
)
```

The plot should look very similar to the plot of annual average water extent.

11.4 Mapping flood extents

The previous section demonstrates how to analyze surface water dynamics using the JRC Global Surface Water datasets. Note that the JRC datasets are not near real-time and are only updated once a year. As of February 2023, the JRC datasets are only available from 1984 to 2021. Therefore, they are not suitable for mapping floods occurring in 2022 and beyond. In this section, you will learn how to map flood extents using Landsat and Sentinel-1 data. We will use the 2022 Pakistan floods as an example. From 14 June to October 2022, floods in Pakistan killed 1,739 people, and caused 3.2 trillion Pakistani Rupees (14.9 billion US Dollars) of damage and 3.3 trillion Pakistani Rupees (15.2 billion US Dollars) of economic losses. See the Wikipedia[219] page for more information about the 2022 Pakistan floods.

Create an interactive map

Specify the map center location [lat, lon] and zoom level:

```
Map = geemap.Map(center=[29.3055, 68.9062], zoom=6)
Map
```

Search datasets

Click on the globe icon in the top left corner of the map to open the search panel. Select the data tab and enter a keyword to search for datasets, e.g., countries. Press Enter to search. The search results will populate the dropdown list. Select the dataset you want to add to the map from the dropdown

---

[219]Wikipedia: tiny.geemap.org/hjot

list, such as the LSIB 2017: Large Scale International Boundary Polygons, Simplified[220]. Click on the
`import` button to add a new code cell in Jupyter Notebook (see Fig. 11.6). Note that Google Colab and
JupyterLab do not support creating new code cells programmatically. You will need to manually add a
new code cell and copy the data sample code to the new cell.

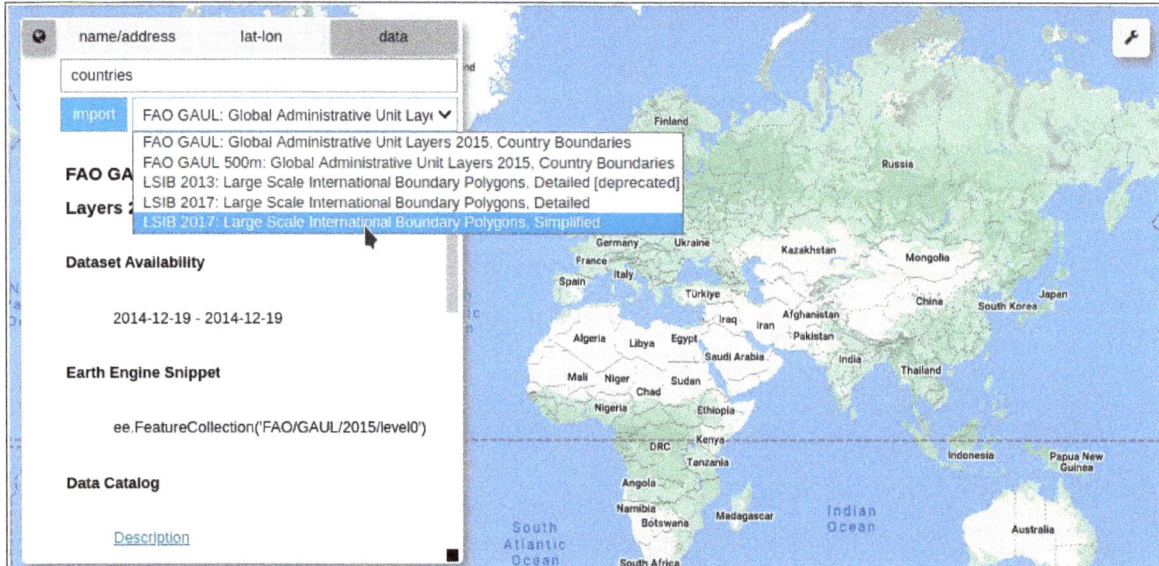

Figure 11.6: Search for countries dataset.

In this example, we will focus on Pakistan, but the code can be easily modified to visualize and analyze
floods in other countries. Modify the `country_name` variable below to specify the country of interest
and set the date range for the flood event. In order to extract the flood extent, we also need to specify
the date range for the pre-flood period.

```
country_name = 'Pakistan'
pre_flood_start_date = '2021-08-01'
pre_flood_end_date = '2021-09-30'
flood_start_date = '2022-08-01'
flood_end_date = '2022-09-30'
```

Visualize datasets

Specify the country of interest and filter the dataset by the country name:

```
country = ee.FeatureCollection('USDOS/LSIB_SIMPLE/2017').filter(
    ee.Filter.eq('country_na', country_name)
)
style = {'color': 'black', 'fillColor': '00000000'}
Map.addLayer(country.style(**style), {}, country_name)
Map.centerObject(country)
Map
```

Create Landsat composites

Create a Landsat 8 composite for the pre-flood period (August 1 to September 30, 2021) using the
USGS Landsat 8 Collection 2 Tier 1 Raw Scenes[221] , which can be used to create a composite image with

---
[220]LSIB 2017: Large Scale International Boundary Polygons, Simplified: tiny.geemap.org/daqa

only one line of code using the `ee.Algorithms.Landsat.simpleComposite()` method. We then clip the composite to the country boundary using the `image.clipToCollection()` method:

```
landsat_col_2021 = (
    ee.ImageCollection('LANDSAT/LC08/C02/T1')
    .filterDate(pre_flood_start_date, pre_flood_end_date)
    .filterBounds(country)
)
landsat_2021 = ee.Algorithms.Landsat.simpleComposite(landsat_col_2021).clipToCollection(
    country
)
vis_params = {'bands': ['B6', 'B5', 'B4'], 'max': 128}
Map.addLayer(landsat_2021, vis_params, 'Landsat 2021')
```

Similarly, create a Landsat 8 composite for the flood period (August 1 to September 30, 2022).

```
landsat_col_2022 = (
    ee.ImageCollection('LANDSAT/LC08/C02/T1')
    .filterDate(flood_start_date, flood_end_date)
    .filterBounds(country)
)
landsat_2022 = ee.Algorithms.Landsat.simpleComposite(landsat_col_2022).clipToCollection(
    country
)
Map.addLayer(landsat_2022, vis_params, 'Landsat 2022')
Map.centerObject(country)
Map
```

The resulting Landsat composites are shown in Fig. 11.7.

Figure 11.7: Landsat composites of Pakistan for the pre-flood (2021) and flood (2022) periods.

---

[221]USGS Landsat 8 Collection 2 Tier 1 Raw Scenes: tiny.geemap.org/bfrb

Compare Landsat composites side by side

You can change the layer opacity interactively to compare the two Landsat composites. Alternatively, you can use the `split_map()` function to compare the two Landsat composites side by side. The `split_map()` function takes two layers as input and displays them side by side. You can also specify the labels for the two layers:

```
Map = geemap.Map()
Map.setCenter(68.4338, 26.4213, 7)

left_layer = geemap.ee_tile_layer(
    landsat_2021, vis_params, 'Landsat 2021'
    )
right_layer = geemap.ee_tile_layer(
    landsat_2022, vis_params, 'Landsat 2022'
)

Map.split_map(
    left_layer, right_layer, left_label='Landsat 2021', right_label='Landsat 2022'
)
Map.addLayer(country.style(**style), {}, country_name)
Map
```

Compute Normalized Difference Water Index (NDWI)

The Normalized Difference Water Index[222] (NDWI) is a commonly used index for detecting water bodies. It is calculated as follows:

```
NDWI = (Green - NIR)/(Green + NIR)
```

Where Green is the green band and NIR is the near-infrared band. The NDWI values range from -1 to 1. The NDWI values usually have a threshold applied to them to a positive number (e.g., 0.1-0.3) to identify water bodies.

Landsat 8 imagery has 11 spectral bands[223] (see Fig. 11.8). The Landsat 8 NDWI is calculated using the green (`B3`) and NIR (`B5`) bands.

Use the `normalizedDifference()` function to compute the NDWI for the pre-flood and flood periods:

```
ndwi_2021 = landsat_2021.normalizedDifference(['B3', 'B5']).rename('NDWI')
ndwi_2022 = landsat_2022.normalizedDifference(['B3', 'B5']).rename('NDWI')
```

Create the NDWI layers for the pre-flood and flood periods and visualize them side by side (see Fig. 11.9).

```
Map = geemap.Map()
Map.setCenter(68.4338, 26.4213, 7)
ndwi_vis = {'min': -1, 'max': 1, 'palette': 'ndwi'}

left_layer = geemap.ee_tile_layer(ndwi_2021, ndwi_vis, 'NDWI 2021')
right_layer = geemap.ee_tile_layer(ndwi_2022, ndwi_vis, 'NDWI 2022')

Map.split_map(left_layer, right_layer, left_label='NDWI 2021', right_label='NDWI 2022')
Map.addLayer(country.style(**style), {}, country_name)
Map
```

---

[222]Normalized Difference Water Index: tiny.geemap.org/bzkx

[223]11 spectral bands: tiny.geemap.org/flje

Name	Pixel Size	Wavelength	Description
B1	30 meters	0.43 - 0.45 µm	Coastal aerosol
B2	30 meters	0.45 - 0.51 µm	Blue
B3	30 meters	0.53 - 0.59 µm	Green
B4	30 meters	0.64 - 0.67 µm	Red
B5	30 meters	0.85 - 0.88 µm	Near infrared
B6	30 meters	1.57 - 1.65 µm	Shortwave infrared 1
B7	30 meters	2.11 - 2.29 µm	Shortwave infrared 2
B8	15 meters	0.52 - 0.90 µm	Band 8 Panchromatic
B9	15 meters	1.36 - 1.38 µm	Cirrus
B10	30 meters	10.60 - 11.19 µm	Thermal infrared 1, resampled from 100m to 30m
B11	30 meters	11.50 - 12.51 µm	Thermal infrared 2, resampled from 100m to 30m

Figure 11.8: The list of Landsat 8 spectral bands.

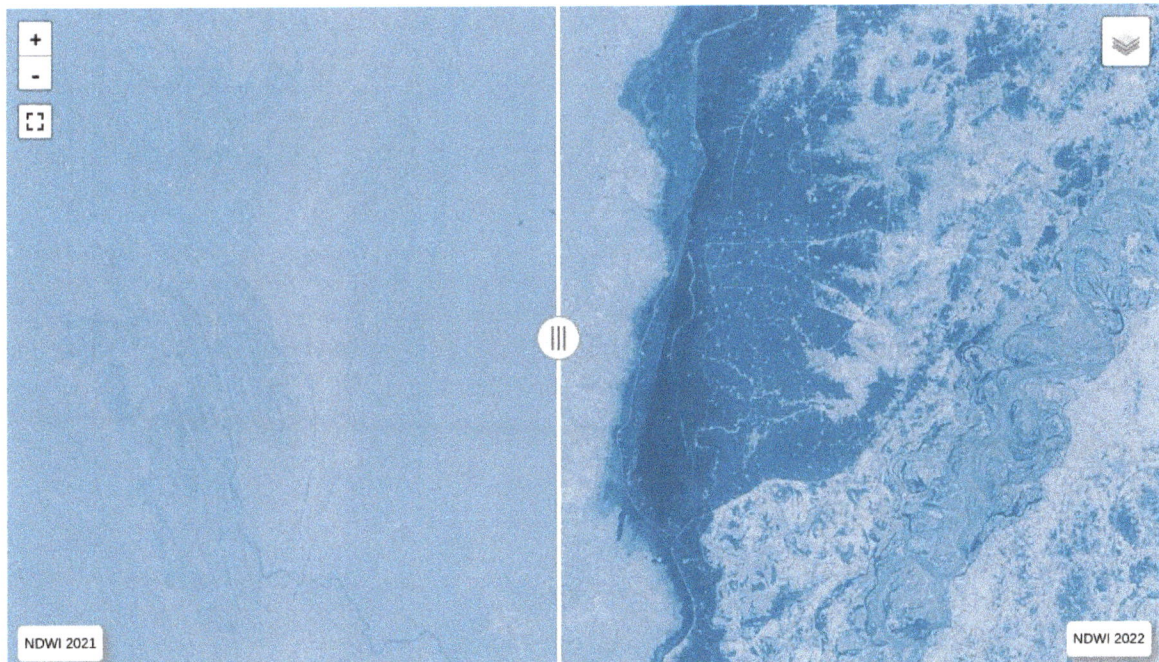

Figure 11.9: The NDWI layers for the pre-flood (left) and flood (right) periods.

Extract Landsat water extent

To extract the water extent, we need to convert the NDWI images to binary images using a threshold value. The threshold value is usually set to a positive value between 0 and 0.3. The smaller the threshold value, the more water bodies will be detected, which may increase the false positive rate. The larger the threshold value, the fewer water bodies will be detected, which may increase the false negative rate. In this example, we set the threshold value to 0.1:

```
threshold = 0.1
water_2021 = ndwi_2021.gt(threshold).selfMask()
water_2022 = ndwi_2022.gt(threshold).selfMask()
```

Compare the pre-flood and surface water extent side by side:

```
Map = geemap.Map()
Map.setCenter(68.4338, 26.4213, 7)

Map.addLayer(landsat_2021, vis_params, 'Landsat 2021', False)
Map.addLayer(landsat_2022, vis_params, 'Landsat 2022', False)

left_layer = geemap.ee_tile_layer(
    water_2021, {'palette': 'blue'}, 'Water 2021'
)
right_layer = geemap.ee_tile_layer(
    water_2022, {'palette': 'red'}, 'Water 2022'
)

Map.split_map(
    left_layer, right_layer, left_label='Water 2021', right_label='Water 2022'
)
Map.addLayer(country.style(**style), {}, country_name)
Map
```

The red pixels in the right panel represent the flooded water extent in 2022. The blue pixels in the left panel represent the pre-flood water extent in 2021.

Extract Landsat flood extent

To extract the flood extent, we need to subtract the pre-flood water extent from the flood water extent. The flood extent is the difference between the flood water extent and the pre-flood water extent. In other words, pixels identified as water in the flood period but not in the pre-flood period are considered as flooded pixels. The selfMask() method is used to mask out the no-data pixels.

```
flood_extent = water_2022.unmask().subtract(water_2021.unmask()).gt(0).selfMask()
```

Add the flood extent layer to the map (see Fig. 11.10).

```
Map = geemap.Map()
Map.setCenter(68.4338, 26.4213, 7)

Map.addLayer(landsat_2021, vis_params, 'Landsat 2021', False)
Map.addLayer(landsat_2022, vis_params, 'Landsat 2022', False)

left_layer = geemap.ee_tile_layer(
    water_2021, {'palette': 'blue'}, 'Water 2021'
)
right_layer = geemap.ee_tile_layer(
    water_2022, {'palette': 'red'}, 'Water 2022'
)

Map.split_map(
```

```
    left_layer, right_layer, left_label='Water 2021', right_label='Water 2022'
)

Map.addLayer(flood_extent, {'palette': 'cyan'}, 'Flood Extent')
Map.addLayer(country.style(**style), {}, country_name)
Map
```

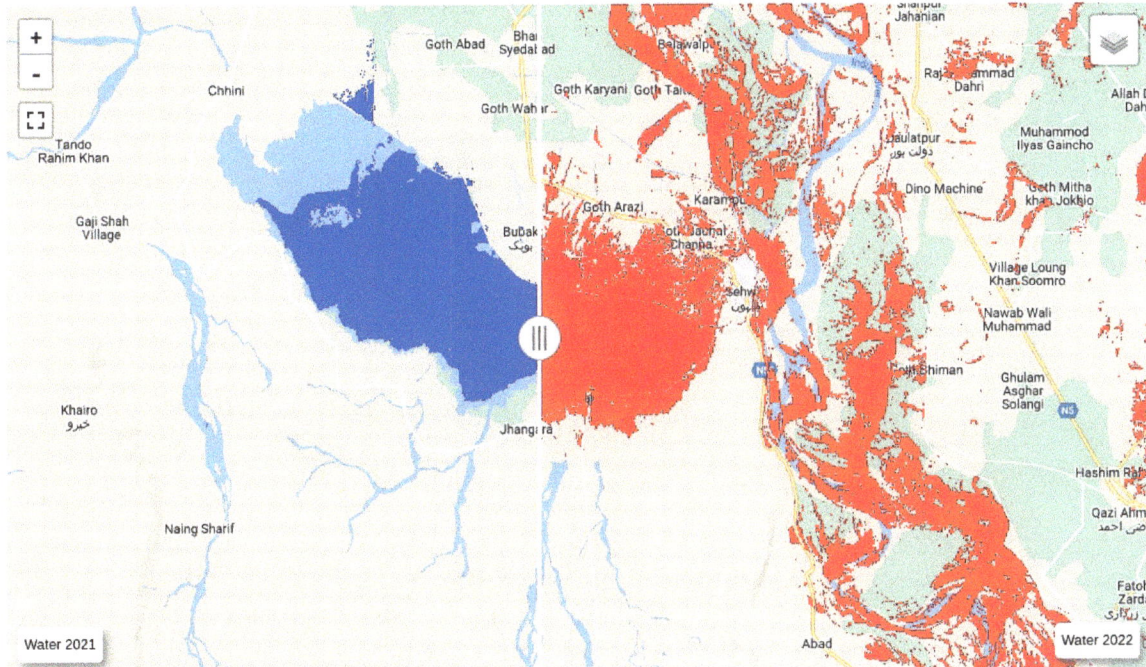

Figure 11.10: The flood extent delineated by Landsat 8 imagery.

Calculate Landsat flood area

To calculate the flood area, we can use the geemap.zonal_stats()[224] function. The required input parameters are the flood extent layer and the country boundary layer. The `scale` parameter can be set to `1000` to specify the spatial resolution of the image to be used for calculating the zonal statistics. The `stats_type` parameter can be set to `SUM` to calculate the total area of the flood extent in square kilometers. Set `return_fc=True` to return the zonal statistics as an `ee.FeatureCollection` object, which can be converted to a Pandas DataFrame.

Calculate the total surface water area in 2021:

```
area_2021 = geemap.zonal_stats(
    water_2021, country, scale=1000, statistics_type='SUM', return_fc=True
)
geemap.ee_to_df(area_2021)
```

sum	wld_rgn	country_na	abbreviati	country_co
4205.678431	S Asia	Pakistan	Pak.	PK

Calculate the total surface water area in 2022:

---

[224]geemap.zonal_stats(): tiny.geemap.org/iztj

```
area_2022 = geemap.zonal_stats(
    water_2022, country, scale=1000, statistics_type='SUM', return_fc=True
)
geemap.ee_to_df(area_2022)
```

sum	wld_rgn	country_na	abbreviati	country_co
13145.027451	S Asia	Pakistan	Pak.	PK

Calculate the total flood area:

```
flood_area = geemap.zonal_stats(
    flood_extent, country, scale=1000, statistics_type='SUM', return_fc=True
)
geemap.ee_to_df(flood_area)
```

sum	wld_rgn	country_na	abbreviati	country_co
11065.72549	S Asia	Pakistan	Pak.	PK

The total area of the flood extent is 11,065 square kilometers based on Landsat 8 images.

Create Sentinel-1 SAR composites

Besides Landsat, we can also use Sentinel-1 Synthetic Aperture Radar (SAR)[225] data to extract flood extents. Radar can collect signals in different polarizations by controlling the analyzed polarization in both the transmit and receive paths. A VH would indicate signals emitted in vertical (V) and received in horizontal (H) polarization. Alternatively, a signal that was emitted in horizontal (H) and received in horizontal (H) would be indicated by HH, and so on. Examining the signal strength from these different polarizations carries information about the structure of the imaged surface. Rough surface scattering, such as that caused by bare soil or water, is most sensitive to VV scattering. Therefore, VV polarization is often used to detect water bodies.

Sentinel-1 operates in four exclusive acquisition modes[226] :

- Stripmap (SM)
- Interferometric Wide swath (IW)
- Extra-Wide swath (EW)
- Wave mode (WV)

The Interferometric Wide swath (IW) mode allows combining a large swath width (250 km) with a moderate geometric resolution (5 m by 20 m). The IW mode is the default acquisition mode over land. In this example, we will use Sentinel-1 IW mode data to extract flood extents.

The Sentinel-1 SAR data[227] are available from 2014 to present. Let's filter the COPERNICUS/S1_GRD dataset by the date range and location:

```
s1_col_2021 = (
    ee.ImageCollection('COPERNICUS/S1_GRD')
    .filter(ee.Filter.listContains('transmitterReceiverPolarisation', 'VV'))
    .filter(ee.Filter.eq('instrumentMode', 'IW'))
```

[225]Synthetic Aperture Radar (SAR): tiny.geemap.org/kgnt
[226]acquisition modes: tiny.geemap.org/rmbx
[227]Sentinel-1 SAR data: tiny.geemap.org/kkad

```
        .filter(ee.Filter.eq('orbitProperties_pass', 'ASCENDING'))
        .filterDate(pre_flood_start_date, pre_flood_end_date)
        .filterBounds(country)
        .select('VV')
    )
    s1_col_2021
```

247 Sentinel-1 IW mode images are available for the pre-flood period in 2021. Similarly, let's find out how many Sentinel-1 IW mode images are available for the flood period in 2022:

```
    s1_col_2022 = (
        ee.ImageCollection('COPERNICUS/S1_GRD')
        .filter(ee.Filter.listContains('transmitterReceiverPolarisation', 'VV'))
        .filter(ee.Filter.eq('instrumentMode', 'IW'))
        .filter(ee.Filter.eq('orbitProperties_pass', 'ASCENDING'))
        .filterDate(flood_start_date, flood_end_date)
        .filterBounds(country)
        .select('VV')
    )
    s1_col_2022
```

250 Sentinel-1 IW mode images are available for the flood period in 2022. Let's visualize the Sentinel-1 IW mode images for the pre-flood and flood periods (see Fig. 11.11).

```
    Map = geemap.Map()
    Map.add_basemap('HYBRID')
    sar_2021 = s1_col_2021.reduce(ee.Reducer.percentile([20])).clipToCollection(country)
    sar_2022 = s1_col_2022.reduce(ee.Reducer.percentile([20])).clipToCollection(country)
    Map.addLayer(sar_2021, {'min': -25, 'max': -5}, 'SAR 2021')
    Map.addLayer(sar_2022, {'min': -25, 'max': -5}, 'SAR 2022')
    Map.centerObject(country)
    Map
```

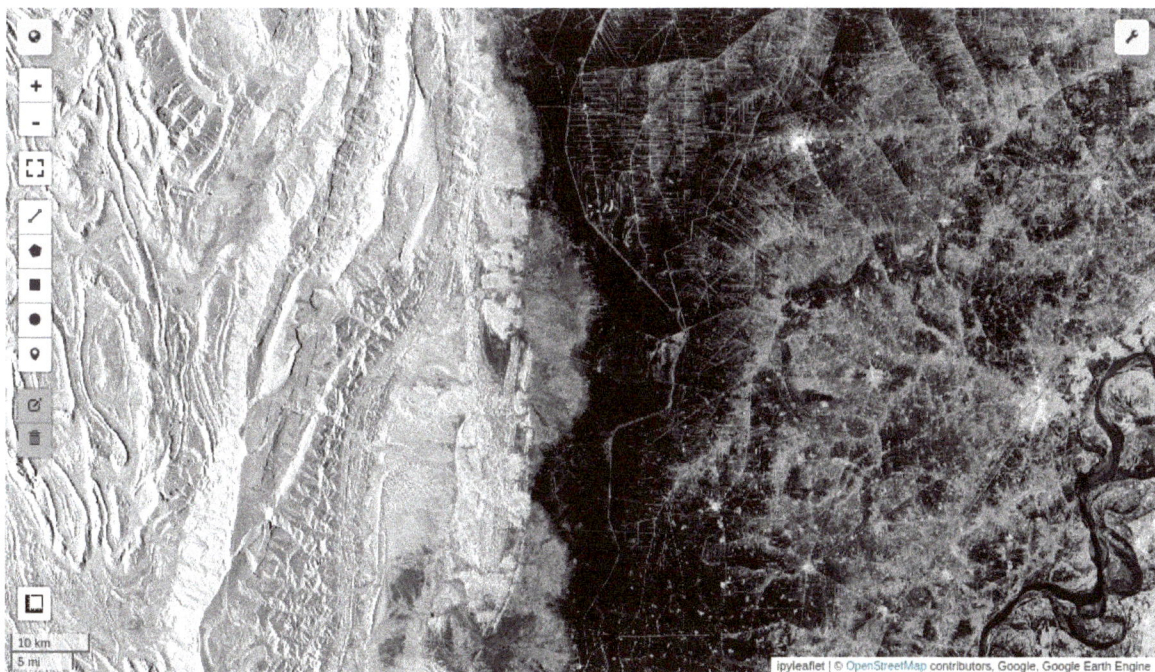

Figure 11.11: Sentinel-1 SAR imagery covering southern Pakistan in 2022.

Apply speckle filtering

Speckle, appearing in synthetic aperture radar (SAR) images as granular noise, is due to the interference of waves reflected from many elementary scatterers. Speckle in SAR images complicates the image interpretation problem by reducing the effectiveness of image segmentation and classification (Lee et al., 1994[228]). Therefore, speckle filtering is often applied to SAR images to reduce the speckle noise. In this example, we apply a morphological speckle filter to the Sentinel-1 SAR images. The morphological speckle filter is a non-linear filter that uses the median value of a pixel and its neighboring pixels to replace the pixel value. The kernel size is set to 100 meters.

```
col_2021 = s1_col_2021.map(lambda img: img.focal_median(100, 'circle', 'meters'))
col_2022 = s1_col_2022.map(lambda img: img.focal_median(100, 'circle', 'meters'))

Map = geemap.Map()
Map.add_basemap('HYBRID')
sar_2021 = col_2021.reduce(ee.Reducer.percentile([20])).clipToCollection(country)
sar_2022 = col_2022.reduce(ee.Reducer.percentile([20])).clipToCollection(country)
Map.addLayer(sar_2021, {'min': -25, 'max': -5}, 'SAR 2021')
Map.addLayer(sar_2022, {'min': -25, 'max': -5}, 'SAR 2022')
Map.centerObject(country)
Map
```

Compare Sentinel-1 SAR composites side by side

Create a split-view map to compare the pre-flood and flood SAR composites side by side:

```
Map = geemap.Map()
Map.setCenter(68.4338, 26.4213, 7)

left_layer = geemap.ee_tile_layer(sar_2021, {'min': -25, 'max': -5}, 'SAR 2021')
right_layer = geemap.ee_tile_layer(sar_2022, {'min': -25, 'max': -5}, 'SAR 2022')

Map.split_map(
    left_layer, right_layer, left_label='Sentinel-1 2021', right_label='Sentinel-1 2022'
)
Map.addLayer(country.style(**style), {}, country_name)
Map
```

Extract SAR water extent

Water usually appears dark in SAR images because radar waves are reflected differently by different surfaces. Water is a smooth, flat surface that does not reflect radar waves very well, so it appears dark in SAR images. Thresholding SAR imagery is one of the most widely used approaches to delineate water extents for its effectiveness and efficiency (Liang and Liu, 2020[229]). Thresholding methods can be generally divided into two categories: global and local. Global thresholding methods use a single threshold value to segment the entire image. Local thresholding methods use a different threshold value for each pixel. In this example, we use a global threshold of -18 dB to extract the water extent:

```
threshold = -18
water_2021 = sar_2021.lt(threshold)
water_2022 = sar_2022.lt(threshold)
```

Create a split-view map to compare the pre-flood and flood water extent side by side (see Fig. 11.12).

```
Map = geemap.Map()
```

[228]Lee et al., 1994: tiny.geemap.org/wntd

[229]Liang and Liu, 2020: tiny.geemap.org/ivvg

```
Map.setCenter(68.4338, 26.4213, 7)

Map.addLayer(sar_2021, {'min': -25, 'max': -5}, 'SAR 2021')
Map.addLayer(sar_2022, {'min': -25, 'max': -5}, 'SAR 2022')

left_layer = geemap.ee_tile_layer(
    water_2021.selfMask(), {'palette': 'blue'}, 'Water 2021'
)
right_layer = geemap.ee_tile_layer(
    water_2022.selfMask(), {'palette': 'red'}, 'Water 2022'
)

Map.split_map(
    left_layer, right_layer, left_label='Water 2021', right_label='Water 2022'
)
Map.addLayer(country.style(**style), {}, country_name)
Map
```

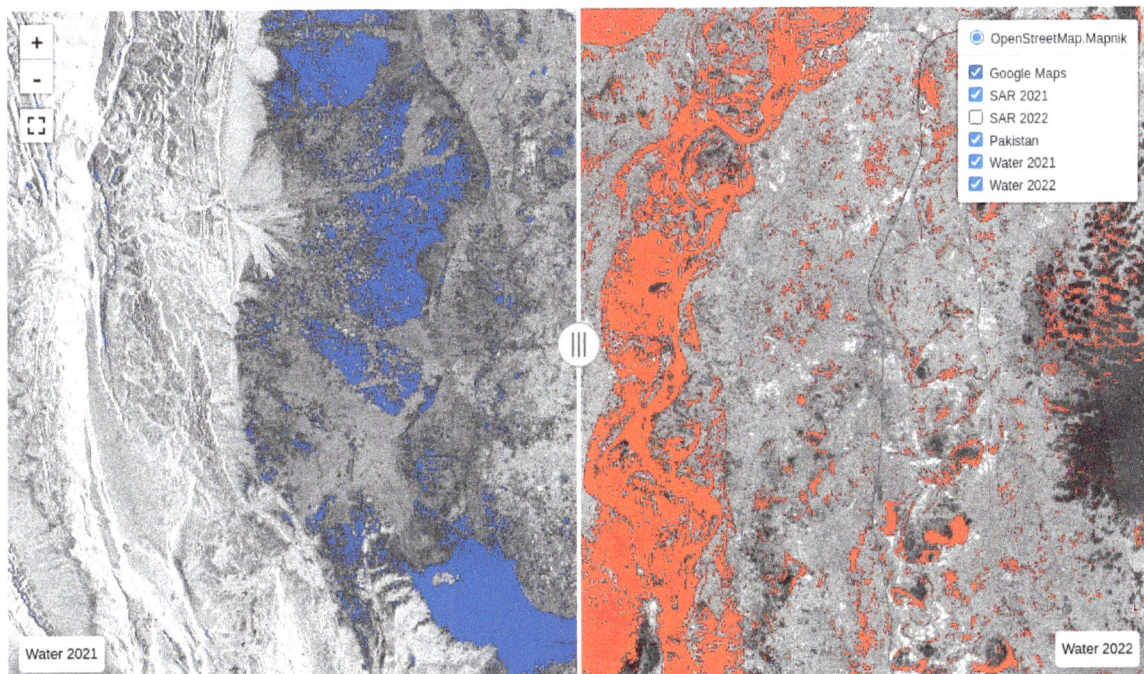

Figure 11.12: Surface water extent extracted from Sentinel-1 SAR imagery covering southern Pakistan in 2021 and 2022.

Extract SAR flood extent

Similar to the Landsat approach, we can subtract the pre-flood water extent from the flood water extent to extract the flood extent.

```
flood_extent = water_2022.unmask().subtract(water_2021.unmask()).gt(0).selfMask()
```

The flood extent is the difference between the flood water extent and the pre-flood water extent. In other words, pixels identified as water in the flood period but not in the pre-flood period are considered as flooded pixels, which are shown in cyan.

```
Map = geemap.Map()
Map.setCenter(68.4338, 26.4213, 7)
```

```
Map.addLayer(sar_2021, {'min': -25, 'max': -5}, 'SAR 2021')
Map.addLayer(sar_2022, {'min': -25, 'max': -5}, 'SAR 2022')

left_layer = geemap.ee_tile_layer(
    water_2021.selfMask(), {'palette': 'blue'}, 'Water 2021'
)
right_layer = geemap.ee_tile_layer(
    water_2022.selfMask(), {'palette': 'red'}, 'Water 2022'
)

Map.split_map(
    left_layer, right_layer, left_label='Water 2021', right_label='Water 2022'
)

Map.addLayer(flood_extent, {'palette': 'cyan'}, 'Flood Extent')
Map.addLayer(country.style(**style), {}, country_name)
Map
```

Calculate SAR flood area

```
area_2021 = geemap.zonal_stats(
    water_2021, country, scale=1000, statistics_type='SUM', return_fc=True
)
geemap.ee_to_df(area_2021)
```

sum	wld_rgn	country_na	abbreviati	country_co
68949.458824	S Asia	Pakistan	Pak.	PK

```
area_2022 = geemap.zonal_stats(
    water_2022, country, scale=1000, statistics_type='SUM', return_fc=True
)
geemap.ee_to_df(area_2022)
```

sum	wld_rgn	country_na	abbreviati	country_co
59224.121569	S Asia	Pakistan	Pak.	PK

```
flood_area = geemap.zonal_stats(
    flood_extent, country, scale=1000, statistics_type='SUM', return_fc=True
)
geemap.ee_to_df(flood_area)
```

sum	wld_rgn	country_na	abbreviati	country_co
12264.835294	S Asia	Pakistan	Pak.	PK

The total area of the flood extent is 12,264 square kilometers based on Sentinel-1 SAR images.

11.5   Forest cover change analysis

Forest cover change analysis is a common application of remote sensing. It is used to monitor forest cover change over time and to estimate forest loss. In this section, we will use the Hansen Global Forest Change[230] dataset to analyze forest cover and forest loss at the global scale. According to the Food and Agriculture Organization (FAO) of the United Nations, forest is defined as land spanning more than

0.5 hectares with trees higher than 5 meters and a canopy cover of more than 10 percent, or trees able to reach these thresholds in situ. The Hansen dataset is a global forest cover change dataset that is updated annually. As of February 2023, the dataset includes forest cover change information from 2000 to 2021 at a spatial resolution of 30 meters.

Forest cover mapping

First, let's load the Hansen dataset and visualize the forest cover along with the multi-spectral Landsat 7 images:

```
dataset = ee.Image('UMD/hansen/global_forest_change_2021_v1_9')
dataset.bandNames()

['treecover2000',
 'loss',
 'gain',
 'lossyear',
 'first_b30',
 'first_b40',
 'first_b50',
 'first_b70',
 'last_b30',
 'last_b40',
 'last_b50',
 'last_b70',
 'datamask']
```

As shown above, the dataset contains several bands, including the tree cover in 2000, the forest loss and gain during the period from 2000 to 2021, the year of forest loss, and the multi-spectral imagery from the first available year (i.e., 2000) and the last available year (i.e., 2021). The tree cover band is an integer value ranging from 0 to 100, where 0 represents no tree cover and 100 represents 100 percent tree cover. The loss and gain bands are boolean values, where 1 represents the presence of loss or gain and 0 represents the absence of loss or gain.

```
Map = geemap.Map()
first_bands = ['first_b50', 'first_b40', 'first_b30']
first_image = dataset.select(first_bands)
Map.addLayer(first_image, {'bands': first_bands, 'gamma': 1.5}, 'Year 2000 Bands 5/4/3')

last_bands = ['last_b50', 'last_b40', 'last_b30']
last_image = dataset.select(last_bands)
Map.addLayer(last_image, {'bands': last_bands, 'gamma': 1.5}, 'Year 2021 Bands 5/4/3')

treecover = dataset.select(['treecover2000'])
treeCoverVisParam = {'min': 0, 'max': 100, 'palette': ['black', 'green']}
name = 'Tree cover (%)'
Map.addLayer(treecover, treeCoverVisParam, name)
Map.add_colorbar(treeCoverVisParam, label=name, layer_name=name)
Map
```

The global tree cover in 2000 is shown in Fig. 11.13.

According to the FAO forest definition of at least 10 percent tree canopy cover, we can convert the tree cover band to a binary image by setting a threshold of 10 percent. Pixels with tree cover greater than or equal to 10 percent are set to 1, and pixels with tree cover less than 10 percent are set to 0. The resulting binary image represents the global forest cover in 2000:

```
threshold = 10
```

---

[230]Hansen Global Forest Change: tiny.geemap.org/aueu

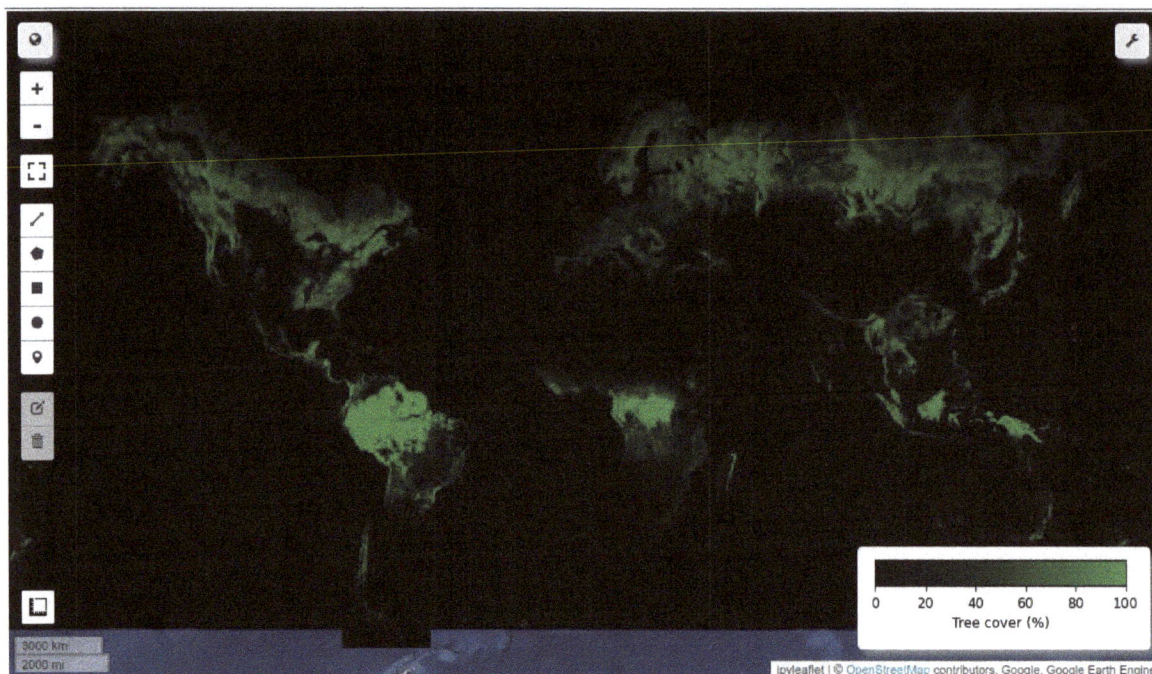

Figure 11.13: Global tree cover in 2000.

```
treecover_bin = treecover.gte(threshold).selfMask()
treeVisParam = {'palette': ['green']}
Map.addLayer(treecover_bin, treeVisParam, 'Tree cover bin')
```

Forest loss and gain mapping

Using the global forest cover in 2000 as a baseline, we can visualize the forest loss and gain during the period from 2000 to 2021. The `lossyear` band represents the year of tree loss. The value ranges from 0 to 21, representing the year of tree loss from 2000 to 2021. A value of 1 indicates that the tree loss occurred in 2001, a value of 2 indicates that the tree loss occurred in 2002, and so on.

```
Map = geemap.Map()
treeloss_year = dataset.select(['lossyear'])
treeLossVisParam = {'min': 0, 'max': 21, 'palette': ['yellow', 'red']}
layer_name = 'Tree loss year'
Map.addLayer(treeloss_year, treeLossVisParam, layer_name)
Map.add_colorbar(treeLossVisParam, label=layer_name, layer_name=layer_name)
Map
```

The resulting tree loss year map is shown in Fig. 11.14.

We can visualize the tree loss and gain using the `loss` and `gain` bands, respectively. The `loss` and `gain` bands are boolean values, where 1 represents the presence of loss or gain from 2000 to 2021.

```
treeloss = dataset.select(['loss']).selfMask()
treegain = dataset.select(['gain']).selfMask()
Map.addLayer(treeloss, {'palette': 'red'}, 'Tree loss')
Map.addLayer(treegain, {'palette': 'yellow'}, 'Tree gain')
Map
```

The resulting tree loss and gain map is shown in Fig. 11.15.

Figure 11.14: Map of the year of tree loss from 2000 to 2021.

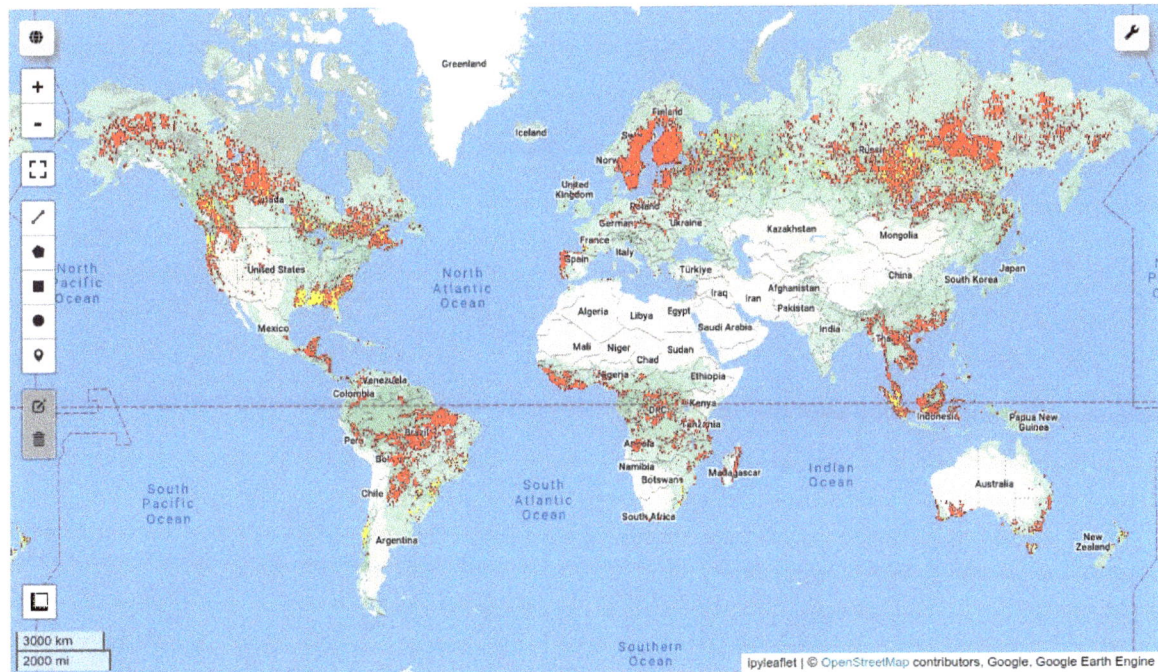

Figure 11.15: Map of tree loss and gain from 2000 to 2021. The red color represents tree loss, and the yellow color represents tree gain.

Zonal statistics by country

The previous sections showed how to visualize the global forest cover in 2000 and the forest loss and gain during the period from 2000 to 2021. We can also calculate the forest cover area by country. First, load the countries feature collection:

```
Map = geemap.Map()
countries = ee.FeatureCollection(geemap.examples.get_ee_path('countries'))
style = {'color': '#000000ff', 'fillColor': '#00000000'}
Map.addLayer(countries.style(**style), {}, 'Countries')
Map
```

Then, use the `zonal_stats_by_group()` function to calculate the forest cover area by country. The function parameters include the `treecover_bin` image, `countries` feature collection, output CSV file name, statistics type, denominator, and scale. Set the statistics type to `SUM` to calculate the total forest cover area. Set the denominator to 1e6 to convert the area from square meters to square kilometers. The output CSV file should contain a `Class_sum` column, which represents the forest cover area in square kilometers.

```
geemap.zonal_stats_by_group(
    treecover_bin,
    countries,
    'forest_cover.csv',
    statistics_type='SUM',
    denominator=1e6,
    scale=1000,
)
```

After the CSV file is generated, we can use the `pie_chart()` and `bar_chart()` functions to visualize the forest cover area by country. The following code creates a pie chart showing the top 20 countries with the largest forest cover area. The `max_rows` parameter specifies the number of countries to show in the chart. Countries falling outside the top 20 are grouped into the `Other` category. The `height` parameter specifies the height of the chart in pixels:

```
geemap.pie_chart(
    'forest_cover.csv', names='NAME', values='Class_sum', max_rows=20, height=400
)
```

We can see that the top five countries with the largest forest cover area are Russia, Brazil, Canada, United States, and China. The top 20 countries account for 74.7% of the global forest cover area (see Fig. 11.16).

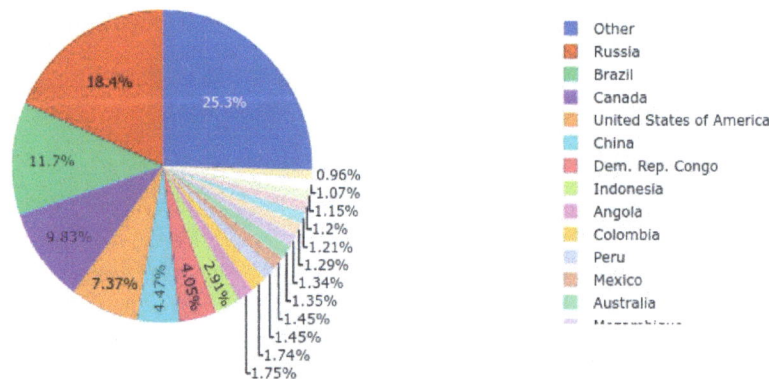

Figure 11.16: A pie chart showing the forest cover area by country.

Below is a bar chart showing the top 20 countries with the largest forest cover area (see Fig. 11.17).

```
geemap.bar_chart(
    'forest_cover.csv',
    x='NAME',
    y='Class_sum',
    max_rows=20,
    x_label='Country',
    y_label='Forest area (km2)',
)
```

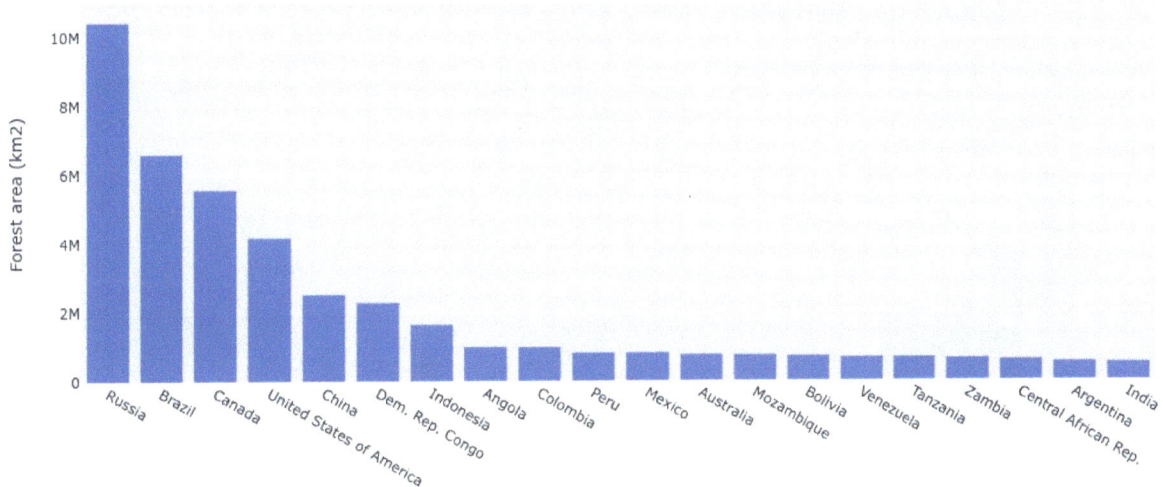

Figure 11.17: A bar chart showing the forest cover area by country.

Similarly, we can calculate the forest loss area by country. The following code creates a pie chart showing the top 20 countries with the largest forest loss area:

```
geemap.zonal_stats_by_group(
    treeloss,
    countries,
    'treeloss.csv',
    statistics_type='SUM',
    denominator=1e6,
    scale=1000,
)

geemap.pie_chart(
    'treeloss.csv', names='NAME', values='Class_sum', max_rows=20, height=600
)
```

The pie chart shows that the top five countries with the largest forest loss area are Russia, Brazil, Indonesia, United States, Canada, and Indonesia. The top 20 countries account for 79.9% of the global forest loss area (see Fig. 11.18).

Below is a bar chart showing the top 20 countries with the largest forest loss area (see Fig. 11.19).

```
geemap.bar_chart(
    'treeloss.csv',
    x='NAME',
    y='Class_sum',
    max_rows=20,
    x_label='Country',
    y_label='Forest loss area (km2)',
)
```

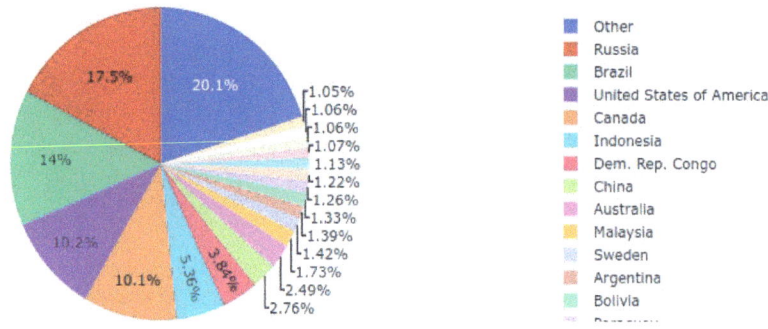

Figure 11.18: A pie chart showing the forest loss area by country.

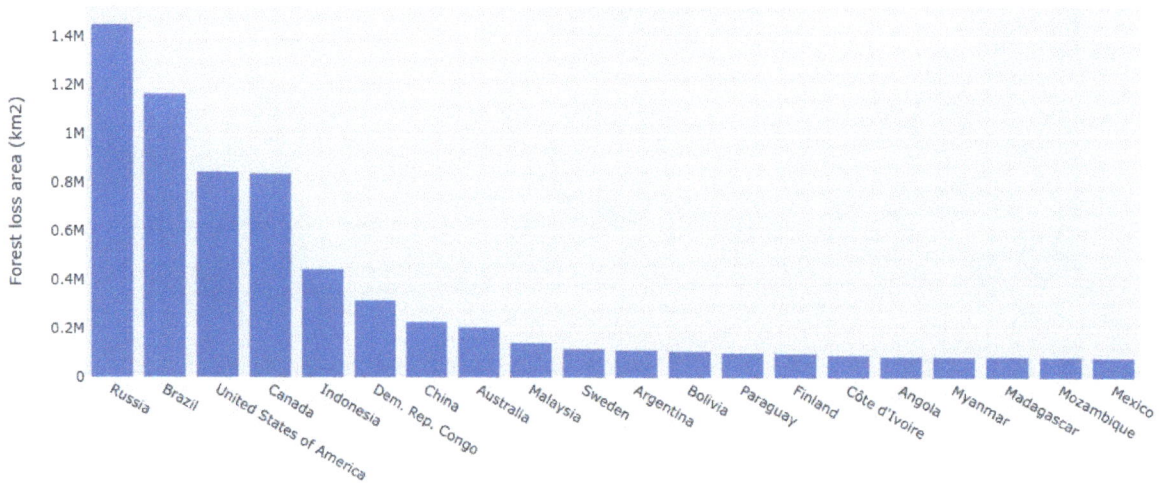

Figure 11.19: A bar chart showing the forest loss area by country.

11.6 Global land cover mapping

Global land cover mapping is a challenging task in remote sensing. In recent years, several global land cover data products have become available thanks to the development of cloud computing and machine learning. The notable global land cover products include the Dynamic World[231] , ESA World-Cover[232] (2020-2021), and Esri annual global land cover[233] (2017-2022). All of these products are provided at a 10-m spatial resolution. The Dynamic World and Esri land cover products are based on Sentinel-2 optical data, while the ESA WorldCover product is based on both Sentinel-1 SAR data and Sentinel-2 optical data. The ESA WorldCover and Esri land cover products are updated annually (i.e., one global land cover map per year), while the Dynamic World is a near-real-time product that provides classification of every Sentinel-2 image with a cloud coverage less than 35%. In this section, we will explore these global land cover products with geemap.

---

[231]Dynamic World: dynamicworld.app

[232]ESA WorldCover: esa-worldcover.org

[233]Esri annual global land cover: tiny.geemap.org/pgut

Dynamic World

The Dynamic World dataset is available in the Earth Engine Data Catalog with the image collection ID of "GOOGLE/DYNAMICWORLD/V1". All Sentinel-2 images with a cloud coverage less than 35% are classified into nine land cover classes, including water, trees, grass, flooded vegetation, crops, shrub and scrub, built up, bare soil, and snow and ice. Each classified image contains 10 bands, where the first nine bands represent the estimated probability of each land cover class, and the last band represents the index of the band with the highest probability. The Dynamic World dataset does not provide a global land cover map. Instead, it provides a classification of every Sentinel-2 image with a cloud coverage less than 35%. To generate a land cover map for a specific region larger than a single Sentinel-2 image, we need to aggregate the classified images into a single image. Luckily, geemap provides a function called `dynamic_world()` to create a land cover map for a specified region within a specified time range. The result can be returned as different data types, including `hillshade`, `class`, `visualize`, and `probability`. The default type is `hillshade`, which returns a hill-shaded image, where pixels with a higher probability of a land cover class are shown as elevated areas and pixels with a lower probability are shown as low-lying areas. The following code creates a hill-shaded land cover map of Madison, Wisconsin, USA, for the year 2021 by aggregating Dynamic World classified images:

```python
Map = geemap.Map()
region = ee.Geometry.BBox(-89.7088, 42.9006, -89.0647, 43.2167)
start_date = '2021-01-01'
end_date = '2022-01-01'

image = geemap.dynamic_world_s2(region, start_date, end_date, clip=True)
vis_params = {'bands': ['B4', 'B3', 'B2'], 'min': 0, 'max': 3000}
Map.addLayer(image, vis_params, 'Sentinel-2 image')

landcover = geemap.dynamic_world(
    region, start_date, end_date, return_type='hillshade', clip=True
)
Map.addLayer(landcover, {}, 'Land Cover')

Map.add_legend(
    title="Dynamic World Land Cover",
    builtin_legend='Dynamic_World',
    layer_name='Land Cover',
)
Map.centerObject(region, 13)
Map
```

The result is shown in Fig. 11.20.

To generate a Sentinel-2 image composite corresponding to the Dynamic World land cover map, we can use the `dynamic_world_s2()` function. The function parameters are similar to the `dynamic_world()` function, such as `region`, `start_date`, and `end_date`. Set `clip` parameter to `True` to clip the Sentinel-2 image composite to the specified bounding box. Note that the hill-shaded land cover map is for visualization only. It is not suitable for further analysis. To generate a classified image that can be used for further analysis (e.g., zonal statistics), use the `dynamic_world()` function with the `return_type` parameter set to `class` and provide visualization parameters to the `addLayer()` method:

```python
classes = geemap.dynamic_world(
    region, start_date, end_date, return_type='class', clip=True
)
vis_params = {
    "min": 0,
    "max": 8,
    "palette": [
        "#419BDF",
        "#397D49",
```

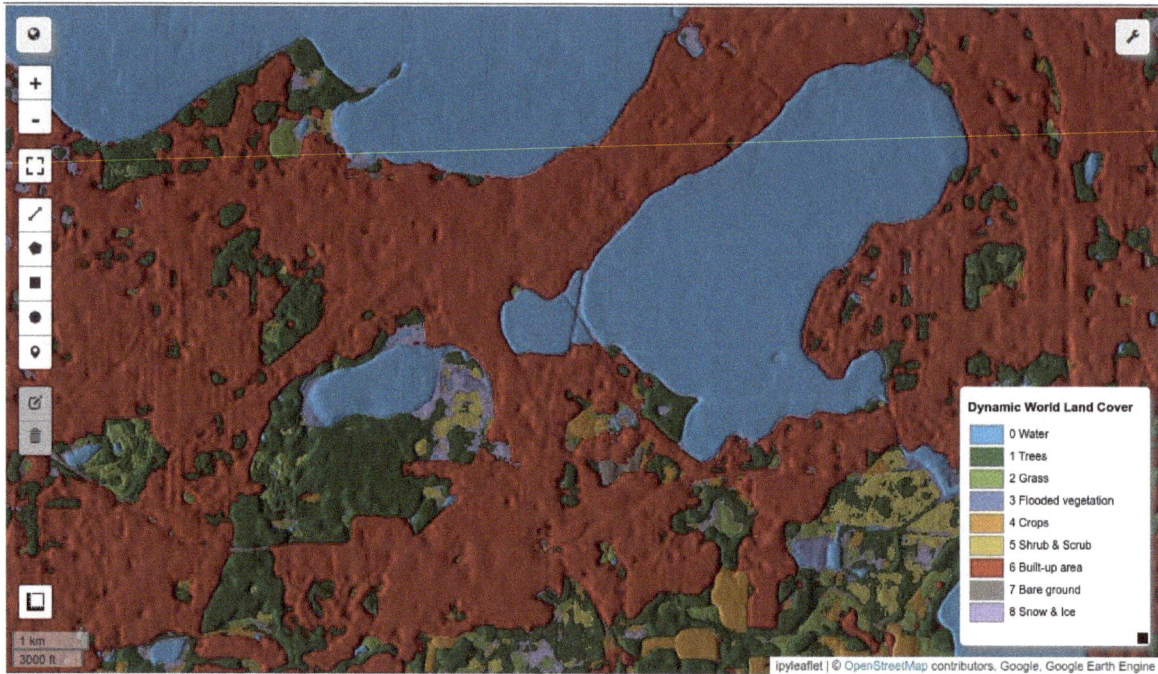

Figure 11.20: A hillshaded image of the Dynamic World land cover map.

```
        "#88B053",
        "#7A87C6",
        "#E49635",
        "#DFC35A",
        "#C4281B",
        "#A59B8F",
        "#B39FE1",
    ],
}
Map.addLayer(classes, vis_params, 'Class')
```

The result is shown in Fig. 11.21.

The classified image should look similar to the hill-shaded image above except that there is no shaded relief effect. The Dynamic World land cover map can also be visualized with the `dynamic_world()` function by setting the `return_type` parameter to `visualize`, which will automatically use the visualization parameters as shown above. In this way, you don't have to specify the visualization parameters manually.

```
probability = geemap.dynamic_world(
    region, start_date, end_date, return_type='visualize', clip=True
)
Map.addLayer(probability, {}, 'Visualize')
```

Again, the image returned by the `dynamic_world()` function with the `return_type` parameter set to `visualize` is not suitable for further analysis. It is for visualization only.

One of the advantages of the Dynamic World land cover map is that it provides a probability image for each land cover class. We can use the `dynamic_world()` function with the `return_type` parameter set to `probability` to generate a probability image, where each pixel represents the probability of a land cover class (see Fig. 11.22).

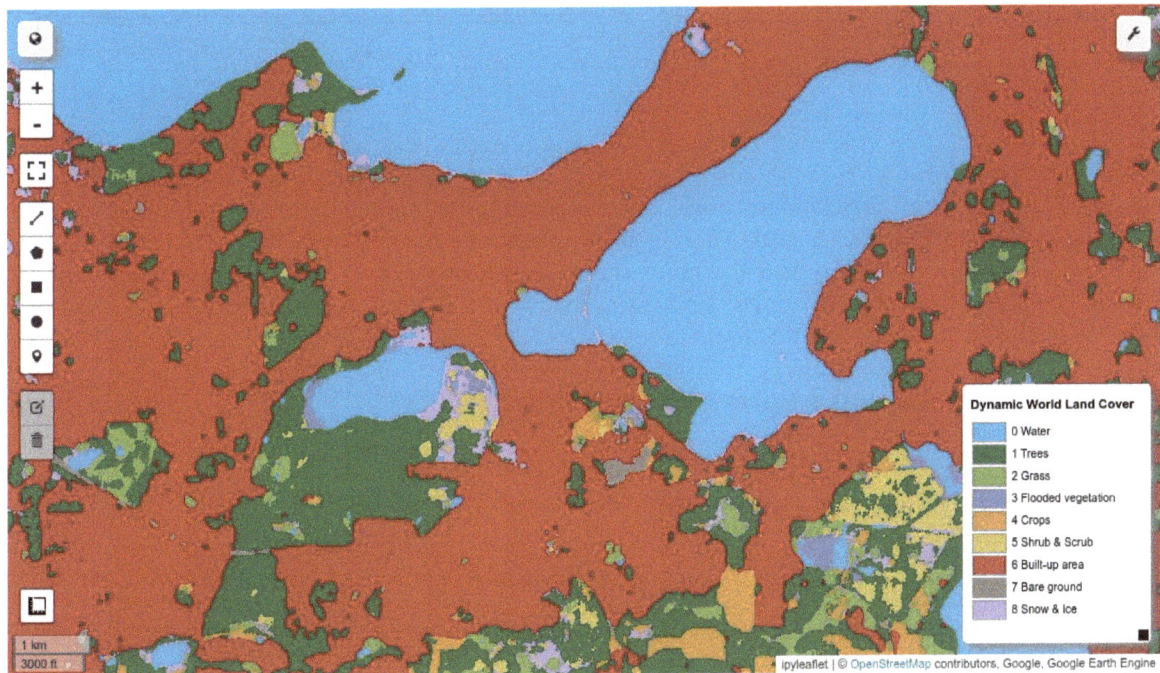

Figure 11.21: A classified image of the Dynamic World land cover map.

```
probability = geemap.dynamic_world(
    region, start_date, end_date, return_type='probability', clip=True
)
Map.addLayer(probability, {}, 'Probability')
```

Pixel values in the probability image range from 0 to 1. Pixels with a higher probability value are shown in brighter colors, while pixels with a lower probability value are shown in darker colors. For example, most water pixels are relatively easy to identify, so they have a higher probability value (brighter color). On the other hand, most mixed pixels along the water edge are relatively difficult to identify, so they have a lower probability value (darker color). The probability image can provide insights into the accuracy of the Dynamic World land cover map.

To calculate the area of each land cover class within a specific region, use the `image_area_by_group()` function. Specify the region, scale, and denominator (e.g., 1e6 for square kilometers). The function will return a Pandas DatFrame with the area and percentage of each land cover class:

```
df = geemap.image_area_by_group(classes, region=region, scale=10, denominator=1e6)
df
```

group	area	percentage
0	93.1509	0.0506
1	414.9350	0.2252
2	111.6175	0.0606
3	3.9708	0.0022
4	783.6755	0.4253
5	12.1400	0.0066
6	363.0357	0.1970
7	6.8808	0.0037

...continued on next page

group	area	percentage
8	53.1881	0.0289

Besides creating a land cover map for a specific region within a specific time range, you can also create land cover time series (e.g., yearly, monthly) for a specific region using the `dynamic_world_timeseries()` function. The function parameters are similar to the `dynamic_world()` function, such as `region`, `start_date`, and `end_date`. The `frequency` parameter can be set to `year` or `month`. Similarly, the `return_type` parameter can be set to `class`, `visualize`, `probability`, or `hillshade`. The following example creates a timeseries of annual land cover maps for the same region as above and visualizes them using the `ts_inspector()` function:

```
Map = geemap.Map()
region = ee.Geometry.BBox(-89.7088, 42.9006, -89.0647, 43.2167)
start_date = '2017-01-01'
end_date = '2022-12-31'
images = geemap.dynamic_world_timeseries(
    region, start_date, end_date, frequency='year', return_type="hillshade"
)
Map.ts_inspector(images, date_format='YYYY')
Map.add_legend(title="Dynamic World Land Cover", builtin_legend='Dynamic_World')
Map.centerObject(region)
Map
```

The result is shown in Fig. 11.23.

ESA WorldCover

The ESA WorldCover maps are provided at 10-m resolution based on Sentinel-1 and Sentinel-2 data. The datasets are updated annually. Currently, there are two time series of ESA WorldCover maps: v100 (2020)[234] and v200 (2021)[235]. Both datasets are available in the Earth Engine Data Catalog. It should be noted that different algorithm versions were used to generate the 2020 and 2021 WorldCover maps. Consequently, changes between the maps include both real changes in land cover and changes due to the algorithms used. The ESA WorldCover Viewer[236] provides tools to compare the 2020 and 2021 WorldCover maps along with Sentinel-1 and Sentinel-2 annual composites. Similarly, we can visualize the 2020 and 2021 ESA WorldCover maps using geemap:

```
Map = geemap.Map()
esa_2020 = ee.ImageCollection("ESA/WorldCover/v100").first()
esa_2021 = ee.ImageCollection("ESA/WorldCover/v200").first()
Map.addLayer(esa_2020, {'bands': ['Map']}, 'ESA 2020')
Map.addLayer(esa_2021, {'bands': ['Map']}, 'ESA 2021')
Map.add_legend(builtin_legend='ESA_WorldCover', title='ESA Land Cover')
Map
```

Pan and zoom the map to an area of interest and use the layer control to change the layer opacity interactively to compare the 2020 and 2021 ESA WorldCover maps. With the ESA WorldCover maps readily available, we can also calculate the area of each land cover class using the `image_area_by_group()` function. The result can be returned as a Pandas DataFrame:

```
df = geemap.image_area_by_group(
    esa_2021, scale=1000, denominator=1e6, decimal_places=4, verbose=True
)
df
```

---

[234]v100 (2020): tiny.geemap.org/ayac

[235]v200 (2021): tiny.geemap.org/hdgj

[236]ESA WorldCover Viewer: tiny.geemap.org/tkhm

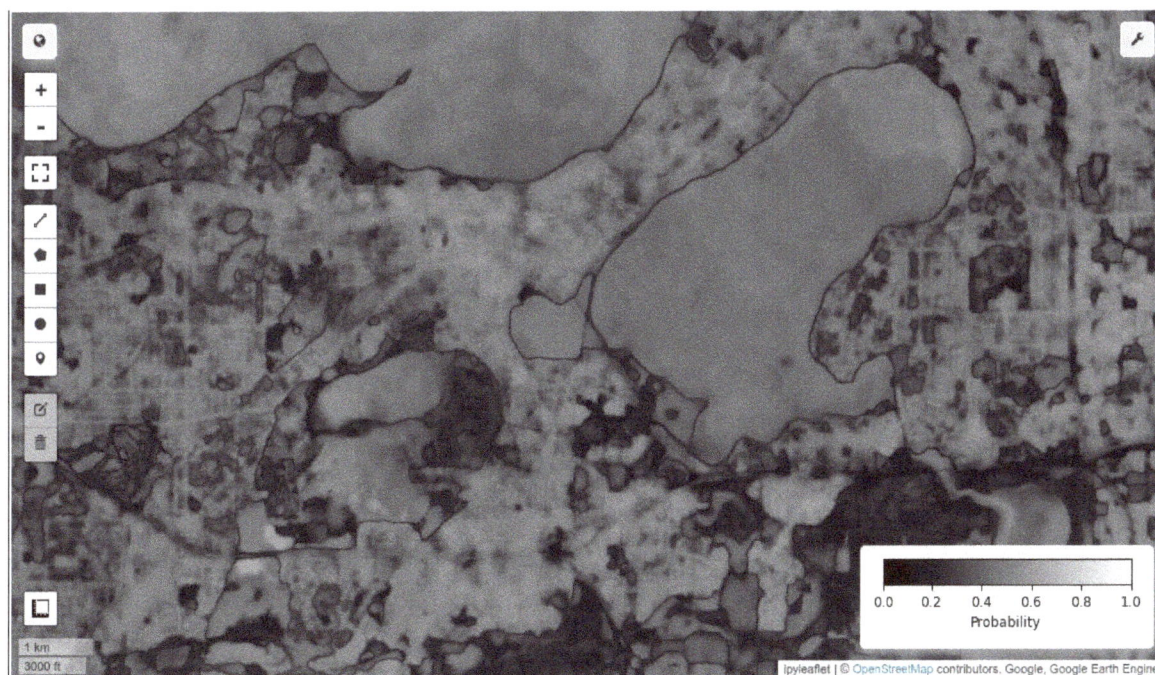

Figure 11.22: A probability image of the Dynamic World land cover map.

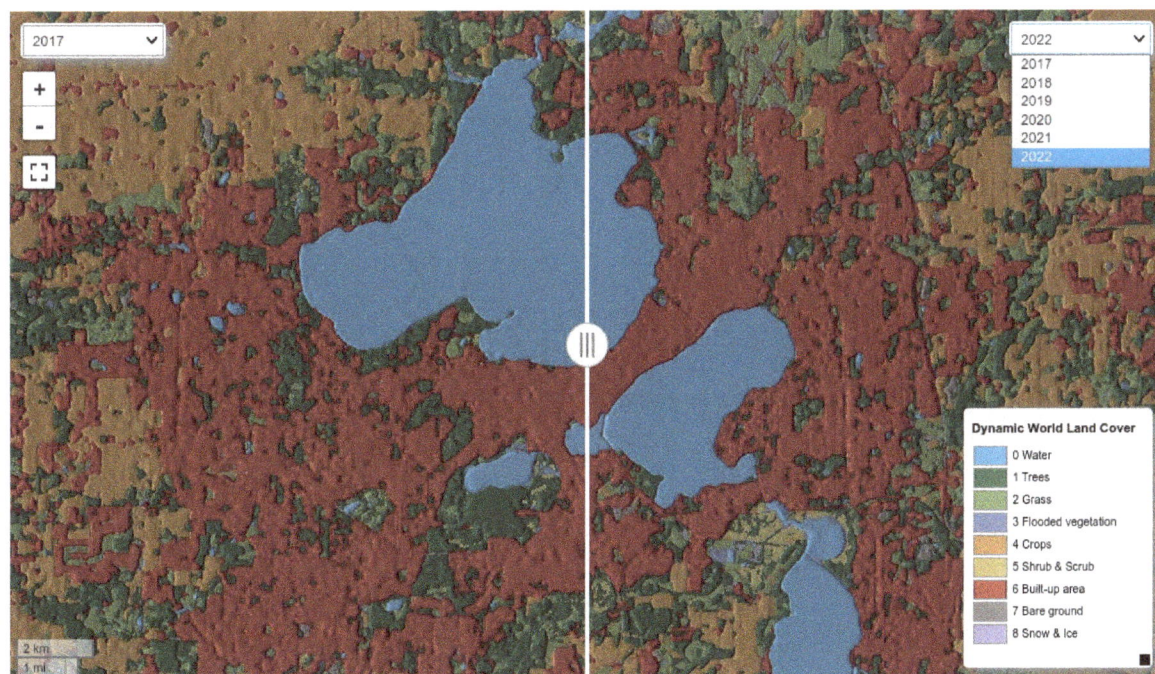

Figure 11.23: Annual land cover maps based on the Dynamic World dataset.

group	area	percentage
10	4.585214e+07	0.3009
20	9.607911e+06	0.0631
30	3.183386e+07	0.2089
40	1.307032e+07	0.0858
50	7.794734e+05	0.0051
60	2.234225e+07	0.1466
70	2.469120e+06	0.0162
80	2.093868e+07	0.1374
90	2.123103e+06	0.0139
95	1.691649e+05	0.0011
100	3.178479e+06	0.0209

From the table above, we can see that tree cover and grassland cover are the dominant land cover types globally, accounting for 30.09% and 20.89% of the total land area, respectively. The built-up land cover type only accounts for 0.51% of the total land area. To save the result as a CSV file, use the `to_csv()` function:

```
df.to_csv('esa_2021.csv')
```

The following example shows how to compare the ESA WorldCover 2021 and Dynamic World Global Land Cover 2021 maps. Note that the Dynamic World does not provide a global map. Therefore, we need to create a global map using the `dynamic_world()` function:

```
Map = geemap.Map(center=[43.0006, -88.9088], zoom=12)
# Create Dynamic World Land Cover 2021
region = ee.Geometry.BBox(-179, -89, 179, 89)
start_date = '2021-01-01'
end_date = '2022-01-01'
dw_2021 = geemap.dynamic_world(region, start_date, end_date, return_type='hillshade')
# Create ESA WorldCover 2021
esa_2021 = ee.ImageCollection("ESA/WorldCover/v200").first()
# Create a split map
left_layer = geemap.ee_tile_layer(esa_2021, {'bands': ['Map']}, "ESA 2021")
right_layer = geemap.ee_tile_layer(dw_2021, {}, "Dynamic World 2021")
Map.split_map(left_layer, right_layer)
# Add legends
Map.add_legend(
    title="ESA WorldCover", builtin_legend='ESA_WorldCover', position='bottomleft'
)
Map.add_legend(
    title="Dynamic World Land Cover",
    builtin_legend='Dynamic_World',
    position='bottomright',
)
Map
```

The split map is shown in Fig. 11.24.

Esri global land cover

The Esri global land cover maps[237] are provided at 10-m resolution based on Sentinel-2 data. The datasets are updated annually. Currently, there are six time series of Esri global land cover maps available from 2017 to 2022. Note that the Esri global land cover maps are not available in the Earth Engine Data Catalog. Luckily, they are available in the Awesome GEE Community Catalog[238]. As the datasets are provided as image tiles, we need to mosaic them into a single image for each year. The following

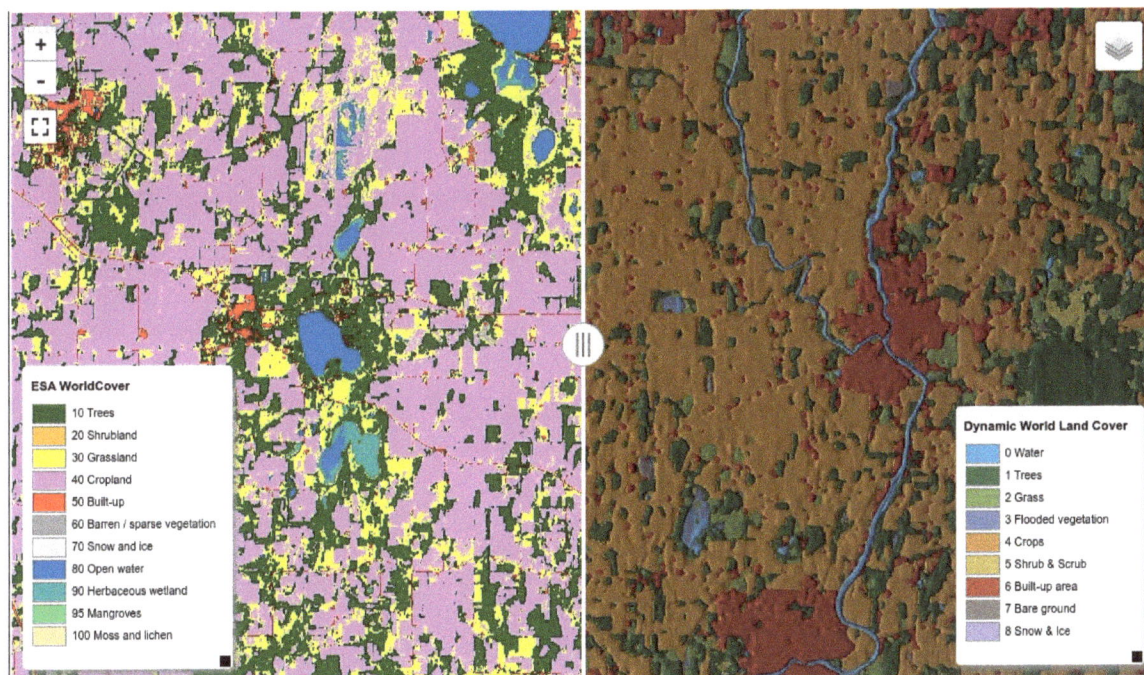

Figure 11.24: Comparing ESA WorldCover 2021 and Dynamic World Global Land Cover 2021.

code shows how to create a function to mosaic the Esri global land cover maps for a given year:

```python
def esri_annual_land_cover(year):
    collection = ee.ImageCollection('projects/sat-io/open-datasets/landcover/ESRI_Global-LULC_10m_TS')
    start_date = ee.Date.fromYMD(year, 1, 1)
    end_date = start_date.advance(1, 'year')
    image = collection.filterDate(start_date, end_date).mosaic()
    return image.set('system:time_start', start_date.millis())
```

With the `esri_annual_land_cover()` function, we can create an image collection for the years 2017 to 2021 by using the `map()` function to loop through the years:

```python
start_year = 2017
end_year = 2021
years = ee.List.sequence(start_year, end_year)
images = ee.ImageCollection(years.map(esri_annual_land_cover))
images
```

Specify the visualization parameters for the Esri global land cover maps:

```python
palette = [
    "#1A5BAB",
    "#358221",
    "#000000",
    "#87D19E",
    "#FFDB5C",
    "#000000",
    "#ED022A",
    "#EDE9E4",
    "#F2FAFF",
    "#C8C8C8",
```

---

[237]Esri global land cover maps: tiny.geemap.org/kixo

[238]Awesome GEE Community Catalog: tiny.geemap.org/zuog

```
    "#C6AD8D",
    ]
vis_params = {"min": 1, "max": 11, "palette": palette}
```

Visualize the Esri global land cover maps using the geemap timeseries inspector:

```
Map = geemap.Map()
Map.ts_inspector(images, left_vis=vis_params, date_format='YYYY')
Map.add_legend(title="Esri Land Cover", builtin_legend='ESRI_LandCover')
Map
```

Lastly, we can compare the Esri global land cover maps and Dynamic World Global Land Cover 2021 maps side by side:

```
Map = geemap.Map(center=[43.0006, -88.9088], zoom=12)
# Create Dynamic World Land Cover 2021
dw_2021 = geemap.dynamic_world(region, start_date, end_date, return_type='hillshade')
# Create Esri Global Land Cover 2021
esri_2021 = esri_annual_land_cover(2021)
# Create a split map
left_layer = geemap.ee_tile_layer(esri_2021, vis_params, "Esri 2021")
right_layer = geemap.ee_tile_layer(dw_2021, {}, "Dynamic World 2021")
Map.split_map(left_layer, right_layer)
# Add legends
Map.add_legend(
    title="Esri Land Cover", builtin_legend='ESRI_LandCover', position='bottomleft'
)
Map.add_legend(
    title="Dynamic World Land Cover",
    builtin_legend='Dynamic_World',
    position='bottomright',
)
Map
```

The split map is shown in Fig. 11.25.

11.7   Concluding remarks

Congratulations! You have reached the end of this book. As we close this journey through Earth Engine and the geemap Python package, we can reflect on the important role these tools have in the geospatial community. From creating interactive maps to analyzing Earth Engine datasets, our exploration underscores the vital role of cloud computing in understanding the world around us. This book has aimed to bridge the knowledge gap in the application of the Earth Engine Python API, presenting comprehensive, hands-on tutorials designed to facilitate user experience in a more interactive and intuitive manner.

We started by introducing GEE and geemap, paving the way for the understanding and application of this cloud computing platform. This led to creating interactive maps, using Earth Engine Data and local geospatial data, and visualizing geospatial data in unique and informative ways. The middle part of the book transitioned into more advanced techniques, such as analyzing geospatial data and exporting Earth Engine data. We then progressed to creating timelapse animations, building interactive web apps, and delving into various Earth Engine applications, including surface water mapping, forest cover change analysis, flood mapping, and global land cover mapping.

The beauty of this exploration is that it is not restricted to a particular group of individuals. Both new and experienced Earth Engine users can find value in this book. I hope that students, researchers, and data scientists who are venturing into geospatial data analysis found the step-by-step guide helpful.

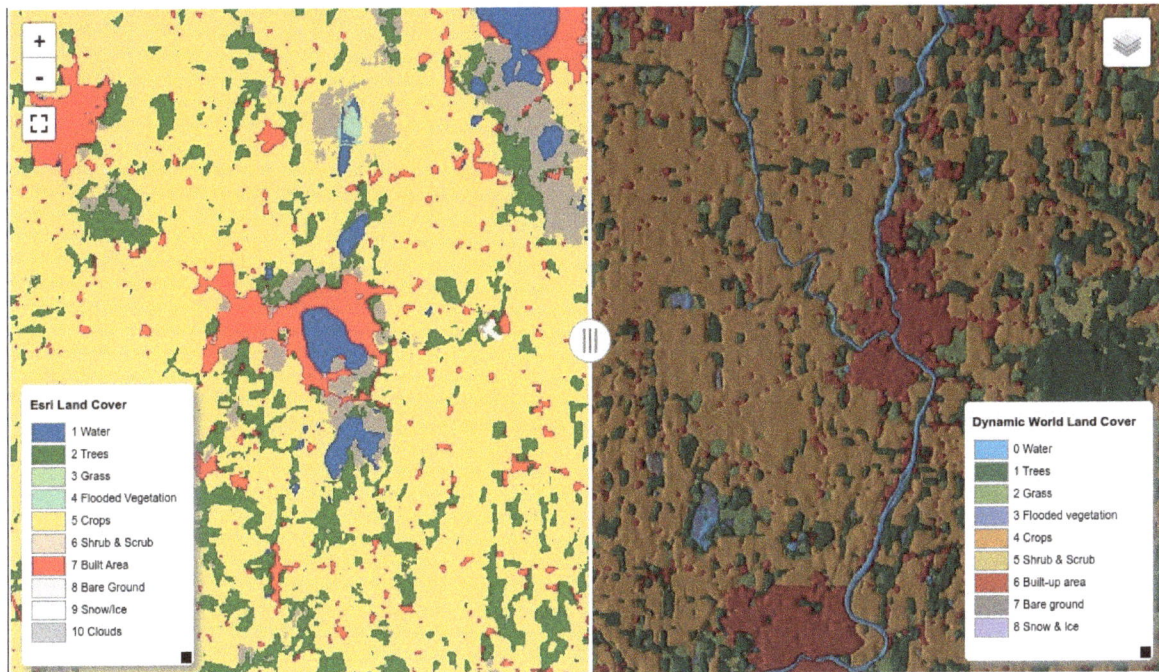

Figure 11.25: Comparing Esri Global Land Cover 2021 and Dynamic World Global Land Cover 2021.

Moreover, this book is not just about providing a technical guide, but also about promoting the open-source philosophy. The practical application of open-source tools such as geemap has the potential to influence research and decision-making processes, with the aim of creating a more sustainable future.

The key to fully benefiting from this book lies in active participation. By typing the code yourself in a Jupyter environment and experimenting with different datasets, you are better equipped to understand the material and eventually, utilize these tools in your own projects. By walking through this process, I hope that the readers gained not only a better understanding of GEE and geemap, but also a sense of accomplishment and curiosity to further their own exploration.

I extend my thanks to all those who made this book possible, from the Earth Engine team to the contributors of the geemap package, and to the reviewers who provided invaluable feedback. Above all, I acknowledge you, the reader, for your interest and commitment to learning.

As the field of geospatial technology continues to evolve, I encourage you to continue exploring, innovating, and making strides in using GEE and geemap. While this book serves as a guide to get started, the possibilities and potential applications of these tools are virtually limitless.

I hope you have learned something new from this book. If you have any questions or suggestions, please ask on the Discussion Forum[239] on GitHub or email me at giswqs@gmail.com. I am always happy to hear from you. Happy learning and coding!

---

[239]Discussion Forum: tiny.geemap.org/rlsk

Books from Locate Press

Be sure to visit http://locatepress.com for information on new and upcoming titles.

Discover QGIS 3.x

SECOND EDITION: EXPLORE THE LATEST LONG TERM RELEASE (LTR) OF QGIS WITH DISCOVER QGIS 3.X!

Discover QGIS 3.x is a comprehensive up-to-date workbook built for both the classroom and professionals looking to build their skills.

Designed to take advantage of the latest QGIS features, this book will guide you in improving your maps and analysis.

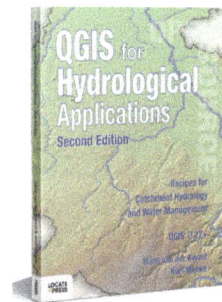

You will find clear learning objectives and a task list at the beginning of each chapter. Of the 31 exercises in this workbook, 7 are new and 8 have seen considerable updates. All exercises are updated to support QGIS 3.26.

The book is a complete resource and includes: lab exercises, challenge exercises, all data, discussion questions, and solutions.

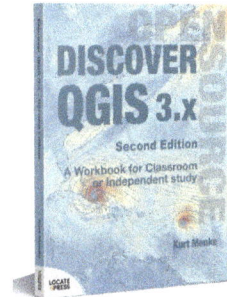

QGIS for Hydrological Applications

SECOND EDITION: RECIPES FOR CATCHMENT HYDROLOGY AND WATER MANAGEMENT.

Now updated - learn even more GIS skills for catchment hydrology and water management with QGIS!

This second edition workbook introduces hydrological topics to professionals in the water sector using state of the art functionality in QGIS. The book is also useful as a beginner's course in GIS concepts, using a problem-based learning approach

Designed to take advantage of the latest QGIS features, this book will guide you in improving your maps and analysis.

Introduction to QGIS

GET STARTED WITH QGIS WITH THIS INTRODUCTION COVERING EVERYTHING NEEDED TO GET YOU GOING USING FREE AND OPEN SOURCE GIS SOFTWARE.

This QGIS tutorial, based on the 3.16 LTR version, introduces you to major concepts and techniques to get you started with viewing data, analysis, and creating maps and reports.

Building on the first edition, the authors take you step-by-step through the process of using the latest map design tools and techniques in QGIS 3. With numerous new map designs and completely overhauled workflows, this second edition brings you up to speed with current cartographic technology and trends.

With this book you'll learn about the QGIS interface, creating, analyzing, and editing vector data, working with raster (image) data, using plugins and the processing toolbox, and more.

Resources for further help and study and all the data you'll need to follow along with each chapter are included.

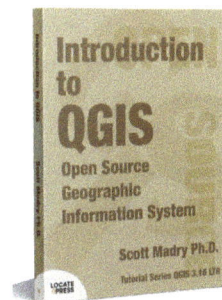

QGIS Map Design - 2nd Edition

LEARN HOW TO USE QGIS 3 TO TAKE YOUR CARTOGRAPHIC PRODUCTS TO THE HIGH-EST LEVEL.

QGIS 3.4 opens up exciting new possibilities for creating beautiful and compelling maps!

Building on the first edition, the authors take you step-by-step through the process of using the latest map design tools and techniques in QGIS 3. With numerous new map designs and completely overhauled workflows, this second edition brings you up to speed with current cartographic technology and trends.

See how QGIS continues to surpass the cartographic capabilities of other geoware available today with its data-driven overrides, flexible expression functions, multitudinous color tools, blend modes, and atlasing capabilities. A prior familiarity with basic QGIS capabilities is assumed. All example data and project files are included.

Get ready to launch into the next generation of map design!

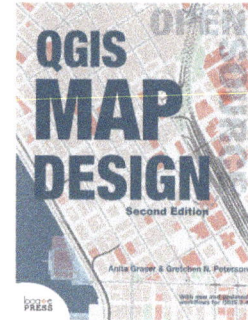

Leaflet Cookbook

COOK UP DYNAMIC WEB MAPS USING THE RECIPES IN THE LEAFLET COOKBOOK.

Leaflet Cookbook will guide you in getting started with Leaflet, the leading open-source JavaScript library for creating interactive maps. You'll move swiftly along from the basics to creating interesting and dynamic web maps.

Even if you aren't an HTML/CSS wizard, this book will get you up to speed in creating dynamic and sophisticated web maps. With sample code and complete examples, you'll find it easy to create your own maps in no time.

A download package containing all the code and data used in the book is available so you can follow along as well as use the code as a starting point for your own web maps.

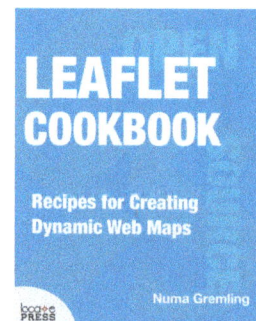

The PyQGIS Programmer's Guide

WELCOME TO THE WORLD OF PYQGIS, THE BLENDING OF QGIS AND PYTHON TO EXTEND AND ENHANCE YOUR OPEN SOURCE GIS TOOLBOX.

With PyQGIS you can write scripts and plugins to implement new features and perform automated tasks.

This book is updated to work with the next generation of QGIS—version 3.x. After a brief introduction to Python 3, you'll learn how to understand the QGIS Application Programmer Interface (API), write scripts, and build a plugin.

The book is designed to allow you to work through the examples as you go along. At the end of each chapter you will find a set of exercises you can do to enhance your learning experience.

The PyQGIS Programmer's Guide is compatible with the version 3.0 API released with QGIS 3.x and will work for the entire 3.x series of releases.

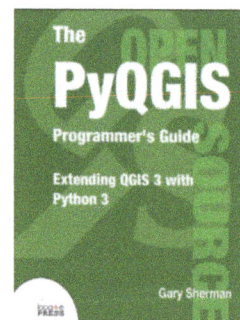

pgRouting: A Practical Guide

WHAT IS PGROUTING?

It's a PostgreSQL extension for developing network routing applications and doing graph analysis.

Interested in pgRouting? If so, chances are you already use PostGIS, the spatial extender for the PostgreSQL database management system.

So when you've got PostGIS, why do you need pgRouting? PostGIS is a great tool for molding geometries and doing proximity analysis, however it falls short when your proximity analysis involves constrained paths such as driving along a road or biking along defined paths.

This book will both get you started with pgRouting and guide you into routing, data fixing and costs, as well as using with QGIS and web applications.

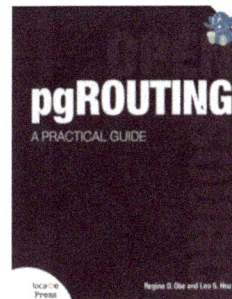

Geospatial Power Tools

EVERYONE LOVES POWER TOOLS!

The GDAL and OGR apps are the power tools of the GIS world—best of all, they're free.

The utilities include tools for examining, converting, transforming, building, and analysing data. This book is a collection of the GDAL and OGR documentation, but also includes new content designed to help guide you in using the utilities to solve your current data problems.

Inside you'll find a quick reference for looking up the right syntax and example usage quickly. The book is divided into three parts: *Workflows and examples*, *GDAL raster utilities*, and *OGR vector utilities*.

Once you get a taste of the power the GDAL/OGR suite provides, you'll wonder how you ever got along without them.

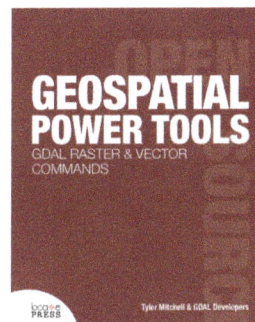

See these books and more at http://locatepress.com